Recent Advances in Machine Learning in Tribology

Recent Advances in Machine Learning in Tribology

Editors

Max Marian
Stephan Tremmel

MDPI

Basel • Beijing • Wuhan • Barcelona • Belgrade • Novi Sad • Cluj • Manchester

Editors

Max Marian
Department of Mechanical and Metallurgical Engineering, School of Engineering Pontificia Universidad Católica de Chile
Santiago
Chile

Stephan Tremmel
Engineering Design and CAD, University of Bayreuth
Bayreuth
Germany

Editorial Office
MDPI AG
Grosspeteranlage 5
4052 Basel, Switzerland

This is a reprint of articles from the Special Issue published online in the open access journal *Lubricants* (ISSN 2075-4442) (available at: https://www.mdpi.com/journal/lubricants/special_issues/L7V01FG2T6).

For citation purposes, cite each article independently as indicated on the article page online and as indicated below:

Lastname, A.A.; Lastname, B.B. Article Title. *Journal Name* **Year**, *Volume Number*, Page Range.

ISBN 978-3-7258-1737-5 (Hbk)
ISBN 978-3-7258-1738-2 (PDF)
doi.org/10.3390/books978-3-7258-1738-2

Contents

About the Editors

Max Marian

Max Marian is Professor and Executive Director of the Institute of Machine Design and Tribology (IMKT) of Leibniz University, Hannover, Germany, and Assistant Professor of Multiscale Engineering Mechanics at the Department of Mechanical and Metallurgical Engineering of Pontificia Universidad Católica de Chile. His research focuses on energy efficiency and sustainability through tribology, with an emphasis on the modification of surfaces. Besides machine elements and engine components, he has expanded his research interests to include biotribology and artificial joints, as well as triboelectric nanogenerators. His research is particularly related to the development of numerical multiscale tribo-simulation and machine learning approaches. He has published more than 50 peer-reviewed publications in reputed journals, has given numerous conference talks and invited talks, and has been awarded with various individual distinctions, as well as best paper and presentation awards. Furthermore, he was listed among the Emerging Leaders 2023 of Surface Topography: Metrology and Properties. Moreover, he is a member of the Editorial Boards of *Frontiers in Chemistry Nanoscience, Industrial Lubrication and Tribology,* and *Lubricants,* as well as *Tribology - Materials, Surfaces & Interfaces,* and is a member of the Society of Tribologists and Lubrication Engineers (STLE) and the German Society for Tribology (GfT).

Stephan Tremmel

Stephan Tremmel is the Chairholder and Head of Engineering Design and CAD at the Faculty of Engineering Science at the University of Bayreuth, Germany. His research focuses on finite element analysis (development, modeling, and simulation) and on design of functionally integrated machine elements, with main applications in drive technology and energy engineering. Furthermore, he has special experience in surface engineering (specifically texturing and PVD coatings) and tribology. He has published more than 100 peer-reviewed journal publications. He is a member of the Editorial Board of *Lubricants* and is a member of the German Research Association for Drive Technology (FVA) and the German Society for Tribology (GfT).

lubricants

MDPI

Editorial

Recent Advances in Machine Learning in Tribology

Max Marian [1,2,*] and Stephan Tremmel [3]

1 Department of Mechanical and Metallurgical Engineering, School of Engineering, Pontificia Universidad Católica de Chile, Vicuña Mackenna 4860, Macul 6904411, Región Metropolitana, Chile
2 Institute of Machine Design and Tribology (IMKT), Leibniz University Hannover, An der Universität 1, 30823 Garbsen, Germany
3 Engineering Design and CAD, University of Bayreuth, Universitätsstraße 30, 95447 Bayreuth, Germany; stephan.tremmel@uni-bayreuth.de
* Correspondence: max.marian@uc.cl

Tribology, the study of friction, wear, and lubrication, has been a subject of interest for researchers exploring the complexities of materials and surfaces. Recently, machine learning has emerged as a valuable tool in this field, offering new avenues for understanding. The second Special Issue in the journal *Lubricants* dedicated to this partnership signifies a step forward in our exploration of these concepts. Machine learning's ability to analyze large datasets and extract patterns has broadened our understanding of tribology. This collaboration between traditional methods and computational techniques has enabled researchers to uncover insights previously inaccessible. From predicting frictional behavior to optimizing lubricant compositions, machine learning's applications in tribology are diverse.

The nine research and two review articles, as well as one technical note, covered in this Special Issue embrace a wide range of topics, from fundamental research on friction mechanisms to practical studies improving industrial machinery performance. Predictive modeling stands out as an area of interest, allowing researchers to forecast tribological properties accurately. This includes predicting material wear rates and optimizing lubricant formulations for specific conditions. Furthermore, machine learning has facilitated the exploration of complex phenomena across different scales, providing a comprehensive understanding of tribological processes. The convergence of tribology and machine learning offers opportunities for synergy and discovery, marking a significant moment in the field's evolution.

The Guest Editors extend their gratitude to all authors and reviewers for their contributions, as well as to the editorial staff of MDPI journal *Lubricants* for their support and guidance.

Conflicts of Interest: The authors declare no conflicts of interest.

Citation: Marian, M.; Tremmel, S. Recent Advances in Machine Learning in Tribology. *Lubricants* **2024**, *12*, 168. https://doi.org/10.3390/lubricants12050168

Received: 4 May 2024
Accepted: 8 May 2024
Published: 9 May 2024

lubricants MDPI

Review

Physics-Informed Machine Learning—An Emerging Trend in Tribology

Max Marian [1,*] and Stephan Tremmel [2]

[1] Department of Mechanical and Metallurgical Engineering, School of Engineering, Pontificia Universidad Católica de Chile, Vicuña Mackenna 4860, Macul 6904411, Chile
[2] Engineering Design and CAD, University of Bayreuth, Universitätsstr. 30, 95447 Bayreuth, Germany; stephan.tremmel@uni-bayreuth.de
* Correspondence: max.marian@uc.cl

Abstract: Physics-informed machine learning (PIML) has gained significant attention in various scientific fields and is now emerging in the area of tribology. By integrating physics-based knowledge into machine learning models, PIML offers a powerful tool for understanding and optimizing phenomena related to friction, wear, and lubrication. Traditional machine learning approaches often rely solely on data-driven techniques, lacking the incorporation of fundamental physics. However, PIML approaches, for example, Physics-Informed Neural Networks (PINNs), leverage the known physical laws and equations to guide the learning process, leading to more accurate, interpretable and transferable models. PIML can be applied to various tribological tasks, such as the prediction of lubrication conditions in hydrodynamic contacts or the prediction of wear or damages in tribo-technical systems. This review primarily aims to introduce and highlight some of the recent advances of employing PIML in tribological research, thus providing a foundation and inspiration for researchers and R&D engineers in the search of artificial intelligence (AI) and machine learning (ML) approaches and strategies for their respective problems and challenges. Furthermore, we consider this review to be of interest for data scientists and AI/ML experts seeking potential areas of applications for their novel and cutting-edge approaches and methods.

Keywords: artificial intelligence; machine learning; tribo-informatics; physics-informed neural network; friction; wear; lubrication

Citation: Marian, M.; Tremmel, S. Physics-Informed Machine Learning—An Emerging Trend in Tribology. *Lubricants* **2023**, *11*, 463. https://doi.org/10.3390/lubricants11110463

Received: 6 September 2023
Revised: 24 September 2023
Accepted: 2 October 2023
Published: 30 October 2023

1. Artificial Intelligence and Machine Learning in Tribology

The complex interactions between surfaces in relative motion or between surfaces and flowing media have substantial impacts on the performance, efficiency, and service life of tribo-technical systems. In recent years, the integration of artificial intelligence (AI) and machine learning (ML) techniques in tribology has opened up new possibilities for improving understanding, prediction, and control of friction, lubrication, and wear phenomena [1,2]. AI refers to the development of intelligent machines that are capable of performing tasks that typically require human intelligence. ML is a subfield of AI (see Figure 1) and focuses on the development of experience-based algorithms that allow for computers to learn and make predictions or decisions (output) based on data (input) without being explicitly programmed [3]. Some notable ML techniques encompass decision trees (tree-like structures that make decisions based on feature values) [3], random forests (combining multiple decision trees to improve prediction accuracy) [4], support vector machines (aiming to find the best decision boundary between classes in a dataset) [5], and neural networks, just to mention a few. Among these techniques, artificial neural networks (ANNs) have gained significant prominence. They consist of interconnected "neurons", organized into layers, whereby each neuron receives an input, performs computations, and passes the result to the next layer. Through training, i.e., adjusting the connections' weights and biases, complex patterns in the data can be captured [3,6,7].

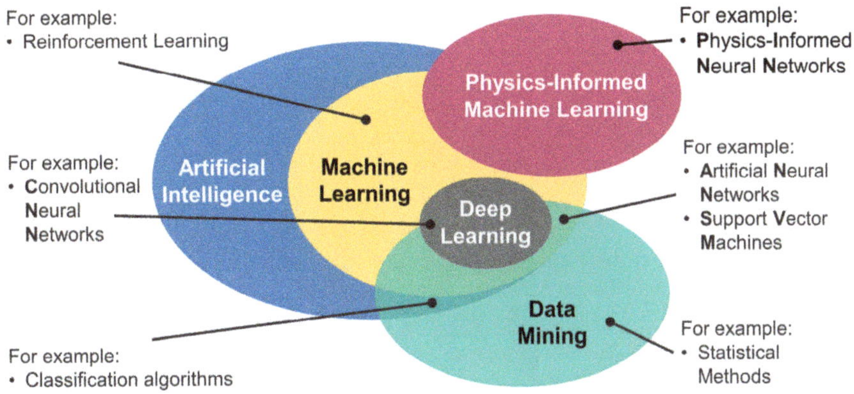

Figure 1. Classification of the terms artificial intelligence, machine learning, deep learning, data mining, and physics-informed machine learning. Redrawn and adapted from [8].

All of these ML/AI approaches possess the potential to revolutionize tribology by enabling more accurate modeling, efficient optimization, and an enhanced control of friction and wear processes [1]. One of the primary applications of AI and ML in tribology is predictive modeling by analyzing large datasets, thus identifying patterns and hidden relationships that may not be apparent through traditional analytical methods [9–12]. Moreover, AI and ML techniques can facilitate condition-based maintenance and real-time monitoring in tribological systems when employing respective integrated sensors and data acquisition systems [13–15]. Furthermore, AI and ML can contribute to designing and optimizing tribo-systems within vast design spaces [16] or can even contribute to discovering novel solutions that may not have been considered previously. All of these aspects may lead to the development of more efficient lubricants [17,18] and materials [19,20], advanced surface modifications [21,22], manufacturing processes [23,24], or innovative tribo-system designs [25,26], not only going beyond mere buzzwords, but actually resulting in improved energy efficiency, reduced emissions, and an enhanced overall system performance [27].

Meanwhile, there is a number of review articles showcasing the usages and many promises of AI and ML within tribology [1,2,28–31]. However, a challenge remains in the training of AI/ML models, which relies heavily on the availability of large amounts of high-quality experimentally [32–38] or numerically [39–42] generated data. Ideally, these data should be FAIR (Findable, Accessible, Interoperable, and Reusable), meaning it should be well documented, easily accessible, compatible with different systems, and suitable for reuse in different contexts [43–45]. However, acquiring such data for scientific or industrial tribology problems can often be challenging, and these data may not always be readily available [46,47]. Also, relying on data alone bears the risks of having misunderstood the scientific problem and not converging towards generalizability.

As an alternative to data-based AI strategies, in situations where there is a scarcity of available data, ML models can be trained using supplementary data derived from the application of physical laws, incorporating mathematical models. This approach, known as physics-informed ML (PIML), thus connects the big data regime, without any knowledge about the underlying physics, with the area of small data and lots of physics [48] (see Figure 2). The employment of PIML in tribology is likewise a comparatively new as well as emerging trend, which has not been covered by other review articles yet. This article therefore seeks to shed some light on the novel trend of physics-informed ML. The concept will be briefly introduced in Section 2, the current state of the art will be discussed in detail in Section 3, and the article will end with some concluding remarks in Section 4.

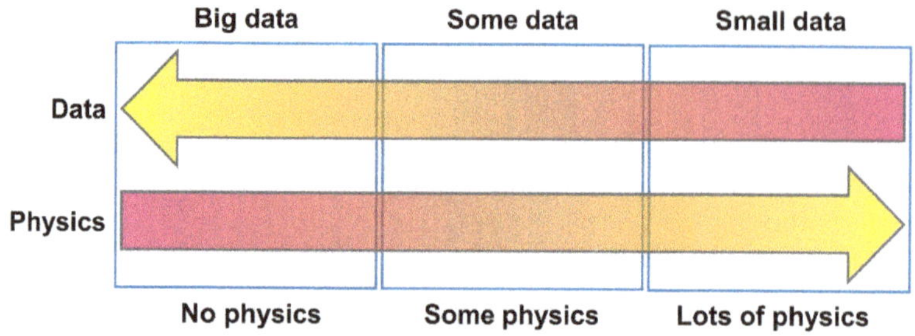

Figure 2. Data and physics scenarios. Redrawn and adapted from [48].

2. Physics-Informed Machine Learning

PIML is an approach that combines ML techniques with the principles and constraints of physics to enhance the accuracy, interpretability, and generalizability of models [48,49]. PIML aims to address the sole dependence on data by incorporating prior knowledge of physics into the learning process, ensuring that the resulting models align with the fundamental principles of the domain [48]. Thus, PIML models can capture the underlying physics, even in cases where the available data are limited, noisy, or incomplete. This integration allows for models that are not only data-driven, but also consistent with the fundamental principles governing the system [48]. By incorporating physics-based knowledge, it is possible to enhance the predictive accuracy compared to conventional, data-driven ML approaches. Furthermore, physics-informed models are often more interpretable, which allows for a better understanding of the underlying mechanisms and optimizing tasks. Thereby, physics-informed models, once properly trained with a solid understanding of the physics involved, can be adapted to various applications and environments with relatively minor adjustments. Finally, by incorporating physical laws, machine learning models are less likely to make predictions that violate fundamental principles, reducing the risk of erroneous or unrealistic results, e.g., predicting a negative film thickness in hydrodynamic contacts, etc. Apart from the observational biases contained in a sufficiently large dataset, as used to train classical ML models, it may consist of inductive biases through a direct intervention into the ML model architecture, for example, in the form of mathematical constraints to be strictly satisfied that are known a priori [48]. Furthermore, learning biases can be incorporated into the training phase through the careful selection of loss functions, constraints, and inference algorithms [48]. These can effectively guide the model towards converging on solutions that align with the fundamental principles of physics [48]. By incorporating soft penalty constraints and fine-tuning them, it becomes possible to approximately satisfy the underlying physical laws, offering a flexible framework to introduce a wide range of physics-based biases, expressed through integrals or differential equations [48]. Observational, inductive, or learning biases are not mutually exclusive and can be combined synergistically to create a diverse set of hybrid approaches to construct PIML systems [48].

Even though a variety of approaches are generally available [50], the most common methodology in PIML is the use of Physics-Informed Neural Networks (PINNs), which combine artificial neural networks with physics-based equations, such as differential equations or conservation laws [49,51]. During the training phase, these equations are incorporated into the loss functions of a neural network to guide the learning process, i.e., there is a data-driven part and a physics-driven part in the loss function. The neural network learns to approximate both the data-driven aspects and the physics-based constraints simultaneously, resulting in models that capture the complex interactions between data and physics [49]. As illustrated in Figure 3, this is achieved by sampling a set of input training data (i.e., spatial coordinates and/or time stamps) and passing it through the neural network. Subsequently, the network's output gradients are computed with respect to its inputs at these locations. These gradients can frequently be analytically obtained via auto-differentiation (AD) and are then used to calculate the residual of the underlying differential equation. The residual is then incorporated as an additional term in the loss function. The aim of including this "physics loss" in the loss function is to guarantee that the solution learned by the network aligns with the established laws of physics.

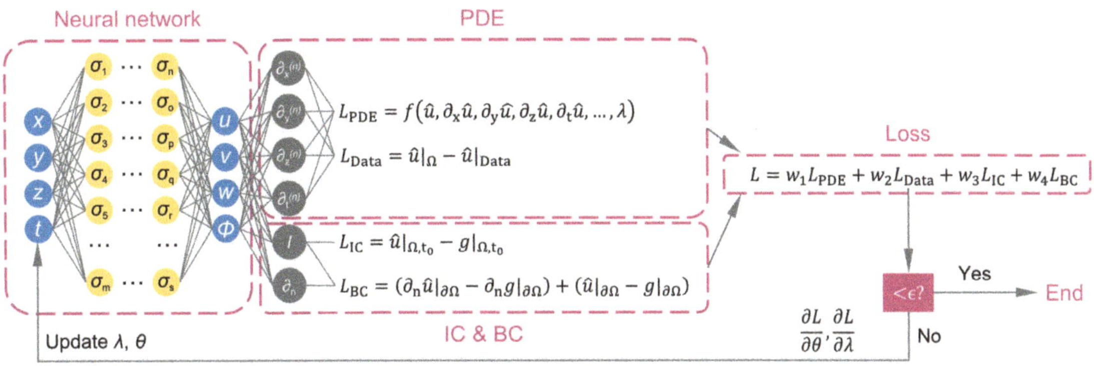

PDE = Partial Differential Equation, IC = Initial Condition, BC = Boundary Condition, x, y, z: cartesian coordinates, t: time, u, v, w: velocities, Φ: angle, $\hat{u} = [u, v, w, \Phi]$, θ: weights/biases, λ: unknown PDE paramters, Ω: domain, g: initial/boundary condition, w_i: weights

Figure 3. Graphical representation of a PINN approach.

Another approach in PIML involves the utilization of probabilistic models, such as Gaussian processes or Bayesian inference, to incorporate physical priors and uncertainties into the learning process [48]. These models enable the quantification of uncertainty and the propagation of physical constraints through the machine learning framework [48].

The applications of PIML are wide-ranging and can be found in various scientific and engineering domains. It has been employed in fluid dynamics for flow prediction and turbulence modeling [52–54], in material science to predict material behavior [55–58] and discover new materials [59], in structural mechanics [60,61], medical imaging [62,63], and many other fields where physical laws play crucial roles. By integrating physics-based knowledge into machine learning models, PIML also offers a powerful tool for understanding and optimizing tribological phenomena and thus represents a very recent and emerging trend in the domain of tribology.

3. Physics-Informed Machine Learning in Tribology

3.1. Lubrication Prediction

PIML can be applied to various tribological tasks, for example, the prediction of lubrication conditions and the optimization of lubrication processes. By considering the governing equations of fluid dynamics and incorporating experimental or simulation data, ML models can learn to predict the lubricant film thickness, pressure, and/or shear stress distribution. As such, Almqvist [64] implemented a PINN in MATHWORKS Matlab to solve the Reynolds boundary value problem (BVP) in a linear slider, assuming a one-dimensional flow of an incompressible and iso-viscous fluid. The rather simple feedforward neural network consisted of one input node (coordinate x), one hidden layer (i.e., a single layer network) with ten neurons, as well as one output node (see Figure 4a), and employed the sigmoid activation function. The Reynolds BVP was described by a second-order ordinary differential equation:

$$\frac{\partial}{\partial x}\left(H^3\frac{\partial p}{\partial x}\right) = \frac{\partial H}{\partial x}, \text{for } 0 < x < 1 \tag{1}$$

with the dimensionless film thickness $H(x)$ and the dimensionless pressure $p(x)$. The pressure at the boundaries was chosen to be zero ($p(0) = 0$, $p(1) = 0$). The Reynolds BVP was then condensed to

$$H^3 p'' + H^{3'} p' - H' = 0 \text{ for } 0 < x < 1\,, \tag{2}$$

$$\begin{bmatrix} p(0) \\ p(0) \end{bmatrix} = 0\,, \tag{3}$$

and the loss function was defined as

$$L = \left[\left(H(x)^3 p'' + H(x)^{3'} p' - H(x)' \right)^2 \right] + p^2(0) + p^2(1) \tag{4}$$

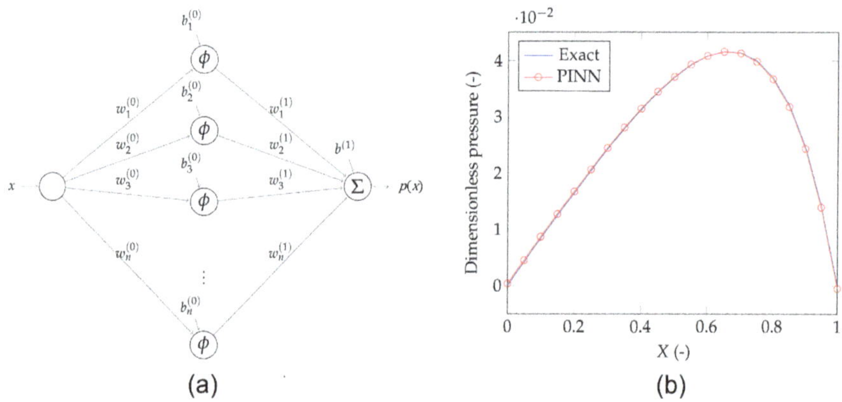

Figure 4. (**a**) Topology of the employed PINN to solve Reynolds BVP and (**b**) comparison of the PINN prediction for a linear converging slider with the exact solution. Reprinted and adapted from [64] with permission from CC BY 4.0.

After establishing the partial derivatives of p'' and $p(1)$ with respect to the weights and bias instead of the commonly employed AD, Almqvist [64] used the PINN approach to solve for the dimensionless pressure in a linear slider with a converging gap of the form $H(x) = 2 - x$ and compared the result to an exact analytical solution (see Figure 4b). Thereby, an overall error of 6.2×10^{-5} as well as errors of 4.1×10^{-4} at $x = 0$ and -4.0×10^{-4} at $x = 1$ were obtained. It is worth noting that this approach does not offer advantages neither with respect to accuracy nor efficiency compared to the established finite difference (FDM) or finite element method-based solutions, but it presents a meshless approach, and not a data-driven approach [64], thus overcoming the "curse of dimension" [65]. Furthermore, cavitation effects were not considered by this formulation, and the study was limited to solving the one-dimensional Reynolds equation for the pressure at a given film thickness profile.

Inspired by the pioneering work from Almqvist [64], several authors have taken up the idea and extended the PINN approach. As such, Zhao et al. [66] solved for the two-dimensional Reynolds equation:

$$\frac{\partial}{\partial x}\left(H^3 \frac{\partial p}{\partial x} \right) = \frac{L}{B} \frac{\partial}{\partial y}\left(H^3 \frac{\partial p}{\partial y} \right) - 6 \frac{\partial H}{\partial x} \tag{5}$$

for a slider bearing with the length L and width B as well as zero-pressure conditions at the edges. The film thickness was described as

$$H(x) = \frac{\theta L}{h_0}(1 - x) + 1 \tag{6}$$

with the inclination of the slider θ and the outlet film thickness h_0. The PINN was programmed in Julia language and followed the examples of [49,67]. The authors studied the influence of the number of training epochs (i.e., the number of complete iterations through the model training process, where the model learns from the available physics-based knowledge, constraints, or equations, making incremental adjustments to its parameters in an effort to improve its performance) as well as the influences of the layer and neuron numbers on the predicted pressure distribution. They reported that the maximum values converged fairly well, while the pressure at the boundaries of the domain as well as the global loss took some more epochs (see Figure 5a). Furthermore, Zhao et al. [66] compared different PINN topologies without hidden layers, with one hidden layer, as well as with two hidden layers with 16 neurons each. As depicted in Figure 5b, while the pressures in the central region were somewhat comparable, the PINN without hidden layers displayed strongly fluctuating pressures at the edges; thus, it strongly diverged from the zero-pressure boundary conditions. In turn, the differences between the PINNs with one hidden layer and two hidden layers were neglectable. Similarly, using fewer neurons in the hidden layers (e.g., four) led to undesired pressure fluctuations at the boundary of the domain, while using either 16 or 32 nodes did not affect the results in a significant way (see Figure 5c). The authors concluded that a PINN topology with 16 neurons in one hidden layer as well as 1000 training epochs allow for a satisfactory solution of the Reynolds equation.

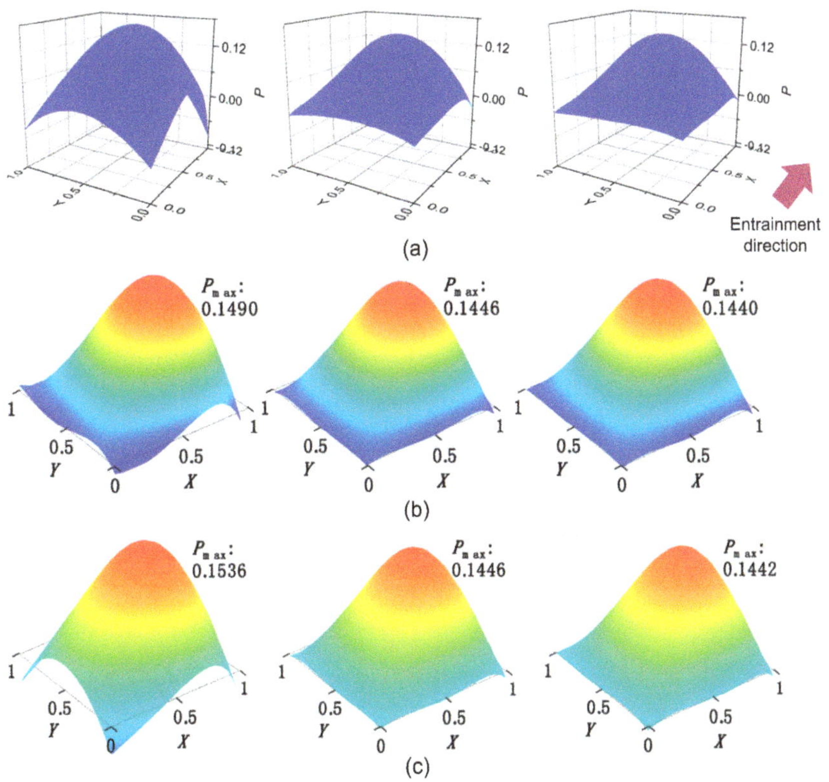

Figure 5. Pressure distribution (**a**) after 100 training epochs (**left**), 500 (**middle**) and 1000 (**right**) training epochs, (**b**) after training without hidden layers (**left**), with one hidden layer (**middle**), and with two (**right**) hidden layers (16 neurons each) as well as (**c**) after training with 4 (**left**), 16 (**middle**), and 32 neurons in one hidden layer. Reprinted and adapted from [66] with permission from CC BY 4.0.

Moreover, Zhao et al. [66] integrated the PINN into an iterative solution process (Figure 6a) for the pressure and film thickness distribution, thus balancing an externally applied load W:

$$\int_{\Omega} p(x,y)dxdy = \frac{Wh_0^2}{\eta UL^2B}, \quad (7)$$

whereby η is the lubricant viscosity and u the sliding velocity. Zhao et al. further verified this developed iterative PINN approach against the results obtained using the finite element method (FEM) as well as the experimentally measured values obtained by means of optical interferometry in a slider-on-disk setup (see Figure 6b). Generally, an excellent agreement was observed. Even though the pressure at the boundaries did not strictly meet the zero-pressure condition in the case of the PINN (deviations up to 3.4%), an excellent correlation between the PINN and FEM prevailed in the majority of the domain (Figure 6c), which was manifested in an overall error of 1.5% between the two.

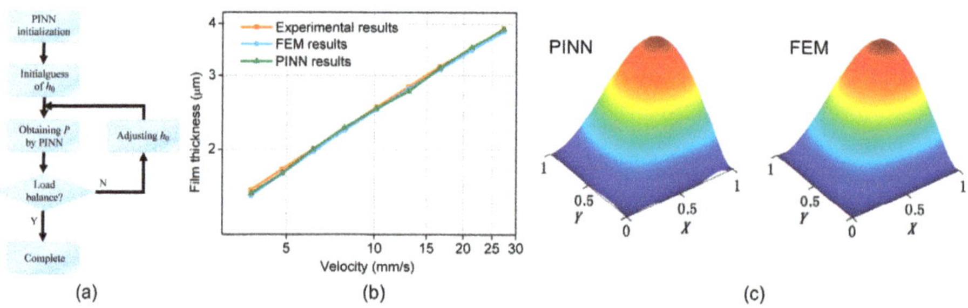

Figure 6. (**a**) Flow chart of the iterative PINN approach for hydrodynamic contact. (**b**) Outlet film thickness at different sliding velocities for the PINN method compared to FEM simulation as well as experimental results. (**c**) Pressure distribution predicted using the PINN (**left**) and the FEM (**right**). Reprinted and adapted from [66] with permission from CC BY 4.0.

Li et al. [68] employed a PINN to solve the Reynolds equation to predict the pressure field and film thickness of a gas-lubricated journal bearing (assuming incompressibility) in order to subsequently calculate the aerodynamic characteristics under variable eccentricity ratio conditions (see Figure 7a,b). The authors compared the results with an FDM solution and reported that the PINN could capture the flow field structure quite well (Figure 7c,d). Thereby, the convergence accuracy was reported to be improved by changing the weight values of different loss items as well as by employing a second-order optimizer to fine-tune the results. Moreover, the authors performed a comprehensive comparison (Figure 7e,f) among three different learning strategies (unsupervised and supervised learning driven by data from FDM, semi-supervised learning with sufficient data, and semi-supervised learning with a small number of noisy data) with respect to the prediction accuracy, i.e., the difference between the predicted results and true physics, and the physics interpretability, which describes the degree to which the results meet the physical equations. It was observed that the data-driven supervised learning method had the best prediction accuracy without a sharp loss increase in the boundary cases, followed by semi-supervised learning, and finally, unsupervised learning. In turn, the supervised learning method did not meet the Reynolds equation and had no interpretability, while the unsupervised and semi-supervised methods satisfied the physics conservation equation with small losses. However, the accuracy of the semi-supervised approach tended to be reduced with noisier data, but not the interpretability. Li et al. [68] concluded that the learning method generally should be chosen based upon the prediction accuracy requirement for the actual application as well as the amount of available data. In situations where there is a lack of experimental or high-precision numerical solution data, the unsupervised learning approach offers a direct solution to approximate the prediction value of the flow field. Thus, it becomes possible to obtain an estimation without relying on specific data or prior knowledge. However, when there is a limited amount of data available, the semi-supervised learning method can be employed to achieve more accurate prediction outcomes. This considers both solution accuracy and physics interpretability, leading to improved results and eliminating the need for simulations in each individual case, which is typically required by conventional numerical methods. In contrast, when complete field physics values are directly provided, the data-driven method can accurately predict the flow field for unknown conditions without possessing physical interpretability.

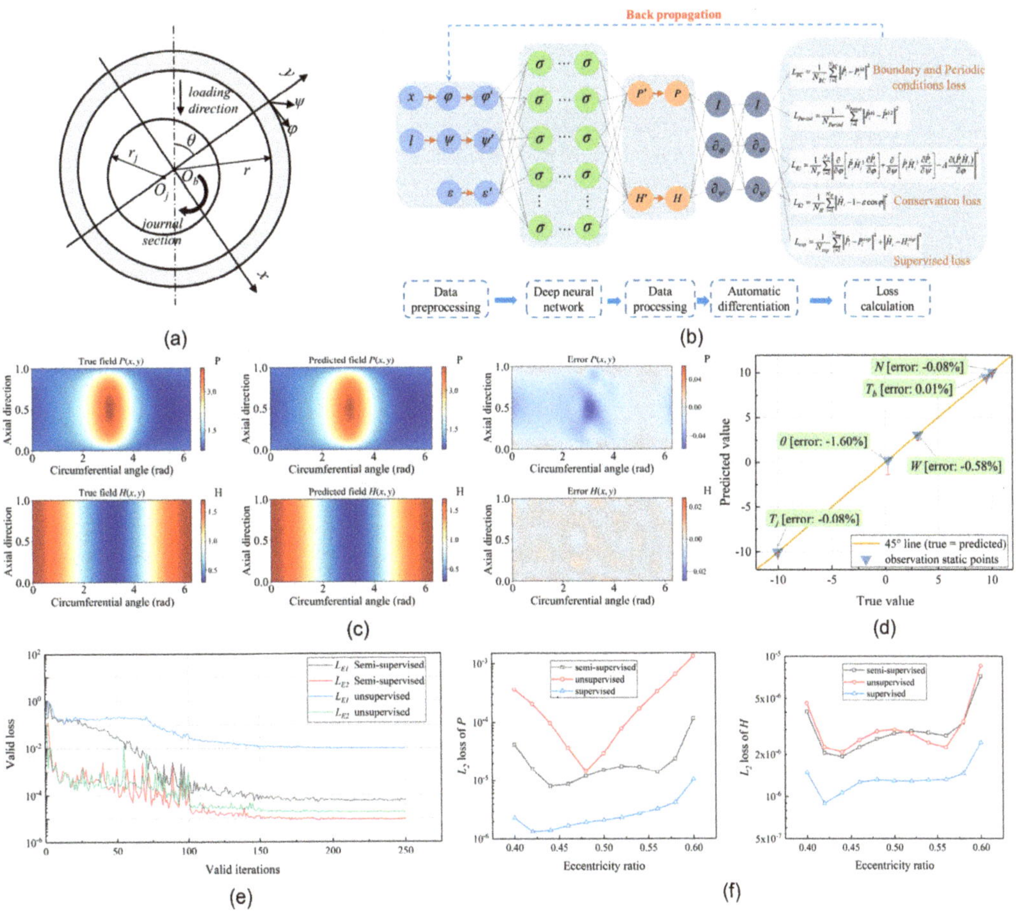

Figure 7. (**a**) Structure of a gas-lubricated journal bearing. (**b**) PINN topology to solve the Reynolds equation. (**c**) Comparison of flow field and (**d**) aerodynamic characteristics between PINN (prediction) and FDM (true). (**e**) Loss function curves against testing data as well as (**f**) L_2 loss comparison for pressure and film thickness at different eccentricities for semi-supervised, unsupervised, and supervised learning methods. Reprinted and adapted from [68] with permission.

Yadav and Thakre [69] also employed a PINN to study the behavior of a fluid-lubricated journal as well as a two-lobe bearing and compared the obtained results against an FEM model. Even though the authors provided few insights and details on the employed model and its implementation, they reported a quite good correlation between the PINN and FEM at various load cases, with errors below 6% and 5% with respect to the predicted eccentricity and friction coefficient.

Xi et al. [70] investigated the application of PINNs to predict the pressure distribution of a finite journal bearing and compared the results when employing soft or hard constraints for the boundary conditions (see Figures 8a and 8b, respectively). The models were implemented in the Python library, DeepXDE, whereby the ANN consisted of three hidden layers with 20 neurons each, and tanh was used as the activation function. The PINN was trained to minimize the loss function using the Gradient Descent Method, and the Adam optimizer was used to obtain the weights. The Dirichlet boundary condition was employed for the Reynolds equation in the case of the soft constraint (Figure 8a). Furthermore, the authors converted the boundary condition into a hard one (Figure 8b) by modifying the neural network, in which the boundary condition could be satisfied. Also, the boundary condition was no longer part of the loss function. Thus, the hard constraint met the pressure boundary condition in a mathematically exact manner and sped up the convergence. The authors compared the developed approaches as well as the FDM results when assuming both constant and variable (temperature-dependent) viscosity, whereby a good agreement was reported.

In the aforementioned studies, the cavitation effects were neglected since they reduce the complexity. Rom [71] extended the idea of using PINNs for lubrication prediction towards the consideration of cavitation by introducing the fractional film content θ to the Reynolds equation,

$$\frac{\partial}{\partial x}\left(H^3\frac{\partial p}{\partial x}\right)+\frac{\partial}{\partial y}\left(H^3\frac{\partial p}{\partial y}\right)=6\eta u\,\frac{\partial H\theta}{\partial x} \tag{8}$$

which was solved with the following underlying constraints:

$$p\geq 0,\ 0\leq\theta\leq 1,\ p(1-\theta)=0. \tag{9}$$

This means that the computational domain was split into two sub-domains, i.e., the full film region with the conventional Reynolds equation ($p > 0$, $\theta = 1$) and the cavitated region ($p = 0$, $\theta < 1$). A priori, the boundary in between the two regions is free and unknown, which makes it complex for conventional algorithms. In turn, strictly dividing both domains is not necessary for PINNs when covered by suitable boundary conditions. Rom [71] specified these problem-/application-specific conditions for the example of journal bearings (see Figure 8a). The author first employed a residual neural network (ResNet) (see Figure 8b), and training was conducted to minimize the error with respect to the mentioned boundary conditions as well as the residual (Reynolds equation divided by H), which was derived via AD. Moreover, the approach was extended to not only develop a PINN for one specific problem (fixed set of parameters), but to account for variable parameters; in this case, the variable eccentricity was the parameter, which was also propagated as the input parameter through the ResNet (extended PINN) (see Figure 8c). This led to a certain generalizability of the model. The loss function consisted of three losses related to the predictions of p and θ on the boundaries as well as three global losses. The neural network parameters were initialized via Glorot initialization and then optimized using a limited-memory Broyden–Fletcher–Goldfarb–Shanno (L-BFGS) algorithm. Tanh was chosen as the activation function and for the output layer. While the fractional film content was between zero and one, this required scaling of the input variables as well as re-scaling of the pressure with an arbitrary chosen upper boundary to obtain dimensional results. Since abrupt jumps and the fractional film content can complicate training, Rom [71] proposed to adaptively add collocation points during the training, i.e., refining the region around the maximum pressure and the boundary between the pressure and cavitation region. The author compared the obtained results for the standard and extended PINNs against the FDM solutions and found a pretty good agreement (see Figure 8d–g). Using a total of 20 neurons in six hidden layers proved to achieve the best results. The errors in between the prediction for the maximum pressure, load carrying capacity, and frictional force at different eccentricities were below 1.6%, 0.3%, and 0.2%, respectively, thus verifying certain generalizability (Figure 8g). However, some minor differences were observed, especially at the transition from the pressure to cavitated region (Figure 8d,e), which were attributed to the high resolution of the FDM region, while the PINN encountered difficulties with the jump in the fractional film content.

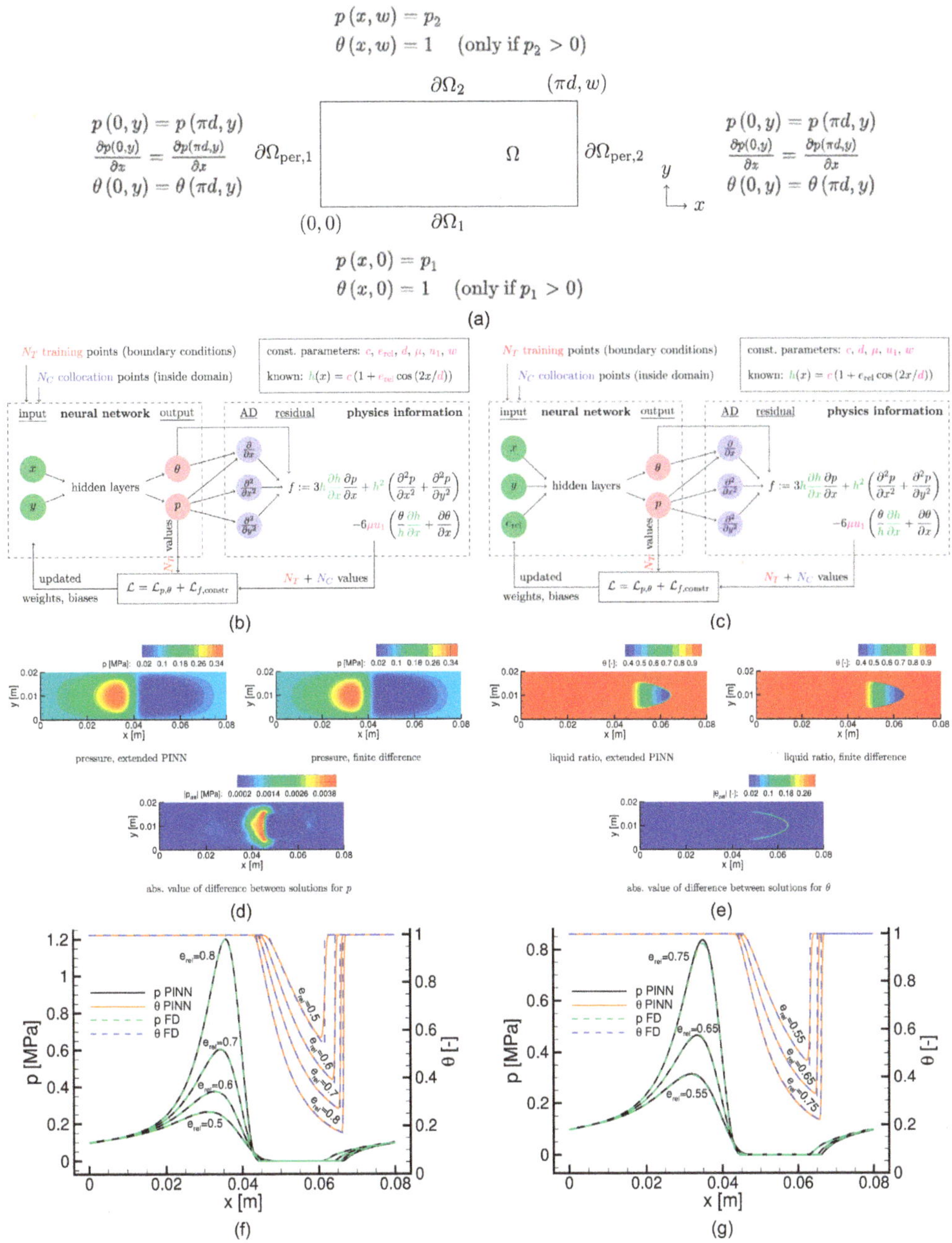

Figure 8. (**a**) Cartesian domain for a journal bearing with respective boundary conditions. (**b**) Standard and (**c**) extended PINN architecture used to solve the Reynolds equation with respective boundary conditions to consider cavitation. Comparison and error between extended PINN and FDM with respect to the (**d**) pressure and (**e**) the fractional film content. Pressure and fractional film content along the contact length for (**f**) the training values of the eccentricity and (**g**) eccentricity values not employed for training. Reprinted and adapted from [71] with permission.

To overcome the manual or computationally expensive initial value threshold selection as well as the weight adjustment/optimization of Rom's approach, Cheng et al. [72] very recently presented a PINN framework for computing the flow field of hydrodynamic lubrication by solving the Reynolds equation while involving cavitation effects by means of the Swift–Stieber model [73,74] as well as the Jakobsson–Floberg–Olsson (JFO) [75,76] model. The authors introduced a penalizing scheme with a residual of non-negativity and an imposing scheme with a continuous differentiable non-negative function to satisfy the non-negativity constraint of the Swift–Stieber approach. To address the complementarity constraint inherent to the JFO theory, the pressure and cavitation fractions were considered as the outputs of the neural network, and the Fischer–Burmeister (FB) equation's residual enforced their complementary relationship. Chen et al. then employed multi-task learning (MTL) techniques (dynamic weight, uncertainty weight, and projecting conflicting gradient method) to strike a balance between optimizing the functions and satisfying the constraints. This was shown to be superior to traditional penalizing schemes. To finally assess the accuracy of their approach, the authors studied the setup of an oil-lubricated 3D journal bearing at a fixed eccentricity with Dirichlet boundary conditions, showing very low errors compared to the respective FEM models.

3.2. Wear and Damage Prediction

Apart from predicting the lubrication phenomena in hydrodynamically or aerodynamically lubricated contacts, PIML has been employed for wear prediction. Haviez et al. [77] suggested the use of a semi-physical neural network when addressing fretting wear and facing scarce datasets due to testing costs and efforts, thus overcoming the drawbacks of purely data-driven ML. To this end, the authors experimentally generated 53 datasets using a fretting wear tester. The two-step semi-PINN was trained without backpropagation or any regularization method simply by introducing (approximate) physical considerations about energy dissipation,

$$\left(\frac{E_\mathrm{d}}{\mu}\right) = \alpha_0 N^{\alpha_1} \delta^{\alpha_2} F^{\alpha_3} \tag{10}$$

and asperity contact to estimate the wear volume,

$$V = \alpha \left(\frac{E_\mathrm{d}}{\mu}\right)^{\beta} \tag{11}$$

according to Archard's law, whereby μ is the coefficient of friction, N is the number of fretting cycles, δ is the sliding amplitude, F is the normal force, and α, α_1, α_2, α_3, α, and β are the fitting parameters to be adjusted according to the input–output relations obtained from the experiments (see Figure 9). Following linearization by taking the logarithmic approach, a single-layer ANN with an exponential activation function and a simple least squares approximation were used to determine the unknown parameters. Despite its simplicity, the authors reported a good generalizability of the suggested approach in terms of the relative quadratic error (RQE) on the new testing data, outperforming conventional ANNs when trained with small data, which might feature overfitting. Yet, it should be considered that fitting an ANN to rather simple analytical functions might be an unnecessary complication compared to other regression methods.

Yucesan and Viana [78] suggested a hybrid PIML approach consisting of a recurrent neural network to develop a cumulative damage model to predict the fatigue of wind turbine main bearings. Thereby, the physics-informed layers were used to model the comparatively well-understood physics, i.e., the bearing lifetime, while a data-driven layer accounted for the aspects, which have so far been beyond the scope of physical modeling, i.e., grease degradation (see Figure 10). The reason was because the input conditions, such as the loads and temperatures, are fully observed over the entire time series, while grease conditions are typically only partially observed at distinct inspection intervals. The model takes the bearing fatigue damage increment.

$$\Delta d_t^{BRG} = \frac{n_t}{\sum \frac{\frac{1}{60 N_i t_i}}{a_1 a_{SKF} \left(\frac{C}{P}\right)^{\frac{10}{3}}}}, \tag{12}$$

where the number of passed cycles is n_t, the total operational hours is t_i, the velocity is N_i, the basic dynamic load rating is C, the equivalent dynamic bearing load is P, and the reliability and life modification factors are a_1 and a_{SKF}. In contrast, the grease damage increment $\Delta d_t{}^{GRS}$, i.e., the degradation of viscosity and increasing contamination, was implemented via a multilayer perceptron. The recurrent neural network then took the wind speed WS_t (mapped to equivalent bearing loads) and bearing temperature T as inputs, thus updating the respective parameters and calculating the cumulative wear. The authors employed their approach to several load cases from real wind turbine data (10 min average operational and monthly grease inspection data for 14 turbines) and demonstrated that the general trends regarding bearing damage and grease degradation could be covered fairly well. Thereby, it was shown that the selection of the initialization of the weights of multilayer perceptron is crucial, and that a set of initial weights that is far away from optimum would not lead to accurate predictions. However, this can be improved by "engineering judgement-based weight initialization" [78], i.e., by performing a sensitivity analysis on the general influence trends of the inputs, thus selecting favorable initial weights.

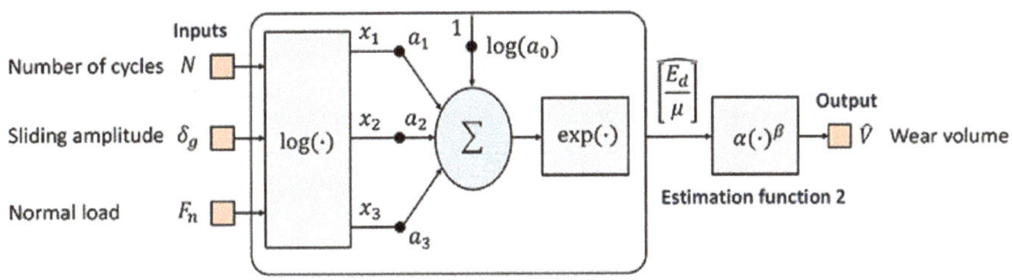

Figure 9. Semi-PINN two-level structure used to predict fretting wear. Reprinted from [31] with permission from CC BY 4.0.

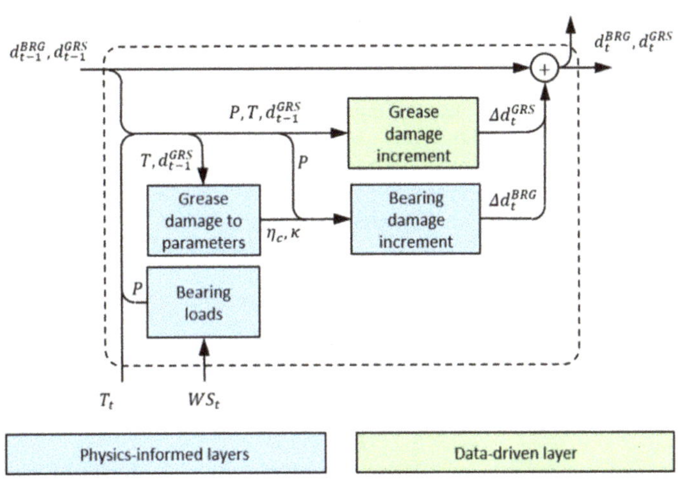

Figure 10. Hybrid PINN for main bearing fatigue and grease degradation. Reprinted from [78] with permission from CC BY 3.0.

Similarly, Shen et al. [79] proposed an approach for bearing fault detection that integrates principles of physics with deep learning methodologies. The approach consisted of two integral components: a straightforward threshold model and a convolutional neural network (CNN). The threshold model initiated the assessment of bearing health statuses by applying established physics principles associated with bearing faults. By following this initial evaluation, the CNN autonomously extracted significant high-level features from the input data, effectively utilizing these features to predict the bearing's health class. To facilitate the incorporation of physics-based knowledge into the deep learning model, the authors developed a loss function that selectively enhanced the influence of the physics-based insights assimilated by the threshold model when embedding this knowledge into the CNN model. To validate the efficacy of their approach, Shen et al. conducted experiments using two distinct datasets. The first dataset comprised data collected from 18 bearings operating in the field of an agricultural machine, while the second dataset contained data from bearings subjected to testing in the laboratory at the Case Western Reserve University (CWRU) Bearing Data Center.

Ni et al. [80] recently presented a physics-informed framework for rolling bearing diagnostics, whereby data were collected from a test rig under varying operating conditions, such as different speeds and loads. The primary difficulties were extracting robust physical information under these diverse conditions and integrating it into the network's architecture. To this end, a first layer was created using cepstrum exponential filtering, emphasizing the modal properties in the signal. The modal properties, being linked to the system characteristics rather than specific operating conditions, offered robustness to varying conditions. The layer served to establish a network that can operate effectively across diverse operating scenarios, including transitions from healthy to faulty states or changes in fault locations. Another layer based on computed order tracking (COT) converted time domain signals into angle domain signals, removing the influence of rotational speed variations and allowing for the extraction of distinctive bearing fault features under conditions of variable or time-varying speeds. Following the initial layers, a parallel bi-channel Physics-Informed Residual Network (PIResNet) architecture was implemented. The processing in the one channel was initiated with the domain conversion layer, followed by the inclusion of a wide kernel CNN layer for the purpose of mitigating high-frequency noise. Subsequently, two residual building blocks (RBBs) and max pooling layers were sequentially introduced. In contrast, the other channel commenced with a modal-property-dominant-generated layer aimed at enhancing the modal properties that were closely tied to the intrinsic characteristics of the system, making them less susceptible to changes in the operating conditions. The remainder of this channel mirrored the configuration of the other with the objective of automatically extracting complex high-dimensional features from the modal-property-dominant signal. Upon completing their respective processes, both channels were flattened and combined. Following this fusion, the fully connected and softmax layers were used for the purpose of classification. The effectiveness of this approach was verified through experiments involving bearings operating under varying speeds, loads, and time-varying speed conditions. Comprehensive comparisons confirmed the excellent performance of the PIResNet in terms of high accuracy, adaptability to different load and speed scenarios, and resilience to noise.

Li et al. [81] presented a PIML framework to predict machining tool wear under varying tool wear rates, consisting of the three modules of piecewise fitting, a hybrid physics-informed data-driven model, and automatic learning (meta-learning) (see Figure 11a). Initially, a piecewise fitting strategy was adopted to estimate the empirical equation parameters and to calculate the tool wear rate in initial, normal, and severe wear states. Subsequently, the physics-informed data-driven (PIDD) model inputs were determined using the parameters derived from the piecewise fitting approach. Utilizing a cross physics–data fusion strategy, i.e., fusing the data and the physical domain, these inputs, along with the local features, were then mapped to the tool wear rate space, thus creating the physics-informed model. Finally, meta-learning was employed to acquire an understanding of the dependable correlations between the tool wear rate and force throughout the tool's lifespan. To enhance interpretability and maintain the physical consistency of the PIML model, a physics-informed loss term was formulated, which served to improve the interpretability of the meta-learning process while ensuring that the PIML model adhered to the governing fundamental physical principles. The authors compared the developed approach for multiple sensory data (vibration, acoustic emission, etc.) and the tool flank wear observations from conducted cutting experiments with various deep learning and conventional machine learning models. Thereby, the proposed PIML framework could relatively accurately predict the tool wear trends and featured a substantially higher accuracy than a bi-directional backward gated recurrent unit (Bi-GRU) neural network, a CNN, long short-term memory (LSTM), and support vector regression (SVR) (see Figure 11b).

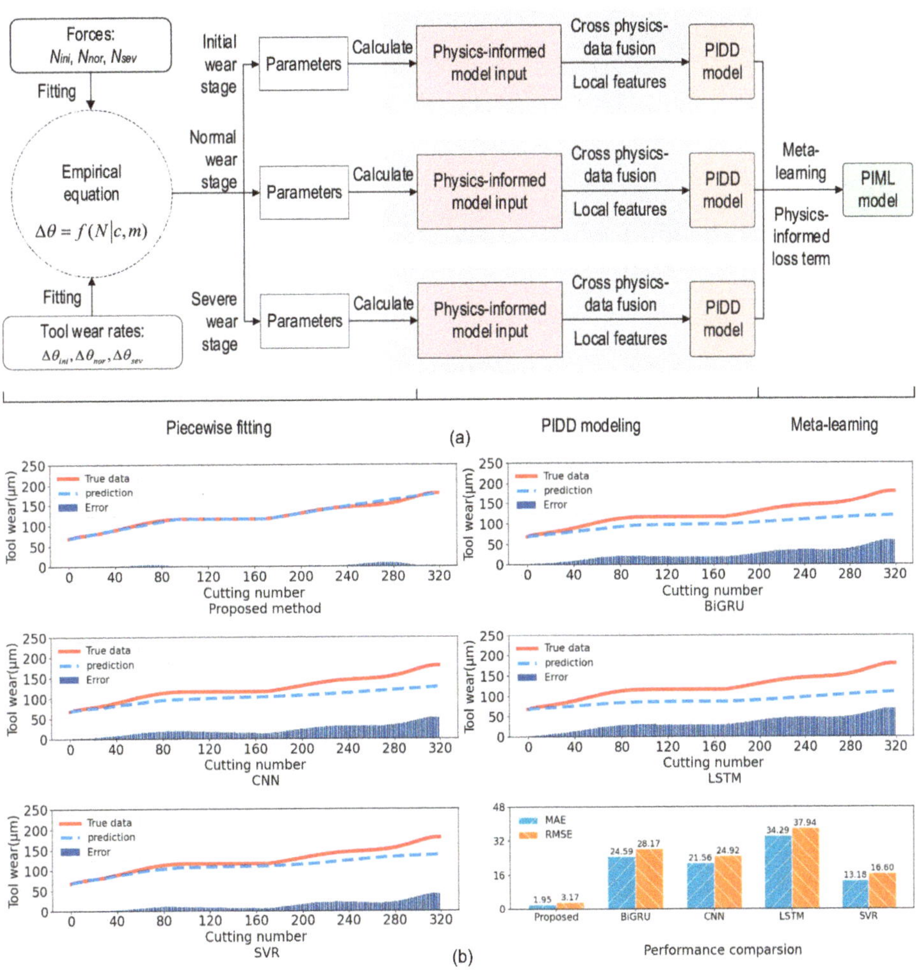

Figure 11. (**a**) Proposed PIML framework and (**b**) predicted tool wear in x-direction of the proposed model compared with various ML approaches. Reprinted and adapted from [81] with permission.

4. Concluding Remarks

To sum up, PIML has gained significant attention in various scientific fields and is now emerging in the area of tribology. By integrating physics-based knowledge into ML, PIML offers potential for understanding and optimizing tribological phenomena, overcoming the drawbacks of traditional ML approaches that rely solely on data-driven techniques. As discussed within Section 3 and summarized in Table 1, PIML can be applied to various tribological tasks.

Table 1. Overview of PIML approaches reported in the literature with their fields of application.

Field of Application	PIML Approach	Year	Reference
Lubrication prediction	Using PINN to solve the 1D Reynolds BVP to predict the pressure distribution in a fluid-lubricated linear converging slider	2021	[64]
	Using PINN to solve the 2D Reynolds equation to predict the pressure and film thickness distribution considering load balance in a fluid-lubricated linear converging slider	2023	[66]
	Using supervised, semi-supervised, and unsupervised PINN to solve the 2D Reynolds equation to predict the pressure and film thickness distribution considering load balance and eccentricity in a gas-lubricated journal bearing	2022	[68]
	Using PINN to solve the 2D Reynolds equation to predict the behavior of fluid-lubricated journal as well as two-lobe bearings	2023	[69]
	Using PINN with soft and hard constraints to solve the 2D Reynolds equation to predict the pressure distribution in fluid-lubricated journal bearings at fixed eccentricity with constant and variable viscosity	2023	[70]
	Using PINN to solve the 2D Reynolds equation to predict the pressure and fractional film content distribution in fluid-lubricated journal bearings at fixed and variable eccentricity considering cavitation	2023	[71]
	Using PINN to solve the 2D Reynolds equation to predict the pressure and fractional film content distribution in fluid-lubricated journal bearings at fixed eccentricity considering cavitation	2023	[72]
Wear and damage prediction	Using semi PINN to find regression fitting parameters for Archard's wear law based upon small data from fretting wear experiments	2015	[77]
	Using hybrid PINN to predict wind turbine bearing fatigue based upon a physics-informed bearing damage model as well as data-driven grease degradation approach	2020	[78]
	Using physics-informed CNN with preceding threshold model for rolling bearing fault detection	2021	[79]
	Using physics-informed residual network for rolling bearing fault detection	2023	[80]
	Using PIML framework consisting of piecewise fitting, a hybrid physics-informed data-driven model, and meta-learning to predict tool wear	2022	[81]

As such, PINNs have been employed for **lubrication prediction** by solving the Reynolds differential equation. Starting with the 1D Reynolds equation for a converging slider, in only two years, the complexity has already been tremendously increased, now covering the 2D Reynolds equation, journal bearings with load balance and variable eccentricity, and cavitation effects. A common limitation of PINNs is that a low loss in terms of the residual of the partial differential equation does not necessarily indicate a small prediction error. Therefore, in the future, it will be crucial to gain experience with these novel techniques to find the most effective algorithms, configurations, and hyperparameters. Future work should also be directed towards expanding the PINN's capabilities by replacing the Reynolds equation with formulations that consider nonstationary flow behavior, lubricant compressibility, or shear-thinning fluids, thus addressing a wider range of application scenarios and obtaining more accurate solutions in various lubrication contexts. Moreover, further input parameters should be incorporated into the Reynolds or film thickness equation. After training, which undoubtedly would be more complex and time-consuming, this would ultimately allow for extensive parameter studies to be conducted for optimization tasks, e.g., of textured surfaces [82], and facilitate faster computation, making it promising for solving elastohydrodynamic problems where the pressure and film thickness need to be computed repeatedly in an iterative procedure until convergence is achieved [71]. Thereby, the computational efficiency and overall accuracy might further be improved by parallel neural networks and extreme learning machines [83,84] as well as advanced adaptive methods, e.g., residual point sampling [85].

With regard to **wear and damage prediction**, semi or hybrid PIML approaches have been employed so far, combining empirical laws and equations with experimentally obtained data. Since testing costs and efforts are generally high or data are simply scarce, these approaches tend to feature advantages compared to purely data-driven ML methods in terms of the prediction accuracy. Since wear processes are inherently strongly statistical and underly scatter, future work might incorporate the Bayesian approach within PIML for uncertainty consideration and quantification. Thereby, a prior distribution is augmented over the model parameters, representing the initial belief about their values. By combining this prior distribution with the observed data, a posterior distribution is obtained, representing the updated beliefs about the parameters given the data. This would ultimately favor the handling of limited and noisy data as well as the ability to quantify uncertainty, providing valuable insights into the reliability of predictions. Furthermore, models used with the aim of predicting damage in real-word tribo-technical systems have so far mainly focused on rolling bearings. Future research should seek to explore the applicability of PIML to other mechanical systems like gears. Such investigations could broaden the scope of the employed method's use towards vibration-based gear and surface wear propagation monitoring.

Author Contributions: Conceptualization, M.M. and S.T.; methodology, formal analysis, and writing—original draft preparation, M.M.; writing—review and editing, S.T.; visualization, M.M. All authors have read and agreed to the published version of the manuscript.

Funding: This work was partially supported by the European Regional Development Fund in Bavaria under the Gate2HPC project.

Data Availability Statement: Not applicable.

Acknowledgments: M.M. greatly acknowledges the financial support from the Vicerrectoría Académica (VRA) of the Pontificia Universidad Católica de Chile within the Programa de Inserción Académica (PIA). S.T. kindly acknowledges the continuous support of the University of Bayreuth, Germany.

Conflicts of Interest: The authors declare no conflict of interest.

References

1. Marian, M.; Tremmel, S. Current Trends and Applications of Machine Learning in Tribology—A Review. *Lubricants* **2021**, *9*, 86. [CrossRef]
2. Rosenkranz, A.; Marian, M.; Profito, F.J.; Aragon, N.; Shah, R. The Use of Artificial Intelligence in Tribology—A Perspective. *Lubricants* **2021**, *9*, 2. [CrossRef]
3. Bell, J. *Machine Learning: Hands-On for Developers and Technical Professionals*; Wiley: Hoboken, NJ, USA, 2014; ISBN 978-1-118-88906-0.
4. Breiman, L. Random Forests. *Mach. Learn.* **2001**, *45*, 5–32. [CrossRef]
5. Schölkopf, B.; Smola, A.J. *Learning with Kernels: Support Vector Machines, Regularization, Optimization, and Beyond*; MIT Press: Cambridge, MA, USA, 2002; ISBN 978-0-262-19475-4.
6. Sarkar, D.; Bali, R.; Sharma, T. *Practical Machine Learning with Python: A Problem-Solver's Guide to Building Real-World Intelligent Systems*; Apress: Berkeley, CA, USA, 2017; ISBN 978-1-4842-3206-4.
7. Kruse, R.; Borgelt, C.; Braune, C.; Klawonn, F.; Moewes, C.; Steinbrecher, M. *Computational Intelligence: Eine Methodische Einführung in Künstliche Neuronale Netze, Evolutionäre Algorithmen, Fuzzy-Systeme und Bayes-Netze*, 2nd ed.; Überarbeitete und Erweiterte Auflage; Springer Vieweg: Wiesbaden, Germany, 2015; ISBN 978-3-658-10904-2.
8. Gerschütz, B.; Sauer, C.; Wallisch, A.; Mehlstäubl, J.; Kormann, A.; Schleich, B.; Alber-Laukant, B.; Paetzold, K.; Rieg, F.; Wartzack, S. Towards Customized Digital Engineering: Challenges and potentials of adapting digital engineering methods for the product development process. In *Stuttgarter Symposium für Produktentwicklung SSP 2021*; Fraunhofer IAO, Ed.; Fraunhofer IAO: Stuttgart, Germany, 2021; pp. 93–104.
9. Kurt, H.I.; Oduncuoglu, M. Application of a Neural Network Model for Prediction of Wear Properties of Ultrahigh Molecular Weight Polyethylene Composites. *Int. J. Polym. Sci.* **2015**, *2015*, 315710. [CrossRef]

10. Vinoth, A.; Datta, S. Design of the ultrahigh molecular weight polyethylene composites with multiple nanoparticles: An artificial intelligence approach. *J. Compos. Mater.* **2020**, *54*, 179–192. [CrossRef]
11. Hasan, M.S.; Kordijazi, A.; Rohatgi, P.K.; Nosonovsky, M. Triboinformatics Approach for Friction and Wear Prediction of Al-Graphite Composites Using Machine Learning Methods. *J. Tribol. Trans. ASME* **2022**, *144*, 011701. [CrossRef]
12. Hasan, M.S.; Kordijazi, A.; Rohatgi, P.K.; Nosonovsky, M. Triboinformatic modeling of dry friction and wear of aluminum base alloys using machine learning algorithms. *Tribol. Int.* **2021**, *161*, 107065. [CrossRef]
13. Kanai, R.A.; Desavale, R.G.; Chavan, S.P. Experimental-Based Fault Diagnosis of Rolling Bearings Using Artificial Neural Network. *J. Tribol. Trans. ASME* **2016**, *138*, 031103. [CrossRef]
14. Prost, J.; Cihak-Bayr, U.; Neacșu, I.A.; Grundtner, R.; Pirker, F.; Vorlaufer, G. Semi-Supervised Classification of the State of Operation in Self-Lubricating Journal Bearings Using a Random Forest Classifier. *Lubricants* **2021**, *9*, 50. [CrossRef]
15. Argatov, I.; Jin, X. Time-delay neural network modeling of the running-in wear process. *Tribol. Int.* **2023**, *178*, 108021. [CrossRef]
16. Marian, M.; Grützmacher, P.; Rosenkranz, A.; Tremmel, S.; Mücklich, F.; Wartzack, S. Designing surface textures for EHL point-contacts—Transient 3D simulations, meta-modeling and experimental validation. *Tribol. Int.* **2019**, *137*, 152–163. [CrossRef]
17. Dai, K.; Gao, X. Estimating antiwear properties of lubricant additives using a quantitative structure tribo-ability relationship model with back propagation neural network. *Wear* **2013**, *306*, 242–247. [CrossRef]
18. Bhaumik, S.; Pathak, S.D.; Dey, S.; Datta, S. Artificial intelligence based design of multiple friction modifiers dispersed castor oil and evaluating its tribological properties. *Tribol. Int.* **2019**, *140*, 105813. [CrossRef]
19. Padhi, P.K.; Satapathy, A. Analysis of Sliding Wear Characteristics of BFS Filled Composites Using an Experimental Design Approach Integrated with ANN. *Tribol. Trans.* **2013**, *56*, 789–796. [CrossRef]
20. Gangwar, S.; Pathak, V.K. Dry sliding wear characteristics evaluation and prediction of vacuum casted marble dust (MD) reinforced ZA-27 alloy composites using hybrid improved bat algorithm and ANN. *Mater. Today Commun.* **2020**, *25*, 101615. [CrossRef]
21. Sahraoui, T.; Guessasma, S.; Fenineche, N.E.; Montavon, G.; Coddet, C. Friction and wear behaviour prediction of HVOF coatings and electroplated hard chromium using neural computation. *Mater. Lett.* **2004**, *58*, 654–660. [CrossRef]
22. Boidi, G.; Rodrigues da Silva, M.; Profito, F.J.J.; Machado, I.F. Using Machine Learning Radial Basis Function (RBF) Method for Predicting Lubricated Friction on Textured and Porous Surfaces. *Surf. Topogr. Metrol. Prop.* **2020**, *8*, 044002. [CrossRef]
23. Gupta, S.K.; Pandey, K.N.; Kumar, R. Artificial intelligence-based modelling and multi-objective optimization of friction stir welding of dissimilar AA5083-O and AA6063-T6 aluminium alloys. *Proc. Inst. Mech. Eng. Part L J. Mater. Des. Appl.* **2018**, *232*, 333–342. [CrossRef]
24. Anand, K.; Shrivastava, R.; Tamilmannan, K.; Sathiya, P. A Comparative Study of Artificial Neural Network and Response Surface Methodology for Optimization of Friction Welding of Incoloy 800 H. *Acta Metall. Sin. (Engl. Lett.)* **2015**, *28*, 892–902. [CrossRef]
25. Francisco, A.; Lavie, T.; Fatu, A.; Villechaise, B. Metamodel-Assisted Optimization of Connecting Rod Big-End Bearings. *J. Tribol. Trans. ASME* **2013**, *135*, 041704. [CrossRef]
26. Zavos, A.; Katsaros, K.P.; Nikolakopoulos, P.G. Optimum Selection of Coated Piston Rings and Thrust Bearings in Mixed Lubrication for Different Lubricants Using Machine Learning. *Coatings* **2022**, *12*, 704. [CrossRef]
27. Tremmel, S.; Marian, M. Machine Learning in Tribology—More than Buzzwords? *Lubricants* **2022**, *10*, 68. [CrossRef]
28. Paturi, U.M.R.; Palakurthy, S.T.; Reddy, N.S. The Role of Machine Learning in Tribology: A Systematic Review. *Arch Comput. Methods Eng* **2023**, *30*, 1345–1397. [CrossRef]
29. Sose, A.T.; Joshi, S.Y.; Kunche, L.K.; Wang, F.; Deshmukh, S.A. A review of recent advances and applications of machine learning in tribology. *Phys. Chem. Chem. Phys.* **2023**, *25*, 4408–4443. [CrossRef]
30. Yin, N.; Xing, Z.; He, K.; Zhang, Z. Tribo-informatics approaches in tribology research: A review. *Friction* **2023**, *11*, 1–22. [CrossRef]
31. Argatov, I. Artificial Neural Networks (ANNs) as a Novel Modeling Technique in Tribology. *Front. Mech. Eng.* **2019**, *5*, 1074. [CrossRef]
32. Boidi, G.; Grützmacher, P.G.; Varga, M.; Da Rodrigues Silva, M.; Gachot, C.; Dini, D.; Profito, F.J.; Machado, I.F. Tribological Performance of Random Sinter Pores vs. Deterministic Laser Surface Textures: An Experimental and Machine Learning Approach. In *Tribology of Machine Elements-Fundamentals and Applications*; IntechOpen: London, UK, 2021. [CrossRef]
33. de La Guerra Ochoa, E.; Otero, J.E.; Tanarro, E.C.; Morgado, P.L.; Lantada, A.D.; Munoz-Guijosa, J.M.; Sanz, J.M. Optimising lubricated friction coefficient by surface texturing. *Proc. Inst. Mech. Eng. Part C J. Mech. Eng. Sci.* **2013**, *227*, 2610–2619. [CrossRef]
34. Gyurova, L.A.; Miniño-Justel, P.; Schlarb, A.K. Modeling the sliding wear and friction properties of polyphenylene sulfide composites using artificial neural networks. *Wear* **2010**, *268*, 708–714. [CrossRef]
35. Thankachan, T.; Soorya Prakash, K.; Kamarthin, M. Optimizing the Tribological Behavior of Hybrid Copper Surface Composites Using Statistical and Machine Learning Techniques. *J. Tribol. Trans. ASME* **2018**, *140*, 031610. [CrossRef]
36. Sadık Ünlü, B.; Durmuş, H.; Meriç, C. Determination of tribological properties at CuSn10 alloy journal bearings by experimental and means of artificial neural networks method. *Ind Lubr. Tribol.* **2012**, *64*, 258–264. [CrossRef]
37. Senatore, A.; D'Agostino, V.; Di Giuda, R.; Petrone, V. Experimental investigation and neural network prediction of brakes and clutch material frictional behaviour considering the sliding acceleration influence. *Tribol. Int.* **2011**, *44*, 1199–1207. [CrossRef]
38. Bhaumik, S.; Mathew, B.R.; Datta, S. Computational intelligence-based design of lubricant with vegetable oil blend and various nano friction modifiers. *Fuel* **2019**, *241*, 733–743. [CrossRef]

39. Schwarz, S.; Grillenberger, H.; Graf-Goller, O.; Bartz, M.; Tremmel, S.; Wartzack, S. Using Machine Learning Methods for Predicting Cage Performance Criteria in an Angular Contact Ball Bearing. *Lubricants* **2022**, *10*, 25. [CrossRef]
40. Marian, M.; Mursak, J.; Bartz, M.; Profito, F.J.; Rosenkranz, A.; Wartzack, S. Predicting EHL film thickness parameters by machine learning approaches. *Friction* **2022**, *11*, 992–1013. [CrossRef]
41. Walker, J.; Questa, H.; Raman, A.; Ahmed, M.; Mohammadpour, M.; Bewsher, S.R.; Offner, G. Application of Tribological Artificial Neural Networks in Machine Elements. *Tribol. Lett.* **2023**, *71*, 3. [CrossRef]
42. Hess, N.; Shang, L. Development of a Machine Learning Model for Elastohydrodynamic Pressure Prediction in Journal Bearings. *J. Tribol. Trans. ASME* **2022**, *144*, 081603. [CrossRef]
43. Garabedian, N.T.; Schreiber, P.J.; Brandt, N.; Zschumme, P.; Blatter, I.L.; Dollmann, A.; Haug, C.; Kümmel, D.; Li, Y.; Meyer, F.; et al. Generating FAIR research data in experimental tribology. *Sci. Data* **2022**, *9*, 315. [CrossRef]
44. Brandt, N.; Garabedian, N.T.; Schoof, E.; Schreiber, P.J.; Zschumme, P.; Greiner, C.; Selzer, M. Managing FAIR Tribological Data Using Kadi4Mat. *Data* **2022**, *7*, 15. [CrossRef]
45. Bagov, I.; Greiner, C.; Garabedian, N. Collaborative Metadata Definition using Controlled Vocabularies, and Ontologies. *RIO* **2022**, *8*, e94931. [CrossRef]
46. Kügler, P.; Marian, M.; Dorsch, R.; Schleich, B.; Wartzack, S. A Semantic Annotation Pipeline towards the Generation of Knowledge Graphs in Tribology. *Lubricants* **2022**, *10*, 18. [CrossRef]
47. Kügler, P.; Marian, M.; Schleich, B.; Tremmel, S.; Wartzack, S. tribAIn—Towards an Explicit Specification of Shared Tribological Understanding. *Appl. Sci.* **2020**, *10*, 4421. [CrossRef]
48. Karniadakis, G.E.; Kevrekidis, I.G.; Lu, L.; Perdikaris, P.; Wang, S.; Yang, L. Physics-informed machine learning. *Nat. Rev. Phys.* **2021**, *3*, 422–440. [CrossRef]
49. Raissi, M.; Perdikaris, P.; Karniadakis, G.E. Physics-informed neural networks: A deep learning framework for solving forward and inverse problems involving nonlinear partial differential equations. *J. Comput. Phys.* **2019**, *378*, 686–707. [CrossRef]
50. Raissi, M.; Karniadakis, G.E. Hidden physics models: Machine learning of nonlinear partial differential equations. *J. Comput. Phys.* **2018**, *357*, 125–141. [CrossRef]
51. Lagaris, I.E.; Likas, A.; Fotiadis, D.I. Artificial neural networks for solving ordinary and partial differential equations. *IEEE Trans. Neural Netw.* **1998**, *9*, 987–1000. [CrossRef]
52. Pioch, F.; Harmening, J.H.; Müller, A.M.; Peitzmann, F.-J.; Schramm, D.; el Moctar, O. Turbulence Modeling for Physics-Informed Neural Networks: Comparison of Different RANS Models for the Backward-Facing Step Flow. *Fluids* **2023**, *8*, 43. [CrossRef]
53. Almajid, M.M.; Abu-Al-Saud, M.O. Prediction of porous media fluid flow using physics informed neural networks. *J. Pet. Sci. Eng.* **2022**, *208*, 109205. [CrossRef]
54. Rudy, S.H.; Brunton, S.L.; Proctor, J.L.; Kutz, J.N. Data-driven discovery of partial differential equations. *Sci. Adv.* **2017**, *3*, e1602614. [CrossRef]
55. Chen, D.; Li, Y.; Liu, K.; Li, Y. A physics-informed neural network approach to fatigue life prediction using small quantity of samples. *Int. J. Fatigue* **2023**, *166*, 107270. [CrossRef]
56. Lee, S.; Popovics, J. Applications of physics-informed neural networks for property characterization of complex materials. *RILEM Tech. Lett.* **2022**, *7*, 178–188. [CrossRef]
57. Taç, V.; Linka, K.; Sahli-Costabal, F.; Kuhl, E.; Tepole, A.B. Benchmarking physics-informed frameworks for data-driven hyperelasticity. *Comput. Mech.* **2023**. [CrossRef]
58. Pun, G.P.P.; Batra, R.; Ramprasad, R.; Mishin, Y. Physically informed artificial neural networks for atomistic modeling of materials. *Nat. Commun.* **2019**, *10*, 2339. [CrossRef] [PubMed]
59. Zhang, Z.; Gu, G.X. Physics-informed deep learning for digital materials. *Theor. Appl. Mech. Lett.* **2021**, *11*, 100220. [CrossRef]
60. Katsikis, D.; Muradova, A.D.; Stavroulakis, G.E. A Gentle Introduction to Physics-Informed Neural Networks, with Applications in Static Rod and Beam Problems. *J. Adv. App. Comput. Math.* **2022**, *9*, 103–128. [CrossRef]
61. Moradi, S.; Duran, B.; Eftekhar Azam, S.; Mofid, M. Novel Physics-Informed Artificial Neural Network Architectures for System and Input Identification of Structural Dynamics PDEs. *Buildings* **2023**, *13*, 650. [CrossRef]
62. van Herten, R.L.M.; Chiribiri, A.; Breeuwer, M.; Veta, M.; Scannell, C.M. Physics-informed neural networks for myocardial perfusion MRI quantification. *Med. Image Anal.* **2022**, *78*, 102399. [CrossRef]
63. Sahli Costabal, F.; Yang, Y.; Perdikaris, P.; Hurtado, D.E.; Kuhl, E. Physics-Informed Neural Networks for Cardiac Activation Mapping. *Front. Phys.* **2020**, *8*, 42. [CrossRef]
64. Almqvist, A. Fundamentals of Physics-Informed Neural Networks Applied to Solve the Reynolds Boundary Value Problem. *Lubricants* **2021**, *9*, 82. [CrossRef]
65. Bach, F. Breaking the Curse of Dimensionality with Convex Neural Networks. *J. Mach. Learn. Res.* **2014**, *18*, 629–681.
66. Zhao, Y.; Guo, L.; Wong, P.P.L. Application of physics-informed neural network in the analysis of hydrodynamic lubrication. *Friction* **2023**, *11*, 1253–1264. [CrossRef]
67. Zubov, K.; McCarthy, Z.; Ma, Y.; Calisto, F.; Pagliarino, V.; Azeglio, S.; Bottero, L.; Luján, E.; Sulzer, V.; Bharambe, A.; et al. NeuralPDE: Automating Physics-Informed Neural Networks (PINNs) with Error Approximations. *arXiv* **2021**, arXiv:2107.09443.
68. Li, L.; Li, Y.; Du, Q.; Liu, T.; Xie, Y. ReF-nets: Physics-informed neural network for Reynolds equation of gas bearing. *Comput. Methods Appl. Mech. Eng.* **2022**, *391*, 114524. [CrossRef]

69. Yadav, S.K.; Thakre, G. Solution of Lubrication Problems with Deep Neural Network. In *Advances in Manufacturing Engineering*; Dikshit, M.K., Soni, A., Davim, J.P., Eds.; Springer Nature Singapore: Singapore, 2023; pp. 471–477, ISBN 978-981-19-4207-5.
70. Xi, Y.; Deng, J.; Li, Y. A solution for finite journal bearings by using physics-informed neural networks with both soft and hard constrains. *Ind Lubr. Tribol.* **2023**, *75*, 560–567. [CrossRef]
71. Rom, M. Physics-informed neural networks for the Reynolds equation with cavitation modeling. *Tribol. Int.* **2023**, *179*, 108141. [CrossRef]
72. Cheng, Y.; He, Q.; Huang, W.; Liu, Y.; Li, Y.; Li, D. HL-nets: Physics-informed neural networks for hydrodynamic lubrication with cavitation. *Tribol. Int.* **2023**, *188*, 108871. [CrossRef]
73. Swift, H.W. The Stability of Lubricating Films in Journal Bearings. *Minutes Proc. Inst. Civ. Eng.* **1932**, *233*, 267–288.
74. Stieber, W. *Hydrodynamische Theorie des Gleitlagers. Das Schwimmlager*; VDI: Berlin, Germany, 1933.
75. Jakobsson, B.; Floberg, L. *The Finite Journal Bearing, Considering Vaporization*; Gumperts: Göteborg, Sweden, 1957.
76. Olsson, K.-O. *Cavitation in Dynamically Loaded Bearings*; Gumperts: Göteborg, Sweden, 1965.
77. Haviez, L.; Toscano, R.; El Youssef, M.; Fouvry, S.; Yantio, G.; Moreau, G. Semi-physical neural network model for fretting wear estimation. *J. Intell. Fuzzy Syst.* **2015**, *28*, 1745–1753. [CrossRef]
78. Yucesan, Y.A.; Viana, F.A.C. A Physics-informed Neural Network for Wind Turbine Main Bearing Fatigue. *Int. J. Progn. Health Manag.* **2020**, *11*. [CrossRef]
79. Shen, S.; Lu, H.; Sadoughi, M.; Hu, C.; Nemani, V.; Thelen, A.; Webster, K.; Darr, M.; Sidon, J.; Kenny, S. A physics-informed deep learning approach for bearing fault detection. *Eng. Appl. Artif. Intell.* **2021**, *103*, 104295. [CrossRef]
80. Ni, Q.; Ji, J.C.; Halkon, B.; Feng, K.; Nandi, A.K. Physics-Informed Residual Network (PIResNet) for rolling element bearing fault diagnostics. *Mech. Syst. Signal Process.* **2023**, *200*, 110544. [CrossRef]
81. Li, Y.; Wang, J.; Huang, Z.; Gao, R.X. Physics-informed meta learning for machining tool wear prediction. *J. Manuf. Syst.* **2022**, *62*, 17–27. [CrossRef]
82. Marian, M.; Almqvist, A.; Rosenkranz, A.; Fillon, M. Numerical micro-texture optimization for lubricated contacts—A critical discussion. *Friction* **2022**, *10*, 1772–1809. [CrossRef]
83. Shukla, K.; Jagtap, A.D.; Karniadakis, G.E. Parallel physics-informed neural networks via domain decomposition. *J. Comput. Phys.* **2021**, *447*, 110683. [CrossRef]
84. Dwivedi, V.; Srinivasan, B. Physics Informed Extreme Learning Machine (PIELM)–A rapid method for the numerical solution of partial differential equations. *Neurocomputing* **2020**, *391*, 96–118. [CrossRef]
85. Wu, C.; Zhu, M.; Tan, Q.; Kartha, Y.; Lu, L. A comprehensive study of non-adaptive and residual-based adaptive sampling for physics-informed neural networks. *Comput. Methods Appl. Mech. Eng.* **2023**, *403*, 115671. [CrossRef]

lubricants

MDPI

Review

Improvement of Generative Adversarial Network and Its Application in Bearing Fault Diagnosis: A Review

Diwang Ruan [1], Xuran Chen [2], Clemens Gühmann [1,*] and Jianping Yan [3,*]

1 Chair of Electronic Measurement and Diagnostic Technology, TU Berlin, 10587 Berlin, Germany
2 School of Electrical Engineering and Computer Science, TU Berlin, 10587 Berlin, Germany
3 School of Aeronautics and Astronautics, Zhejiang University, Hangzhou 310027, China
* Correspondence: clemens.guehmann@tu-berlin.de (C.G.); jianping.yan@zju.edu.cn (J.Y.)

Abstract: A small sample size and unbalanced sample distribution are two main problems when data-driven methods are applied for fault diagnosis in practical engineering. Technically, sample generation and data augmentation have proven to be effective methods to solve this problem. The generative adversarial network (GAN) has been widely used in recent years as a representative generative model. Besides the general GAN, many variants have recently been reported to address its inherent problems such as mode collapse and slow convergence. In addition, many new techniques are being proposed to increase the sample generation quality. Therefore, a systematic review of GAN, especially its application in fault diagnosis, is necessary. In this paper, the theory and structure of GAN and variants such as ACGAN, VAEGAN, DCGAN, WGAN, et al. are presented first. Then, the literature on GANs is mainly categorized and analyzed from two aspects: improvements in GAN's structure and loss function. Specifically, the improvements in the structure are classified into three types: information-based, input-based, and layer-based. Regarding the modification of the loss function, it is sorted into two aspects: metric-based and regularization-based. Afterwards, the evaluation metrics of the generated samples are summarized and compared. Finally, the typical applications of GAN in the bearing fault diagnosis field are listed, and the challenges for further research are also discussed.

Keywords: generative adversarial network (GAN); bearing fault diagnosis; data augmentation; loss function modification; GAN structure improvement; GAN review

Citation: Ruan, D.; Chen, X.; Gühmann, C.; Yan, J. Improvement of Generative Adversarial Network and Its Application in Bearing Fault Diagnosis: A Review. *Lubricants* **2023**, *11*, 74. https://doi.org/10.3390/lubricants11020074

Received: 25 December 2022
Revised: 7 February 2023
Accepted: 9 February 2023
Published: 10 February 2023

1. Introduction

Rotating machinery has many applications in practical engineering, in which bearing is one of the critical components [1–3]. Since bearings usually work in an extremely harsh environment, they are prone to wear, cracks, and other defects, affecting the equipment's normal operation and even leading to huge economic losses and casualties. Therefore, detecting and diagnosing the bearing fault in time is very important.

Bearing fault diagnosis means determining the health status of the bearing based on monitoring data. Commonly used monitoring data include vibrations signal [1], temperature signal [4], current signal [5], stray flux [6], acoustic emission [7], and oil film condition [8]. Among them, the vibration signal is the most widely used in bearing fault diagnosis as it has many advantages, such as low cost, high sensitivity, good robustness, almost no response lag, and it is easy to install [1]. Traditional bearing fault diagnosis is a knowledge-driven approach in which experienced engineers use signal processing techniques to analyze vibration signals and determine the health status of bearings. Therefore, traditional methods are entirely human-dependent and challenging for online fault diagnosis. The advent of the industrial Internet has made massive data monitoring a reality. As a result, data-driven fault diagnosis methods ensue. Many researchers have successfully applied machine learning (ML) theory to bearing fault diagnosis and established diagnostic models to realize the automatic detection and identification of bearing faults. This field is also known as intelligent fault diagnosis [9]. When using traditional ML methods such as k-nearest neighbor (kNN), artificial neural network (ANN), and support vector machine (SVM) for bearing fault diagnosis, the diagnostic model can establish a link between the bearing fault characteristics and bearing health status, thereby automatically identifying the health status of the bearing by calculating the fault characteristics of the input data [10]. Technically, the traditional ML methods still require the manual extraction and selection of valid fault features from the collected data. Deep learning (DL), a branch of ML, enables automatic feature extraction from the collected data, linking the raw monitoring data directly to the health status of the bearing. Commonly used DL networks include convolutional neural network (CNN), stacked autoencoder (AE), long short-term memory (LSTM), deep belief network (DBN), and recurrent neural networks (RNNs). To date, DL has been massively studied in prognostics and health management (PHM) [11–13]. The success of the aforementioned data-driven fault diagnosis approaches is based on the premise that there are sufficient labeled data to train the diagnostic model. However, this assumption is usually unrealistic in practical engineering scenarios. For example, bearings operate under normal conditions for most of their life cycle, with a small percentage of fault conditions. Therefore, most bearing monitoring data are health data. The lack of fault data leads to two main problems. The first is the small sample problem, which refers to the small sample size of the fault data. The second is the data imbalance problem, which means the imbalanced distribution of sample size among measurement data from different bearing health states. Both of these two problems will lead to low diagnostic accuracy. Therefore, bearing fault diagnosis under small samples and imbalanced datasets is a very significant and promising research topic.

Data augmentation is an effective solution to address the small sample problem and the data imbalance problem. Commonly used bearing fault data augmentation methods are divided into oversampling techniques, data transformations, and generative models. As a generative model, GAN is one of the most popular methods for fault data augmentation. This paper will review the aforementioned fault data augmentation methods focusing on GANs. GANs are initially utilized to generate images in the field of computer vision. Liu et al. [14] first introduced GAN to bearing fault diagnosis. In recent years, many researchers improved the training technique and evaluation method of GAN to better apply it to bearing fault data augmentation. Based on our review of existing literature and our experience, we divide these improvements into three categories: improvements in the network structure, improvements in the loss function, and improvements in the evaluation of generated data.

Although there have been several review papers published related to data-driven machinery fault diagnosis, they focus on the whole artificial intelligence technology in mechanical fault diagnosis [1,9,11]. These papers cover both traditional machine learning methods and deep learning and have a wide range of study objects, including bearings, gearboxes, induction motors, and wind turbines. Furthermore, they focus on the improvements in the diagnostic model. As one of the key techniques to improve the accuracy of fault diagnostic models, data augmentation, especially data synthesis using GAN, has developed rapidly in recent years. Therefore, it is necessary to review the research in the field of bearing fault data generation, summarize the existing outcomes, and give possible prospects for future exploration.

The motivation of the study is to provide a systematic review of GAN, including theory, development, problems, and prospect. As presented in Figure 1, the rest of the review is organized. The research methodology and initial analysis are described in Section 2. Section 3 introduces three common methods for data augmentation. Section 4 focuses on the improvements and applications of GAN in the field of bearing fault diagnosis. Specifically, the improvements in GAN are categorized into structure improvement and loss function improvement. The evaluation metrics for the sample generation quality of

GAN are also discussed in this section. Finally, the conclusions and prospects are given in Section 5.

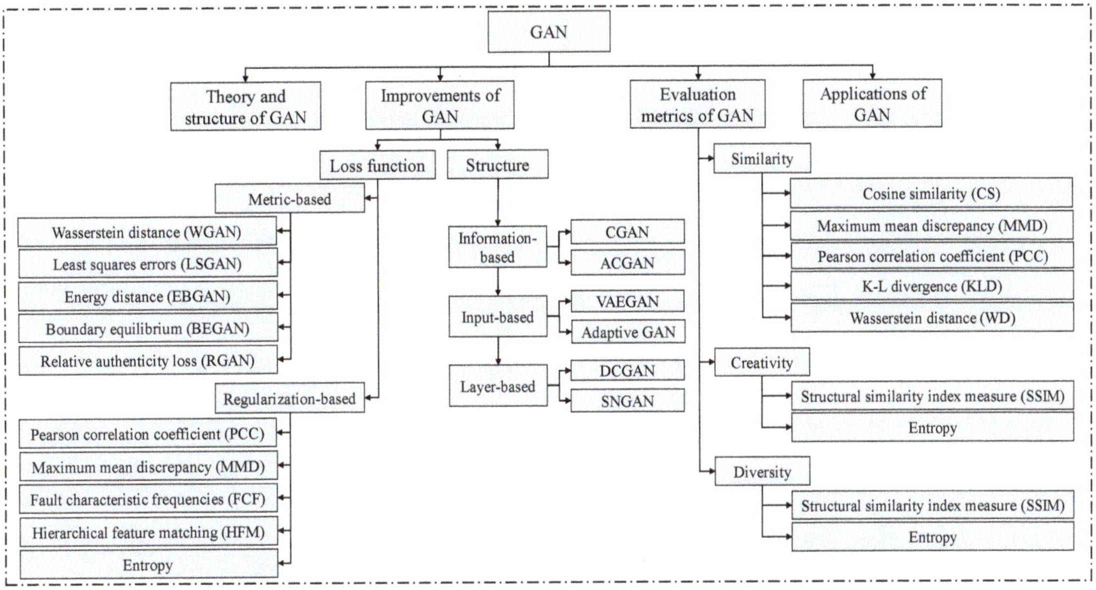

Figure 1. Structure of review analysis.

2. Research Methodology and Initial Analysis

2.1. Research Methodology

To ensure the quality of the literature, the Web of Science Core Collection database was selected for the literature search in this paper. Using the topic keywords "bearing fault diagnosis AND (data augmentation OR data synthesis OR data generation)", we initially obtained a total of 160 English journal and conference articles [15], as shown in Figure 2a. The search results include research articles published up to October 2022. To collect the literature as comprehensively as possible, the topic keywords "bearing fault diagnosis AND oversampling" and "bearing fault diagnosis AND generative adversarial network" were adopted to supplement our search results. The search results [16] of the latter are shown in Figure 2b. In addition, several relevant articles were found and included in our analysis after citation analysis. We first skimmed all the articles for the literature analysis to filter out the irrelevant ones. The remaining articles were further analyzed and categorized for study.

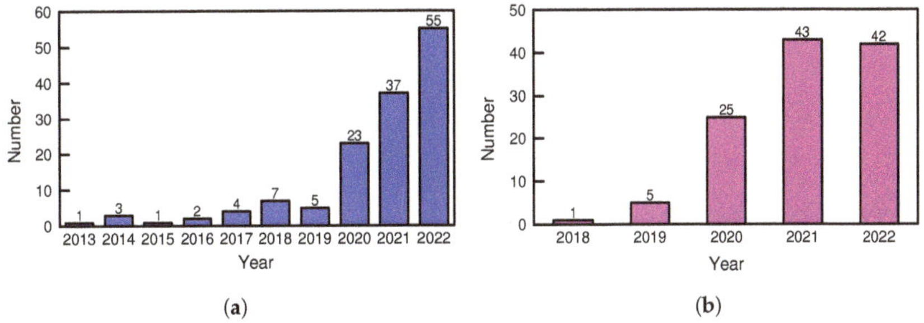

Figure 2. Publications related to data augmentation and GANs (from 2013 to 2022). (**a**) Publications related to bearing data augmentation. (**b**) Publications related to bearing fault diagnosis and GANs.

2.2. Initial Literature Analysis

Figure 2a shows that there has been an increasing number of studies on data augmentation for bearing fault diagnosis in the last decade. This reflects the fact that there is a lack of fault data in practice and the necessity to address this problem. According to Figure 2b, research on bearing fault diagnosis and GANs started in 2018 and rapidly became a research hotspot. From 2018 to 2022, the number of publications per year has grown substantially. Keyword co-occurrence analysis was performed using VOSviewer [17]. As shown in Figure 3, the initial research hotspot for GAN is the combination with CNN. At this time, the popular DCGAN was applied to bearing fault diagnosis. On the other hand, CNNs were commonly used as fault classification models. The next hot topic was the application of GAN as a data augmentation technique to generate fault data to address the small sample and imbalanced data problems, with fault classification problems being the most studied application scenario. Another preferred research direction was improving the training process for GANs. In recent years, transfer learning (TL) has become a popular research issue related to GAN.

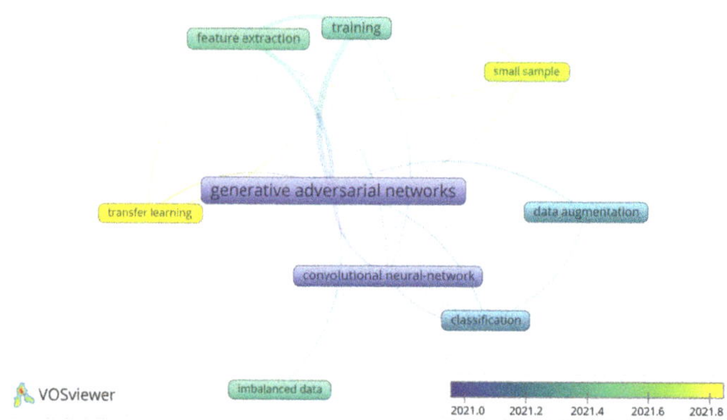

Figure 3. Results of keywords' co-occurrence analysis.

3. Data Augmentation Methods for Bearing Fault Diagnosis

Training bearing fault diagnosis models require a large amount of fault data. However, fault data are usually lacking in practical engineering. Using data augmentation techniques to generate fault data is an effective solution. Data augmentation is the process of creating new similar samples for the original dataset, which can discover the unexplored space of the input data. This helps reduce overfitting when training a machine learning or deep learning model and enhances the generalization performance. Based on our analysis of the existing literature, the data augmentation methods for bearing fault diagnosis are divided into oversampling techniques, data transformations, and GANs. In the following, the introduction of these three data augmentation methods will be presented.

3.1. Data Augmentation Using Oversampling Techniques

Oversampling is a simple and effective method for data augmentation. The most basic oversampling method is random oversampling [18], in which new samples are generated by randomly replicating the samples of the minority class. However, this method does not increase the amount of information in the dataset and may increase the risk of overfitting. To overcome this problem, Chawla et al. [19] further proposed the synthetic minority over-sampling technique (SMOTE), which generates new samples by linear interpolation between two original samples. However, this method does not consider the probability distribution of the original data. Therefore, adding generated samples to the original dataset may lead to a change in its distribution. In addition, the new dataset may not involve real fault information. Although the two above methods can generate samples of the minority class, the synthetic samples cannot provide more fault information. Consequently, they are not feasible in bearing fault diagnosis. Usually, researchers use the two methods as benchmarks to demonstrate the superiority of their new methods [20].

SMOTE is a pioneer oversampling method, based on which many new oversampling techniques have been proposed and successfully improved bearing fault diagnosis accuracy. Jian et al. [21] presented a novel sample information-based synthetic minority oversampling technique (SI-SMOTE). It evaluates the sample information based on the Mahalanobis distance, thereby identifying informative minority samples. The original SMOTE is merely utilized to generate new samples of informative minority classes. Hao et al. [22] proposed the K-means synthetic minority oversampling technique (K-means SMOTE) based on the clustering distribution. This uses the K-means algorithm to filter out target clusters. As a result, only the samples of selected clusters are synthesized.

In addition, researchers have developed other oversampling methods that have proven effective in bearing fault diagnosis. For example, Razavi-Far et al. [23] developed a novel imputation-based oversampling technique to generate new synthetic samples of the minority class. Their approach generates a set of incomplete samples representative of the minor classes and uses the expectation maximization (EM) algorithm to produce new synthetic samples of the minor classes. To overcome the problem of multi-class imbalanced fault diagnosis, Wei et al. [24] proposed the sample-characteristic oversampling technique (SCOTE). It transforms the problem into multiple binary imbalanced problems.

3.2. Data Augmentation Using Data Transformations

The data transformation methods are inspired by data augmentation techniques in computer vision, in which image transformations such as flipping and cropping are often utilized to obtain new samples to enrich the training set. For example, when using vibration signals for the intelligent fault diagnosis of bearings, there are usually two types of input data. The first one is the original vibration signals, which can be directly fed into the machine learning or deep learning model, and the model learns the features of the time series. The other one is images. The vibration signals are first converted into images. This not only enables the utilization of the feature extraction capability of the deep neural network such as CNN for images but also introduces commonly used image augmentation techniques to the field of bearing fault diagnosis.

Raw vibration signals are one-dimensional time series. To construct datasets, it is necessary first to clip time series using the overlapping segmentation method. With the length of the sample and the length of overlap defined, a large number of samples can be obtained. Zhang et al. [25] first proposed this method and verified that the augmented dataset could improve the fault diagnosis accuracy. Kong et al. [26] proposed a novel sparse classification approach to diagnose planetary bearings in which overlapping segmentation is embedded to augment the vibration data. Inspired by image data augmentation, researchers also use similar tricks to enhance the obtained dataset. The most intuitive way is to add Gaussian white noise to the samples. Based on the analysis of the retrieved literature, most of them used this method. Qian et al. [27] first sliced the vibration signal to form a dataset, 25% of which was added with Gaussian noise. Subsequently, the samples were mixed to train their model. Faysal et al. [28] went one step further by proposing a noise-assisted ensemble augmentation technique for 1D time series data. Other commonly used image transformation methods have also been proven effective on time series data, such as translation, rotation, scaling, truncation, and various flipping operations [29–34]. Considering the inherent characteristics of vibration signals, it is also an effective method to rearrange the data points of samples. For example, the samples can be equally divided into two parts to form two groups. New samples can subsequently be obtained by randomly recombining the data from the two groups [35]. Ruan et al. [36] proposed a method called signal concatenation to further increase the number of samples. The original samples are divided into several parts, which are augmented, respectively, and concatenated to form new samples finally.

Some researchers also convert vibration signals into images to diagnose bearing faults. One option is to rearrange the time series into a two-dimensional form and represent them as images. Subsequently, commonly used image augmentation techniques such as flipping can be utilized to double the size of the dataset [37]. Another common option is to use signal processing techniques to transform the vibration signal into a time–frequency spectrogram. For example, Yang et al. [38] introduced the image segmentation theory to augment planetary gearbox-bearing fault spectrogram data fed to the subsequent fault diagnostic model. Specifically, the researchers proposed wavelet transform coefficients cyclic demodulation to obtain a 2D spectrogram of the original vibration signal. They divided the spectrogram into small blocks and defined the overlapping length. This generates smaller spectrograms to compose balanced datasets.

3.3. Data Augmentation Using GANs

According to the purpose of the task, ML/DL models can be generally classified into two categories: discriminative and generative models. Typical discriminative tasks include regression and classification, whereas generative models are widely used to synthesize data. GAN is a kind of generative model. Since it was proposed by Goodfellow et al. [39] in 2014, it has become the most popular method for data augmentation. In contrast to other generative models, such as variational autoencoder (VAE), the idea of adversarial training was introduced in GAN. It consists of two neural networks called discriminator and generator. The structure of a general GAN is shown in Figure 4a.

Generator G is used to generate realistic samples from random noise z. The discriminator D aims to distinguish between real samples x and generated samples $G(z)$. The adversarial learning of GAN is like a zero-sum game. In the beginning, the discriminator can easily distinguish fake samples from real samples because the samples generated from random noise are also random. However, if the GAN is well trained, the discriminator will no longer be able to judge the authenticity of the samples, and the generator can be used to synthesize realistic samples. Essentially, two data distributions are mapped here, from the distribution of random noise to that of real samples. In the training process, all losses are calculated based on the output of the discriminator. Since the task of the discriminator is to judge the authenticity of the input, it can be regarded as a binary classification problem. Therefore, the binary cross-entropy is used as the loss function. First, the discriminator needs to be optimized while the generator is fixed. If 1 denotes true and 0 denotes false, the optimization objective of the discriminator can be formulated as Equation (1), which means to judge the real samples as true and the generated samples as false. After the discriminator is optimized, the discriminator is fixed and the generator needs to be optimized. The optimization goal of the generator is that the discriminator judges the generated samples as true, which can be formulated as Equation (2).

$$\max_{\mathrm{D}} \mathrm{L(D)} = \mathbb{E}_{\mathrm{x} \sim \mathrm{p_{data}}}[\log \mathrm{D(x)}] + \mathbb{E}_{\mathrm{z} \sim \mathrm{p_z}}[\log(1 - \mathrm{D(G(z))})] \tag{1}$$

$$\min_{G} \mathrm{L(G)} = \mathbb{E}_{\mathrm{z} \sim \mathrm{p_z}}[\log(1 - \mathrm{D(G(z))})] \tag{2}$$

where $D(x)$ denotes the probability that an original sample is judged to be real data and $D(G(z))$ is the probability that a generated sample is judged to be fake data.

GAN is initially applied to computer vision to augment the image data [40]. However, as mentioned in Section 2.2, GAN was first introduced to bearing fault diagnosis in 2018 [14] and has become a popular research topic in recent years. Wang et al. [41] then used GAN to generate mechanical fault signals to improve the diagnosis accuracy. Section 4 will introduce the improvements and applications of GANs in bearing fault diagnosis in detail.

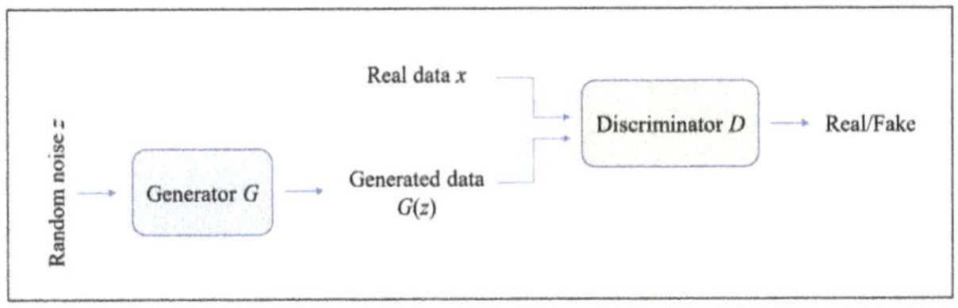

(**a**) Structure of the general GAN

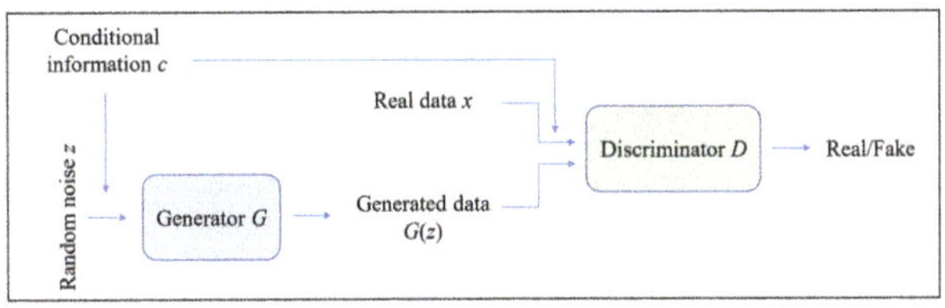

(**b**) Structure of the CGAN

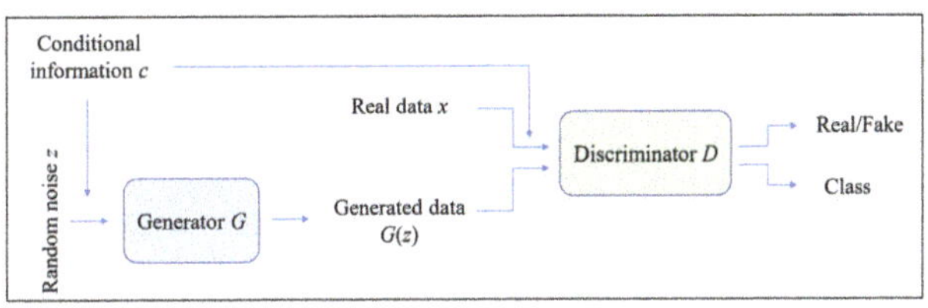

(**c**) Structure of the ACGAN

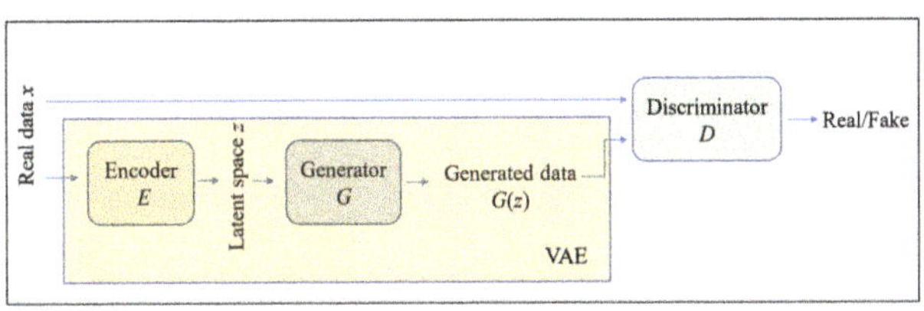

(**d**) Structure of the VAEGAN

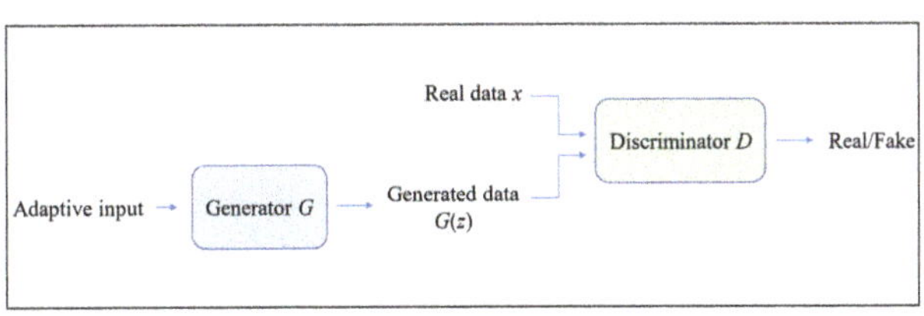

(**e**) Structure of GAN with adaptive input

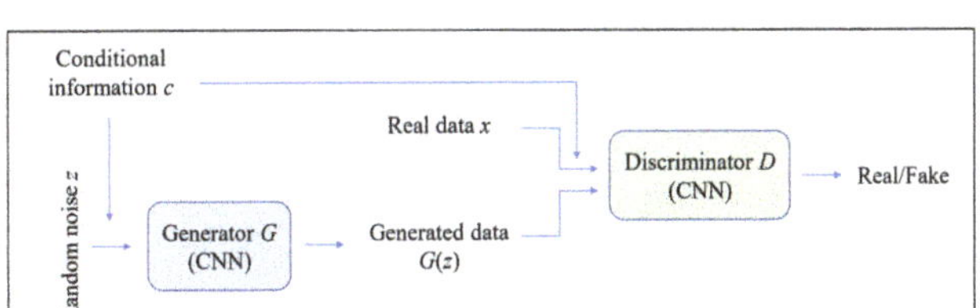

(**f**) Structure of the C-DCGAN

Figure 4. Structure of general GAN and variants.

4. Improvements and Applications of GANs in Bearing Fault Diagnosis

The original GAN has three primary problems: unstable convergence, model collapse, and vanishing gradient. To overcome these problems and enhance the quality of sample generation, many variants of GAN have been proposed in recent years. We classify them into two categories: network structure-based improvements and loss function-based improvements. Apart from this, the quality evaluation of the generated samples is a meaningful topic. At the end of this section, the applications of GANs in bearing fault diagnosis are summarized.

4.1. Improvements in the Network Structure

According to different improvement ideas, we further classify the network structure-based improvements into three categories: information-based improvements, input-based improvements, and layer-based improvements.

4.1.1. Information-Based Improvements

The input to the generator of a general GAN is random noise, which can easily lead to mode collapse. When the mode collapse happens, the GAN's generator can only produce one or a small subset of the different outputs. To address this problem, Mirza et al. [42] proposed the conditional GAN (CGAN). CGAN adds conditional information to the discriminator and generator of the original GAN. The input to CGAN will be a stitching of conditional information with the original input. This additional information such as category labels can control and stabilize the data generation process. By setting different conditional inputs, the samples of different categories can be generated. The other idea is to improve the discriminator so that it can judge the not only authenticity but also output the class of the samples like a classifier. Auxiliary classifier GAN (ACGAN) introduces an auxiliary classifier to the discriminator, which can not only judge the authenticity of the data but also output the class of the data, thereby improving the stability of the training and the quality of the generated samples [43]. The role of the auxiliary classifier is to predict the category of a sample and pass it to the generator as additional conditional information. ACGAN enables a more stable generation of the realistic samples of a specified category. Both CGAN and ACGAN enhance the performance of the general GAN by providing more information. That is why they are regarded as an information-based structural improvement in this paper. Their structures are presented in Figure 4b,c. The original CGAN and ACGAN were successfully applied to bearing fault diagnosis. Wang et al. [44] utilized CGAN to generate the spectrum samples of vibration signals. The use of category labels as condition information to generate the samples of various categories of bearing faults proved to be effective. In [45], ACGAN was directly utilized to generate 1D vibration signals. Experimental results revealed that generated vibration samples improved the accuracy of the bearing fault diagnostic model from 95% to 98%. Some researchers were inspired by the idea of providing more information by the addition of classifiers or other modules. Zhang et al. [46] designed a multi-modules gradient penalized GAN. A classifier as an additional module was added to the Wasserstein GAN with a gradient penalty (WGAN-GP). In [47] and the generator was integrated with a self-modulation (SM) module, which enables the parameter updating based on both the input data and the discriminator. This makes the convergence of the training faster. These papers demonstrate that the idea of designing and integrating more modules concerning the structure of GAN with the goal of providing more useful information is feasible.

4.1.2. Input-Based Improvements

In the general GAN, random noise is fed into the generator to synthesize realistic samples. This may not be reasonable for specific data distributions. Some researchers have made innovative improvements to the structure of the input of the generator, thereby improving the quality of the generated samples. Larsen et al. [48] combined the VAE and GAN and proposed VAEGAN. VAE is an earlier proposed generative model consisting of an encoder and a decoder. The encoder maps the input data to points in the latent space, which are converted back into points in the original space by the decoder. By learning a latent variable model, VAE can be used to generate more data. In a VAEGAN model, the encoder is used to encode existing data, and the encoded latent vectors are used as input to the generator or decoder instead of the random noise. VAEGAN utilizes the latent variable model of VAE to generate the data and uses the discriminator of GAN to evaluate the authenticity of the generated samples. The advantage of VAEGAN is that it can generate high-quality samples and can operate in the latent variable space, such as performing sample interpolation and other modifications. Figure 4d shows the structure of the VAEGAN. Rathore et al. [49] applied VAEGAN to generate time–frequency spectrograms and balanced the bearing fault dataset. The experiment verified that the generated samples are more reasonable and of higher quality. There are a lot of other alternatives to random noise out there. For example, Zhang et al. [50] proposed an adaptive learning method to update the latent vector instead of sampling from Gaussian distribution, realizing adaptive input instead of random noise, as shown in Figure 4e. By using different distributions to generate the latent vector's digits, a better combination effect can be produced. Improving the input structure of the discriminator is, likewise, a good starting point. In [51], the input of the discriminator was changed from real data to latent encoding by the encoder. The mutual information between real data and latent encoding was constrained by the proposed variational information technique, which limited the gradient of the discriminator and ensured a more stable training process.

4.1.3. Layer-Based Improvements

Considering CNN's powerful feature extraction capability, convolutional layers were introduced into a GAN called deep convolutional GAN (DCGAN) and applied to image augmentation [52,53]. The original generator of DCGAN is shown in Figure 5. DCGANs have also proven to be effective in vibration signal augmentation. Luo et al. [54] integrated CGAN and DCGAN into C-DCGAN, as shown in Figure 4f. The augmented data successfully improved the accuracy of the bearing fault diagnosis. Based on the DCGAN, a multi-scale progress GAN (MS-PGAN) framework was designed in [55]. This concatenates multi-DCGANs, which share one generator. Through progressive training, high-scale samples can be generated from low-scale samples. Imposing the spectral normalization (SN) on the layers is another useful trick. Tong et al. [56] proposed a novel auxiliary classifier GAN with spectral normalization (ACGAN-SN) to synthesize the bearing fault data, in which spectral normalization was added to the output of each layer of the discriminator. The introduction of the spectral normalization technique makes the training process more stable.

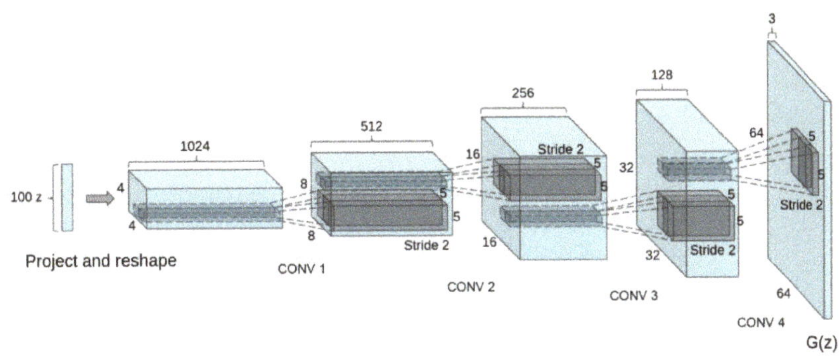

Figure 5. Structure of the original DCGAN generator [53].

The three above cases of layer-based improvements reveal that: (1) convolutional neural network can improve the performance of GAN and produce good results in bearing fault data generation; (2) concatenating multiple neural networks can generate high-scale samples of high quality; and (3) proposed layer normalization methods such as spectral normalization are worth trying.

4.2. Improvements in the Loss Function

4.2.1. Metric-Based Improvements

The original GAN uses J-S divergence to measure the distance between real and generated data distributions. However, it has a drawback that J-S divergence is a fixed value if the distance between distributions is too far, and thereby cannot measure how close two distributions are. This causes vanishing gradient in the training process [57]. To solve this problem, Arjovsky et al. [58] proposed Wasserstein GAN, in which J-S divergence was replaced by Wasserstein distance. As a result, the loss function of the generator and the discriminator can be formulated as follows:

$$L_{WGAN}^{G} = -\mathbb{E}_{p_{G(z)}}[D(G(z))] \tag{3}$$

$$L_{WGAN}^{D} = -\mathbb{E}_{p_x}[D(x)] + \mathbb{E}_{p_{G(z)}}[D(G(z))] \tag{4}$$

where x and $G(z)$ represent the real and generated data, respectively. $D(.)$ is the probability that the data are judged to be real. Compared to the loss functions of the original GAN, the implementation of the Wasserstein distance discards the operation of logarithms in the loss functions [59].

WGAN has been proven effective in many bearing fault diagnosis studies. Zhang et al. [60] proposed an attention-based feature fusion net using WGAN as the data augmentation part. The experimental results verified the feasibility of the scheme under small sample conditions. In [61], a novel imbalance domain adaption network was presented for rolling bearing fault diagnosis, in which WGAN was embedded. The data imbalance between domains and between fault classes in the target domain was considered. WGAN was used to enhance the target domain datasets. However, the performance of WGAN is still limited because of weight clipping. To overcome this problem, Gulrajani et al. [62] combined WGAN with the gradient penalty strategy (WGAN-GP), which is successful in image augmentation. The difference between WGAN-GP and WGAN is that a regularization term is added to the loss function of the discriminator. The loss function of the generator remains the same.

$$L_{WGAN-GP}^{G} = L_{WGAN}^{G} \tag{5}$$

$$L_{WGAN-GP}^{D} = L_{WGAN}^{D} + \lambda L_{gp} \tag{6}$$

Both WGAN and WGAN-GP use the Wasserstein distance to assess the difference between the generated samples and the training samples, which is superior to the J-S divergence, and WGAN-GP adds a gradient penalty on top of WGAN to eliminate the problem of gradient explosion in the network. The discriminator's loss function incorporates a gradient penalty in addition to the judgment of real and fake samples, smoothing the generator and decreasing the risk of mode collapse.

Apart from the Wasserstein distance, Mao et al. [63] proposed the least squares GAN (LSGAN), which uses the least squares error to measure the distance between the generated and real samples. The objective functions of the discriminator and generator are as follows:

$$\min_{D} L(D) = \mathbb{E}_{x \sim p_x}(D(x) - b)^2 + \mathbb{E}_{z \sim p_z}(D(G(z)) - a)^2 \tag{7}$$

$$\min_{G} L(G) = \mathbb{E}_{z \sim p_z}(D(G(z)) - c)^2 \tag{8}$$

Since the discriminator network's goal is to distinguish between real and fake samples, the generated and real samples are encoded as a and b, respectively. The objective function of the generator replaces a with c, indicating that the discriminator treats the generated samples as real samples. It has been proven that the objective function is equivalent to the Pearson χ^2 divergence in a particular case. In [64], LSGAN was used to generate traffic signal images. The results of the comparison experiments show that LSGAN outperforms WGAN and DCGAN in such an application scenario. In [65], Anas et al. reported on a new CT volume registration method, in which LSGAN was employed to learn the 3D dense motion field between two CT scans. After extensive trials and assessments, LSGAN shows higher accuracy than the general GAN in estimating the motion field. LSGAN can alleviate the problem of vanishing gradient during training and generate higher-quality images compared to the general GAN. However, based on our literature research, its application in the field of bearing fault diagnosis was not prevalent. This may be due to the fact that it is not suitable for the generation of bearing fault data. For example, LSGAN shows a worse performance than DCGAN in [66].

Energy-based GAN (EBGAN) [67] introduces an energy function into the discriminator and trains the generator and discriminator by optimizing the energy distance. The discriminator assigns low energy to the real samples and high energy to the fake samples. Usually, the discriminator is a well-trained autoencoder. Instead of judging the authenticity of the input sample, the discriminator calculates its reconstruction score. The loss functions of the discriminator and the generator can be formulated as follows:

$$\mathrm{L}(D) = \mathrm{E}_{x \sim P_{\text{data}}} \left[D(x) + [m - D(G(z))]^{+} \right] \tag{9}$$

$$\mathrm{L}(G) = \mathrm{E}_{z \sim P_z} [D(G(z))] \tag{10}$$

where m is a positive margin used for the selection of energy functions and $[\cdot]^{+} = \max(0, \cdot)$. Yang et al. [68] combined EBGAN and ACGAN in their proposed bearing fault diagnosis method under imbalanced data and obtained good sample generation and classification performance.

Boundary equilibrium GAN (BEGAN) [69] is a further improvement on EBGAN. The main contribution is the introduction of the ratio between the autoencoder reconstruction error and the degree of boundary balance of the generator and discriminator to the loss function. The new loss function balances the competition between the generator and the discriminator, resulting in more realistic generated samples. The loss functions of BEGAN can be formulated as follows:

$$\mathrm{L}(D) = \mathrm{E}_{x \sim P_{\text{data}}} [D(x) - k_t D(G(z))] \tag{11}$$

$$\mathrm{L}(G) = \mathrm{E}_{x \sim P_z} [(1 - D(G(z)))] \tag{12}$$

where k_t is a weighting coefficient to balance the performance of the generator and the discriminator.

Relativistic GAN (RGAN) [70] is another well-known variant of GAN, whose primary idea is to turn the discriminator's output into relative authenticity, i.e., the degree to which the discriminator finds the generated samples to be more realistic than the real ones. RGAN optimizes the model using the relative authenticity loss function and has been demonstrated to converge more easily and be more effective in creating high-quality images. However, RGAN still lacks relevant research in the field of bearing fault diagnosis.

This subsection examines various well-known metric-based improvements for the loss function and their effective applications, particularly in bearing fault diagnosis. The loss functions of the aforementioned GAN variants, as well as the general GAN, are all comparable in that they calculate a certain distance between two distributions, with the optimization goal of minimizing this distance. Much of the literature has verified the validity of the Wasserstein distance in the field of bearing fault diagnosis. However, there are still a number of alternatives that require further research.

4.2.2. Regularization-Based Improvements

Directly applying WGAN-GP to bearing fault diagnosis rarely yields satisfactory results. However, the idea of adding regularizations to the loss function has been proven effective in bearing fault diagnosis. In [71], a new GAN named parallel classification Wasserstein GAN with gradient penalty (PCWGAN-GP) is presented, in which the Pearson loss function was introduced to enhance the performance of the GAN. It can generate the faulty samples of bearings with healthy samples as input. The maximum mean discrepancy (MMD) is a commonly used metric to measure the similarity between domains in transfer learning. Inspired by this, Zheng et al. [55] introduced the MMD to the loss function of WGAN-GP as a new penalty. The experimental results verified the effectiveness of this method in bearing fault sample augmentation. Ruan et al. [72] added the error of fault characteristic frequencies and the results of the fault classifier to the loss function. The improvement in sample quality is evident in the envelope spectrum. In [50,51], the proposed reconstruction module or representation matching module maps the distribution between real and generated data. The calculated difference is sensitive to the data class and can provide additional constraints on the generator. The collected regularizations to the loss function are listed in Table 1.

Table 1. Regularization to the loss function of various GANs.

Number	Loss	Formulation	Source
1	PCWGAN-GP	$\mathrm{PC}\left(\bar{x}_j^k, \bar{x}^k\right) = \frac{1}{n}\sum_{j=1}^{n} \frac{\sum_{m=1}^{M}\left(\bar{x}_{j,m}^k - \bar{x}_{j,m}^k\right)\left(\bar{x}_m^k - \bar{x}_m^k\right)}{\sqrt{\sum_{m=1}^{M}\left(\bar{x}_{j,m}^k - \bar{x}_{j,m}^k\right)^2}\sqrt{\sum_{m=1}^{M}\left(\bar{x}_m^k - \bar{x}_m^k\right)^2}}$	[71]
2	MS-PGAN	$MMD^2[D_X, D_Y] = \left\Vert \frac{1}{x}\sum_{i=1}^{x} f(x_i) - \frac{1}{y}\sum_{j=1}^{y} f(y_i) \right\Vert^2$	[56]
3	FCFE	$L_{\text{frequency}} = \sum_{i=1}^{N}\left(\left\vert M_{\text{real}}^i - M_{\text{fake}}^i \right\vert + \left\vert F_{\text{real}}^i - F_{\text{fake}}^i \right\vert\right)$	[72]
4	WCGAN-HFM	$HFM = \sum_l \omega_l \cdot \vert D_l(x \mid y) - D_l(G(z \mid y))\vert$	[73]
5	Entropy	$H(G(z)) = E_{z\sim P_z}\left[\Vert z - E_n(G(z))\Vert^2\right]$	[74]

(1) M is the dimension of generated samples, $\bar{x}_{j,m}^k$ denotes the mth element in the j-th sample with the category of k, and $\bar{x}$ represents the mean value of x. (2) Maximum mean discrepancy (MMD) measures the similarity between two distributions in transfer learning. The value of MMD^2 was used as the MMD penalty between the source domain D_x and the target domain D_y. (3) N denotes the maximum order of FCF. M stands for the i-th order FCF amplitude from the real and generated sample. F represents the i-th order FCF frequency from the real and generated sample. (4) ω_l is the weighting factor of the l-th layer loss. Hierarchical feature matching (HFM) provides additional information from the perspective of differences between classes. (5) E_n is an encoder with parameters and $E_n(G(z))$ denotes the intermediate layer feature of the generated sample output by the discriminator. The entropy reflects the diversity of generated samples.

Adding regularizations to the loss function usually provides more information and constraints, which helps to stabilize the GAN training and improve the quality of the generated samples. On the other hand, knowledge of physics, such as bearing fault mechanisms, can be combined with the loss function of the general GAN, which can not only refine the quality of the generated samples but also make them more interpretable.

4.2.3. Summary

Based on our analysis, there are two kinds of methods to improve the loss function: metric-based improvements and regularization-based improvements. The former is to adopt a new metric to replace the original J-S divergence, thereby more efficiently measuring the similarity between data distributions. The Wasserstein distance is such an excellent example. WGAN and its variants have been used a lot in bearing fault diagnosis. Other GAN variants, such as LSGAN, EBGAN, BEGAN, and RGAN, require more investigation in the field of bearing fault diagnosis. However, proposing entirely new metrics requires advanced mathematical knowledge, which is a challenging work. The improvements in the loss function by adding regularization terms are more popular. Introducing more constraints can effectively stabilize the training of GANs and enable the generation of high-quality samples. The introduction of physical knowledge as a regularization term into GAN has also been shown to be feasible and deserves more research.

4.3. Evaluation of Generated Samples

The samples generated by GANs are not really collected from mechanical equipment. Therefore, to ensure their feasibility as training data, it is necessary to evaluate the quality of the generated samples, which can be considered in three aspects: similarity, creativity, and diversity.

High similarity means that the generated and real data have the most similar distributions possible. This is the most essential requirement for generated data. Based on our analysis of the existing literature, the evaluation methods concerning their similarity can be divided into two categories: qualitative methods and quantitative metrics.

Qualitative methods refer to the comparison of data visualizations, including the time and frequency domains. This method enables an initial evaluation of the similarity between samples. In the time domain, the most intuitive evaluation method is to compare the waveforms of the generated signal and the real signal. Amplitude and peaks should be noticed. In the frequency domain, it is valuable to check the fault characteristic frequencies (FCFs), which are crucial for bearing fault diagnosis [71,72]. In addition, the features extracted from real and generated samples can be compared using the t-distributed stochastic neighbor embedding (t-SNE) technique as a qualitative approach to validate the usability of the generated samples [44,57,71].

To further accurately quantify the similarity, some indicators have been proposed. Cosine similarity (CS) can measure the similarity between two sequences. In [72], cosine similarity was adopted as the time domain similarity metric to evaluate the quality of the generated bearing fault samples. However, a relatively small cosine similarity value can be obtained if the samples are too long. The maximum mean discrepancy (MMD) was initially used to measure the similarity between domains in transfer learning. In [55], it is introduced to measure the similarity between the generated and real samples. In [56,60,71], the correlation between the spectra of real samples and those of generated samples was calculated by the Pearson correlation coefficient (PCC). The K-L divergence and the Wasserstein distance calculate the similarity between data distributions, which can also be used to quantitatively characterize the quality of the generated samples [47,49,56]. Some bearing fault diagnosis schemes first use signal processing techniques such as short-time Fourier transform (STFT) to convert the original vibration signal into a time–frequency spectrogram, and the features are extracted from the spectrograms for subsequent fault diagnosis. Since the GAN is used to directly generate images, it is reasonable to assess the quality of the generated images. In [49], the peak signal-to-noise ratio (PSNR) and structural similarity index measure (SSIM) were utilized to investigate the quality of the generated samples. Furthermore, the GAN-test can be conducted to measure the feasibility of the generated data [71]. The real and generated data can be treated as the training and test sets, respectively. The accuracy of the diagnostic network shows the variation between real and generated data. The collected metrics for similarity evaluation are listed in Table 2.

Table 2. Evaluation metrics for the sample generation quality of GAN.

Number	Metric	Formulation	Source
1	CS	$\cos\theta = \frac{\vec{m}\cdot\vec{n}}{\lvert\vec{m}\rvert\cdot\lvert\vec{n}\rvert}$	[72]
2	MMD	$MMD[F,p,q] = \sup_{f\in F}(E_{x\sim p}[f(x)] - E_{x\sim q}[f(x)])$	[56]
3	PCC	$PCC_{X,Y} = \frac{\mathrm{cov}(X,Y)}{\sigma_X\sigma_Y}$	[57,60,71]
4	KLD	$D_{\mathrm{KL}}(P\Vert Q) = \sum_{x\in\mathcal{X}} P(x)\log\left(\frac{P(x)}{Q(x)}\right)$	[51]
5	WD	$\mathrm{WD}(\mathrm{P}_1,\mathrm{P}_2) = \inf_{\gamma\sim(\mathrm{P}_1,\mathrm{P}_2)} \mathbb{E}_{(\mathrm{x,y})\sim\gamma}[\Vert\mathrm{x}-\mathrm{y}\Vert]$	[57]
6	PSNR	$\mathrm{PSNR} = 10\log_{10}\left(\frac{\mathrm{MAX}_I^2}{\mathrm{MSE}^2}\right)$	[51]
7	SSIM	$\mathrm{SSIM}(x,y) = \frac{(2\mu_x\mu_y + C_1)(2\sigma_{xy}+C_2)}{(\mu_x^2+\mu_y^2+C_1)(\sigma_x^2+\sigma_y^2+C_2)}$	[51]

(1) m and n stand for two time series. (2) F denotes a given set of functions, p and q are two independent distributions, x and y obey p and q, respectively, sup denotes an upper bound, and $f()$ denotes a function mapping. (3) X and Y are two variables, σ_X and σ_Y are the standard deviations of X and Y, respectively. (4) P and Q are two probability distributions in the same probability space, and X is the relative entropy from Q to P. (5) P_1 and P_2 are two probability distributions, and γ is a joint probability distribution. (6) MAX_I represents the image with the maximum valid value of the pixel in the image, and MSE is the mean squared error estimated over two images.. (7) μ_x, μ_y, σ_x, σ_y, σ_{xy} are the mean, standard deviation, and cross-covariances of the x and y. C_1, C_2, and C_3 are the regularization constants.

Creativity and diversity are further requirements for generated data. The former means that the generated signals are not duplicates of the real signals, and the latter requires that the generated signals are not duplicates of each other. In [73], the SSIM and entropy were adopted to quantify the creativity and the diversity of the generated images. Specifically, the SSIM was used to cluster similar generated samples, and the entropy of these clusters reflects their diversity. The entropy can be formulated as follows:

$$\text{Entropy} = -\sum_{i}^{m} p_i \log p_i \tag{13}$$

where m is the number of clusters and p_i denotes the probability that the i-th cluster belongs to non-replicated clusters. The duplication occurs when the SSIM is equal to or greater than 0.8. Greater cluster entropy indicates that the generated signals are more diverse. However, there is still a lack of studies evaluating the creativity and diversity of bearing vibration signals.

4.4. Applications of GAN in Bearing Fault Diagnosis

Small sample and data imbalance are two main challenges encountered in data-driven bearing fault diagnosis. In practical engineering, the collected fault data are usually insufficient. On the one hand, machinery and its components are in a healthy status under normal production conditions. On the other hand, they cannot remain faulty for long. Therefore, it is expensive or even impractical to obtain sufficient fault data for the training of diagnostic models. Meanwhile, the probability of various faults, including inner ring faults, outer ring faults, and many other faults, varies due to the inherent characteristics of bearings and different working environments. Therefore, the collected fault data are also unbalanced. The two problems restrict the performance of various ML/DL models and lead to relatively low diagnosis accuracy. An intuitive and widely used solution is to synthesize samples artificially, resulting in a sufficient and balanced dataset. The commonly used data augmentation approaches in bearing fault diagnosis, including traditional oversampling methods and data transformation methods, were covered in previous sections of this paper. As a generative model, one of the most fundamental and important applications of GAN is data augmentation. It is a very promising alternative method for generating bearing fault data. Bearing fault data can be classified into two types based on data dimensions, namely one-dimensional fault data and two-dimensional fault data. The raw vibration signal is one-dimensional time series. GANs are able to directly generate one-dimensional vibration data [45,75]. As the frequency domain of the vibration signal contains a wealth of fault information, in many cases, the raw vibration data are converted from the time domain into the frequency domain. GAN can also be used to generate one-dimensional spectrum data [76,77]. As GAN has its origins in image processing from computer vision, there is no doubt that GAN can also be used to synthesize two-dimensional fault data. One option is to reshape one-dimensional fault data into two-dimensional data [72], while another is to utilize GAN to generate the two-dimensional time–frequency spectrograms of vibration signals [78].

The variety of working conditions is another key issue for data-driven bearing fault diagnosis. Differences in equipment and operational conditions have an impact on the diagnostic model's generalization performance. GAN can also be applied to transfer learning. Transfer learning refers to the application of a previously trained model to a new task to achieve better performance [79,80]. GAN or the idea of adversarial learning can be integrated into a general transfer learning method to improve the performance of the transfer learning method [27,36,81]. For example, Pei et al. [82] combined WGAN-GP and transfer learning in their proposed rolling bearing fault diagnosis method. Using fault data from only one working condition as the source domain, the fault diagnosis of the target domain under different working conditions is achieved. On the other hand, GAN enables the data transfer between the source and target domains. In [61], Zhu et al. applied adversarial learning to achieve a balance between the data distributions of source and target domain.

From the perspective of application scenarios, promising experimental results have been demonstrated in the two main tasks of bearing fault diagnosis: fault classification and remaining useful life (RUL) prediction. Fault classification is the basic task of fault diagnosis, including the classification of different fault types [74] and the classification of faults of different severity levels [78,83]. To improve the accuracy of bearing RUL prediction, there have also been some studies on the generation of bearing aging data using GANs [84–87].

In summary, starting from the challenges encountered in practical engineering, GAN can not only be used as a data augmentation technique to address the problem of small sample and data imbalance problems, but can also be applied to transfer learning to improve the ability of models in across-domain diagnosis. Starting from the application scenarios of bearing fault diagnosis, GAN contributes to two major tasks: fault classification and RUL prediction.

5. Conclusions

5.1. Summary

The small sample and data imbalance problems seriously hinder the deployment of DL-based techniques in bearing fault diagnosis. Apart from traditional data augmentation techniques such as oversampling and data transformation, GAN is the most promising method to enable the artificial synthesis of high-quality samples. This paper first reviewed the development of traditional data augmentation methods for bearing fault diagnosis. Subsequently, the recent advances of GANs in bearing fault diagnosis are introduced in detail. Firstly, we divide the improvements of GANs into two primary categories: the improvements in the network structure and the improvements in the loss function. For the former, we further summarized them into three types: information-based, input-based, and layer-based improvements. Likewise, the improvements of loss function are divided into two categories: metric-based improvements and regularization-based improvements. Additionally, we also reviewed the commonly used evaluation methods for generated samples. Finally, we work through the applications of GANs in bearing fault diagnosis. To give an overview of the comparison, Table 3 summarizes the advantages and disadvantages of typical GANs, which can be used to guide the choice of GANs under different application scenarios.

Table 3. Advantages and disadvantages of typical GANs.

Number	Type	Advantages	Disadvantages
1	CGAN	To generate samples with specific attributes such as specific categories.	A large amount of training data with labels are required.
2	ACGAN	With auxiliary classifier, different classes of samples can be generated.	(1) Complex training; (2) Limited quality of generated samples.
3	VAEGAN	The generated samples can be controlled by the autoencoder.	The training is relatively more difficult.
4	DCGAN	The powerful feature extraction capability of CNN is exploited.	More computational resources are required for the training.
5	SNGAN	Exploding and vanishing gradient can be solved effectively.	(1) Slow training speed; (2) Limited diversity of generated samples.
6	WGAN	Wasserstein distance provides a better measure of the difference between distributions.	The training is not stable enough.
7	WGAN-GP	With gradient penalty integrated into WGAN, the stability is improved.	More training time and computational resources.
8	LSGAN	Effectively solves the problems of exploding gradient and vanishing gradient.	Excessive penalization of outliers may lead to a reduction in the diversity of samples being generated.
9	EBGAN	(1) Energy-based loss function allows better interpretability; (2) Improved stability and diversity of sample generation.	(1) Quite complex to implement and train; (2) Prone to mode collapse.
10	BEGAN	Mode collapse can be effectively alleviated.	(1) A relatively complex architecture; (2) Sensitive to hyperparameters.
11	RGAN	With relativistic loss, the quality of sample generation is improved and mode collapse is reduced.	(1) A relatively complex architecture; (2) The relativistic loss is difficult to interpret.

5.2. Outlook

- Explainability from physics
  Due to the black-box properties of DL models, the generated samples lack physical interpretability. Based on our literature research, most studies do not take physical knowledge into account in their models. Although there is a large body of literature on physics-guided neural networks [88,89], there is still a lack of research on introducing physical knowledge into GANs. From our point of view, physics-guided GAN can be studied from two perspectives in the field of bearing fault diagnosis. Based on the taxonomy of improvements of GAN in this paper, the first idea belongs to the improvement of the network structure. For example, the bearing fault mechanism model can be integrated into GAN. The second idea aims to improve the loss function by adding physically interpretable regularization terms to the original loss function.

- Advanced evaluation metrics
  To date, the evaluation of the generated samples is not comprehensive. Almost all of the literature we researched only considered the similarity of the generated samples to the real samples. Apart from similarity, the creativity and diversity of the generated samples should be taken into account to achieve a more comprehensive evaluation. More appropriate evaluation metrics deserve further investigation.
- Application for RUL prediction
  Based on our collation of the literature, there are still a number of promising variants or improvements in GAN that have not yet been applied to bearing fault diagnosis, which deserve further research. For the application in bearing fault diagnosis, the majority of reported GAN variants possess the potential to achieve satisfying results, even under imbalanced or small datasets through sample generation. However, concerning RUL prediction, it is quite another matter. In contrast to fault samples, which have obvious features such as different fault characteristic frequencies for different fault types, samples in the aging period do not have such distinct one-to-one features. Therefore, generating aging samples for bearing during the degradation process with GAN remains an open question. Improving the GAN to generate aging samples for RUL prediction under a dataset with limited run-to-failure trajectories is a challenging but rewarding research topic.

Author Contributions: Conceptualization, D.R.; methodology, D.R. and X.C.; software, D.R. and X.C.; validation, X.C. and D.R.; formal analysis, X.C.; investigation, D.R. and X.C.; resources, D.R. and X.C.; data curation, X.C.; writing—original draft preparation, X.C. and D.R.; writing—review and editing, C.G. and J.Y.; visualization, D.R.; supervision, C.G. and J.Y.; project administration, C.G. and J.Y.; funding acquisition, D.R. All authors have read and agreed to the published version of the manuscript.

Funding: This research is supported by CSC doctoral scholarship (201806250024) and Zhejiang Lab's International Talent Fund for Young Professionals (ZJ2020XT002).

Institutional Review Board Statement: Not applicable.

Informed Consent Statement: Not applicable.

Data Availability Statement: Not applicable.

Conflicts of Interest: The authors declare no conflict of interest.

Abbreviations

The following abbreviations are used in this manuscript:

ACGAN	Auxiliary classifier GAN
ACGAN-SN	Auxiliary classifier GAN with spectral normalization
AE	Stacked autoencoder
ANN	Artificial neural network
BEGAN	Boundary equilibrium GAN
CNN	Convolutional neural networks
CS	Cosine similarity
CGAN	Conditional GAN
C-DCGAN	DCGAN integrated with CGAN
DBN	Deep belief network
DCGAN	Deep convolutional GAN
DL	Deep learning
EM	Expectation maximization
EBGAN	Energy-based GAN
FCFs	Fault characteristic frequencies
GAN	Generative adversarial network
KLD	K-L divergence
LSTM	Long short-term memory

LSGAN	Least squares GAN
ML	Machine learning
MMD	Maximum mean discrepancy
MS-PGAN	Multi-scale progress GAN
PCC	Pearson correlation coefficient
PCWAN-GP	Parallel classification WGAN with gradient penalty
PHM	Prognostics and health management
PSNR	Peak signal-to-noise ratio
RGAN	Relativistic GAN
RNN	Recurrent neural network
RUL	Remaining useful life
SSIM	Structural similarity index measure
STFT	Short-time Fourier transform
SCOTE	Sample-characteristic oversampling technique
SI-SMOTE	Sample information-based SMOTE
SM	Self-modulation
SMOTE	Synthetic minority over-sampling technique
SNGAN	Spectral normalization GAN
SVM	Support vector machine
TL	Transfer learning
VAE	Variational autoencoder
VAEGAN	GAN combined with VAE
WGAN	Wasserstein GAN
WGAN-GP	WGAN with the gradient penalty
kNN	k-nearest neighbor
t-SNE	t-distributed stochastic neighbor embedding

References

1. Hakim, M.; Omran, A.A.B.; Ahmed, A.N.; Al-Waily, M.; Abdellatif, A. A systematic review of rolling bearing fault diagnoses based on deep learning and transfer learning: Taxonomy, overview, application, open challenges, weaknesses and recommendations. *Ain Shams Eng. J.* **2022**, *14*, 101945. [CrossRef]
2. Nandi, S.; Toliyat, H.A.; Li, X. Condition monitoring and fault diagnosis of electrical motors—A review. *IEEE Trans. Energy Convers.* **2005**, *20*, 719–729. [CrossRef]
3. Henao, H.; Capolino, G.A.; Fernandez-Cabanas, M.; Filippetti, F.; Bruzzese, C.; Strangas, E.; Pusca, R.; Estima, J.; Riera-Guasp, M.; Hedayati-Kia, S. Trends in fault diagnosis for electrical machines: A review of diagnostic techniques. *IEEE Ind. Electron. Mag.* **2014**, *8*, 31–42. [CrossRef]
4. Choudhary, A.; Shimi, S.; Akula, A. Bearing fault diagnosis of induction motor using thermal imaging. In Proceedings of the 2018 international conference on computing, power and communication technologies (GUCON), Greater Noida, India, 28–29 September 2018; pp. 950–955.
5. Barcelos, A.S.; Cardoso, A.J.M. Current-based bearing fault diagnosis using deep learning algorithms. *Energies* **2021**, *14*, 2509. [CrossRef]
6. Harlişca, C.; Szabó, L.; Frosini, L.; Albini, A. Bearing faults detection in induction machines based on statistical processing of the stray fluxes measurements. In Proceedings of the 2013 9th IEEE International Symposium on Diagnostics for Electric Machines, Power Electronics and Drives (SDEMPED), Valencia, Spain, 27–30 August 2013; pp. 371–376.
7. Wang, X.; Mao, D.; Li, X. Bearing fault diagnosis based on vibro-acoustic data fusion and 1D-CNN network. *Measurement* **2021**, *173*, 108518. [CrossRef]
8. Inturi, V.; Sabareesh, G.; Supradeepan, K.; Penumakala, P. Integrated condition monitoring scheme for bearing fault diagnosis of a wind turbine gearbox. *J. Vib. Control* **2019**, *25*, 1852–1865. [CrossRef]
9. Zhang, T.; Chen, J.; Li, F.; Zhang, K.; Lv, H.; He, S.; Xu, E. Intelligent fault diagnosis of machines with small & imbalanced data: A state-of-the-art review and possible extensions. *ISA Trans.* **2022**, *119*, 152–171.
10. Moosavian, A.; Ahmadi, H.; Tabatabaeefar, A.; Khazaee, M. Comparison of two classifiers; K-nearest neighbor and artificial neural network, for fault diagnosis on a main engine journal-bearing. *Shock Vib.* **2013**, *20*, 263–272. [CrossRef]
11. Rezaeianjouybari, B.; Shang, Y. Deep learning for prognostics and health management: State of the art, challenges, and opportunities. *Measurement* **2020**, *163*, 107929. [CrossRef]
12. Zhao, R.; Yan, R.; Chen, Z.; Mao, K.; Wang, P.; Gao, R.X. Deep learning and its applications to machine health monitoring. *Mech. Syst. Signal Process.* **2019**, *115*, 213–237. [CrossRef]
13. Ruan, D.; Wu, Y.; Yan, J. Remaining Useful Life Prediction for Aero-Engine Based on LSTM and CNN. In Proceedings of the 2021 33rd Chinese Control and Decision Conference (CCDC), Kunming, China, 22–24 May 2021; pp. 6706–6712.

14. Liu, H.; Zhou, J.; Xu, Y.; Zheng, Y.; Peng, X.; Jiang, W. Unsupervised fault diagnosis of rolling bearings using a deep neural network based on generative adversarial networks. *Neurocomputing* **2018**, *315*, 412–424. [CrossRef]
15. Clarivate. Citation Report of Bearing Fault Diagnosis and Data Augmentation- Web of Science Core Collection. Available online: https://www.webofscience.com/wos/woscc/citation-report/4bc631de-ab77-494c-a5bf-f8fa0b0176e0-588287e2 (accessed on 24 October 2022).
16. Clarivate. Citation Report of Bearing Fault Diagnosis and GAN—Web of Science Core Collection. Available online: https://www.webofscience.com/wos/woscc/citation-report/44ace094-c782-4554-8934-ab2fe0af70e8-5882d97f (accessed on 24 October 2022).
17. VOSViewer. VOSviewer—Visualizing Scientific Landscapes. Available online: https://www.vosviewer.com (accessed on 24 October 2022).
18. Zhang, H.; Li, M. RWO-Sampling: A random walk over-sampling approach to imbalanced data classification. *Inf. Fusion* **2014**, *20*, 99–116. [CrossRef]
19. Chawla, N.V.; Bowyer, K.W.; Hall, L.O.; Kegelmeyer, W.P. SMOTE: Synthetic minority over-sampling technique. *J. Artif. Intell. Res.* **2002**, *16*, 321–357. [CrossRef]
20. Mao, W.; Liu, Y.; Ding, L.; Li, Y. Imbalanced fault diagnosis of rolling bearing based on generative adversarial network: A comparative study. *IEEE Access* **2019**, *7*, 9515–9530. [CrossRef]
21. Jian, C.; Ao, Y. Imbalanced fault diagnosis based on semi-supervised ensemble learning. *J. Intell. Manuf.* **2022**, 1–16. [CrossRef]
22. Hao, W.; Liu, F. Imbalanced data fault diagnosis based on an evolutionary online sequential extreme learning machine. *Symmetry* **2020**, *12*, 1204. [CrossRef]
23. Razavi-Far, R.; Farajzadeh-Zanjani, M.; Saif, M. An integrated class-imbalanced learning scheme for diagnosing bearing defects in induction motors. *IEEE Trans. Ind. Inform.* **2017**, *13*, 2758–2769. [CrossRef]
24. Wei, J.; Huang, H.; Yao, L.; Hu, Y.; Fan, Q.; Huang, D. New imbalanced bearing fault diagnosis method based on Sample-characteristic Oversampling TechniquE (SCOTE) and multi-class LS-SVM. *Appl. Soft Comput.* **2021**, *101*, 107043. [CrossRef]
25. Zhang, W.; Peng, G.; Li, C.; Chen, Y.; Zhang, Z. A new deep learning model for fault diagnosis with good anti-noise and domain adaptation ability on raw vibration signals. *Sensors* **2017**, *17*, 425. [CrossRef]
26. Kong, Y.; Qin, Z.; Han, Q.; Wang, T.; Chu, F. Enhanced dictionary learning based sparse classification approach with applications to planetary bearing fault diagnosis. *Appl. Acoust.* **2022**, *196*, 108870. [CrossRef]
27. Qian, W.; Li, S.; Yi, P.; Zhang, K. A novel transfer learning method for robust fault diagnosis of rotating machines under variable working conditions. *Measurement* **2019**, *138*, 514–525. [CrossRef]
28. Faysal, A.; Keng, N.W.; Lim, M.H. Ensemble Augmentation for Deep Neural Networks Using 1-D Time Series Vibration Data. *arXiv* **2021**, arXiv:2108.03288.
29. Yan, Z.; Liu, H. SMoCo: A Powerful and Efficient Method Based on Self-Supervised Learning for Fault Diagnosis of Aero-Engine Bearing under Limited Data. *Mathematics* **2022**, *10*, 2796. [CrossRef]
30. Wan, W.; Chen, J.; Zhou, Z.; Shi, Z. Self-Supervised Simple Siamese Framework for Fault Diagnosis of Rotating Machinery With Unlabeled Samples. *IEEE Trans. Neural Netw. Learn. Syst.* **2022**. [CrossRef]
31. Peng, T.; Shen, C.; Sun, S.; Wang, D. Fault Feature Extractor Based on Bootstrap Your Own Latent and Data Augmentation Algorithm for Unlabeled Vibration Signals. *IEEE Trans. Ind. Electron.* **2021**, *69*, 9547–9555. [CrossRef]
32. Ding, Y.; Zhuang, J.; Ding, P.; Jia, M. Self-supervised pretraining via contrast learning for intelligent incipient fault detection of bearings. *Reliab. Eng. Syst. Saf.* **2022**, *218*, 108126. [CrossRef]
33. Zhao, J.; Yang, S.; Li, Q.; Liu, Y.; Gu, X.; Liu, W. A new bearing fault diagnosis method based on signal-to-image mapping and convolutional neural network. *Measurement* **2021**, *176*, 109088. [CrossRef]
34. Yu, K.; Lin, T.R.; Ma, H.; Li, X.; Li, X. A multi-stage semi-supervised learning approach for intelligent fault diagnosis of rolling bearing using data augmentation and metric learning. *Mech. Syst. Signal Process.* **2021**, *146*, 107043. [CrossRef]
35. Meng, Z.; Guo, X.; Pan, Z.; Sun, D.; Liu, S. Data segmentation and augmentation methods based on raw data using deep neural networks approach for rotating machinery fault diagnosis. *IEEE Access* **2019**, *7*, 79510–79522. [CrossRef]
36. Ruan, D.; Zhang, F.; Yan, J. Transfer Learning Between Different Working Conditions on Bearing Fault Diagnosis Based on Data Augmentation. *IFAC-PapersOnLine* **2021**, *54*, 1193–1199. [CrossRef]
37. Neupane, D.; Seok, J. Deep learning-based bearing fault detection using 2-D illustration of time sequence. In Proceedings of the 2020 International Conference on Information and Communication Technology Convergence (ICTC), Jeju Island, Korea, 21–23 October 2020; pp. 562–566.
38. Yang, R.; An, Z.; Huang, W.; Wang, R. Data Augmentation in 2D Feature Space for Intelligent Weak Fault Diagnosis of Planetary Gearbox Bearing. *Appl. Sci.* **2022**, *12*, 8414. [CrossRef]
39. Goodfellow, I.; Pouget-Abadie, J.; Mirza, M.; Xu, B.; Warde-Farley, D.; Ozair, S.; Courville, A.; Bengio, Y. Generative adversarial networks. *Commun. ACM* **2020**, *63*, 139–144. [CrossRef]
40. Wang, L.; Chen, W.; Yang, W.; Bi, F.; Yu, F.R. A state-of-the-art review on image synthesis with generative adversarial networks. *IEEE Access* **2020**, *8*, 63514–63537. [CrossRef]
41. Wang, Z.; Wang, J.; Wang, Y. An intelligent diagnosis scheme based on generative adversarial learning deep neural networks and its application to planetary gearbox fault pattern recognition. *Neurocomputing* **2018**, *310*, 213–222. [CrossRef]
42. Mirza, M.; Osindero, S. Conditional generative adversarial nets. *arXiv* **2014**, arXiv:1411.1784.

43. Odena, A.; Olah, C.; Shlens, J. Conditional image synthesis with auxiliary classifier gans. In Proceedings of the International Conference on Machine Learning, PMLR, Sydney, Australia, 6–11 August 2017; pp. 2642–2651.
44. Wang, J.; Han, B.; Bao, H.; Wang, M.; Chu, Z.; Shen, Y. Data augment method for machine fault diagnosis using conditional generative adversarial networks. *Proc. Inst. Mech. Eng. Part D J. Automob. Eng.* **2020**, *234*, 2719–2727. [CrossRef]
45. Guo, Q.; Li, Y.; Song, Y.; Wang, D.; Chen, W. Intelligent fault diagnosis method based on full 1-D convolutional generative adversarial network. *IEEE Trans. Ind. Inform.* **2019**, *16*, 2044–2053. [CrossRef]
46. Zhang, T.; Chen, J.; Li, F.; Pan, T.; He, S. A small sample focused intelligent fault diagnosis scheme of machines via multimodules learning with gradient penalized generative adversarial networks. *IEEE Trans. Ind. Electron.* **2020**, *68*, 10130–10141. [CrossRef]
47. Liu, Y.; Jiang, H.; Wang, Y.; Wu, Z.; Liu, S. A conditional variational autoencoding generative adversarial networks with self-modulation for rolling bearing fault diagnosis. *Measurement* **2022**, *192*, 110888. [CrossRef]
48. Larsen, A.B.L.; Sønderby, S.K.; Larochelle, H.; Winther, O. Autoencoding beyond pixels using a learned similarity metric. In Proceedings of the International Conference on Machine Learning PMLR, New York, NY, USA, 20–22 June 2016; pp. 1558–1566.
49. Rathore, M.S.; Harsha, S. Non-linear Vibration Response Analysis of Rolling Bearing for Data Augmentation and Characterization. *J. Vib. Eng. Technol.* **2022**. [CrossRef]
50. Zhang, K.; Chen, Q.; Chen, J.; He, S.; Li, F.; Zhou, Z. A multi-module generative adversarial network augmented with adaptive decoupling strategy for intelligent fault diagnosis of machines with small sample. *Knowl.-Based Syst.* **2022**, *239*, 107980. [CrossRef]
51. Liu, S.; Jiang, H.; Wu, Z.; Liu, Y.; Zhu, K. Machine fault diagnosis with small sample based on variational information constrained generative adversarial network. *Adv. Eng. Inform.* **2022**, *54*, 101762. [CrossRef]
52. Dewi, C.; Chen, R.C.; Liu, Y.T.; Tai, S.K. Synthetic Data generation using DCGAN for improved traffic sign recognition. *Neural Comput. Appl.* **2022**, *34*, 21465–21480. [CrossRef]
53. Radford, A.; Metz, L.; Chintala, S. Unsupervised representation learning with deep convolutional generative adversarial networks. *arXiv* **2015**, arXiv:1511.06434.
54. Luo, J.; Huang, J.; Li, H. A case study of conditional deep convolutional generative adversarial networks in machine fault diagnosis. *J. Intell. Manuf.* **2021**, *32*, 407–425. [CrossRef]
55. Zheng, M.; Chang, Q.; Man, J.; Liu, Y.; Shen, Y. Two-Stage Multi-Scale Fault Diagnosis Method for Rolling Bearings with Imbalanced Data. *Machines* **2022**, *10*, 336. [CrossRef]
56. Tong, Q.; Lu, F.; Feng, Z.; Wan, Q.; An, G.; Cao, J.; Guo, T. A novel method for fault diagnosis of bearings with small and imbalanced data based on generative adversarial networks. *Appl. Sci.* **2022**, *12*, 7346. [CrossRef]
57. Liu, S.; Jiang, H.; Wu, Z.; Li, X. Data synthesis using deep feature enhanced generative adversarial networks for rolling bearing imbalanced fault diagnosis. *Mech. Syst. Signal Process.* **2022**, *163*, 108139. [CrossRef]
58. Arjovsky, M.; Chintala, S.; Bottou, L. Wasserstein generative adversarial networks. In Proceedings of the International Conference on Machine Learning. PMLR, Sydney, Australia, 6–11 August 2017; pp. 214–223.
59. Zhang, Y.; Ai, Q.; Xiao, F.; Hao, R.; Lu, T. Typical wind power scenario generation for multiple wind farms using conditional improved Wasserstein generative adversarial network. *Int. J. Electr. Power Energy Syst.* **2020**, *114*, 105388. [CrossRef]
60. Zhang, T.; He, S.; Chen, J.; Pan, T.; Zhou, Z. Towards Small Sample Challenge in Intelligent Fault Diagnosis: Attention Weighted Multi-depth Feature Fusion Net with Signals Augmentation. *IEEE Trans. Instrum. Meas.* **2022**, *71*, 1–13. [CrossRef]
61. Zhu, H.; Huang, Z.; Lu, B.; Cheng, F.; Zhou, C. Imbalance domain adaptation network with adversarial learning for fault diagnosis of rolling bearing. *Signal Image Video Process.* **2022**, *16*, 2249–2257. [CrossRef]
62. Gulrajani, I.; Ahmed, F.; Arjovsky, M.; Dumoulin, V.; Courville, A.C. Improved training of wasserstein gans. *Adv. Neural Inf. Process. Syst.* **2017**, *30*, 5769–5779.
63. Mao, X.; Li, Q.; Xie, H.; Lau, R.Y.; Wang, Z.; Paul Smolley, S. Least squares generative adversarial networks. In Proceedings of the IEEE International Conference on Computer Vision, Venice, Italy, 22–29 October 2017; pp. 2794–2802.
64. Dewi, C.; Chen, R.C.; Liu, Y.T.; Yu, H. Various generative adversarial networks model for synthetic prohibitory sign image generation. *Appl. Sci.* **2021**, *11*, 2913. [CrossRef]
65. Anas, E.R.; Onsy, A.; Matuszewski, B.J. Ct scan registration with 3d dense motion field estimation using lsgan. In Proceedings of the Medical Image Understanding and Analysis: 24th Annual Conference, MIUA 2020, Oxford, UK, 15–17 July 2020; pp. 195–207.
66. Wang, R.; Zhang, S.; Chen, Z.; Li, W. Enhanced generative adversarial network for extremely imbalanced fault diagnosis of rotating machine. *Measurement* **2021**, *180*, 109467. [CrossRef]
67. Zhao, J.; Mathieu, M.; LeCun, Y. Energy-based generative adversarial network. *arXiv* **2016**, arXiv:1609.03126.
68. Yang, J.; Yin, S.; Gao, T. An efficient method for imbalanced fault diagnosis of rotating machinery. *Meas. Sci. Technol.* **2021**, *32*, 115025. [CrossRef]
69. Berthelot, D.; Schumm, T.; Metz, L. Began: Boundary equilibrium generative adversarial networks. *arXiv* **2017**, arXiv:1703.10717.
70. Jolicoeur-Martineau, A. The relativistic discriminator: A key element missing from standard GAN. *arXiv* **2018**, arXiv:1807.00734.
71. Yu, Y.; Guo, L.; Gao, H.; Liu, Y. PCWGAN-GP: A New Method for Imbalanced Fault Diagnosis of Machines. *IEEE Trans. Instrum. Meas.* **2022**, *71*, 1–11. [CrossRef]
72. Ruan, D.; Song, X.; Gühmann, C.; Yan, J. Collaborative Optimization of CNN and GAN for Bearing Fault Diagnosis under Unbalanced Datasets. *Lubricants* **2021**, *9*, 105. [CrossRef]

73. Guan, S.; Loew, M. Evaluation of generative adversarial network performance based on direct analysis of generated images. In Proceedings of the 2019 IEEE Applied Imagery Pattern Recognition Workshop (AIPR), Washington, DC, USA, 15–17 October 2019; pp. 1–5.
74. Hang, Q.; Yang, J.; Xing, L. Diagnosis of rolling bearing based on classification for high dimensional unbalanced data. *IEEE Access* **2019**, *7*, 79159–79172. [CrossRef]
75. Zhang, W.; Li, X.; Jia, X.D.; Ma, H.; Luo, Z.; Li, X. Machinery fault diagnosis with imbalanced data using deep generative adversarial networks. *Measurement* **2020**, *152*, 107377. [CrossRef]
76. Li, Z.; Zheng, T.; Wang, Y.; Cao, Z.; Guo, Z.; Fu, H. A novel method for imbalanced fault diagnosis of rotating machinery based on generative adversarial networks. *IEEE Trans. Instrum. Meas.* **2020**, *70*, 1–17. [CrossRef]
77. Peng, Y.; Wang, Y.; Shao, Y. A novel bearing imbalance Fault-diagnosis method based on a Wasserstein conditional generative adversarial network. *Measurement* **2022**, *192*, 110924. [CrossRef]
78. Pham, M.T.; Kim, J.M.; Kim, C.H. Rolling Bearing Fault Diagnosis Based on Improved GAN and 2-D Representation of Acoustic Emission Signals. *IEEE Access* **2022**, *10*, 78056–78069. [CrossRef]
79. Ruan, D.; Chen, Y.; Gühmann, C.; Yan, J.; Li, Z. Dynamics Modeling of Bearing with Defect in Modelica and Application in Direct Transfer Learning from Simulation to Test Bench for Bearing Fault Diagnosis. *Electronics* **2022**, *11*, 622. [CrossRef]
80. Ruan, D.; Wu, Y.; Yan, J.; Gühmann, C. Fuzzy-Membership-Based Framework for Task Transfer Learning Between Fault Diagnosis and RUL Prediction. *IEEE Trans. Reliab.* **2022**. [CrossRef]
81. Deng, Y.; Huang, D.; Du, S.; Li, G.; Zhao, C.; Lv, J. A double-layer attention based adversarial network for partial transfer learning in machinery fault diagnosis. *Comput. Ind.* **2021**, *127*, 103399. [CrossRef]
82. Pei, X.; Su, S.; Jiang, L.; Chu, C.; Gong, L.; Yuan, Y. Research on Rolling Bearing Fault Diagnosis Method Based on Generative Adversarial and Transfer Learning. *Processes* **2022**, *10*, 1443. [CrossRef]
83. Akhenia, P.; Bhavsar, K.; Panchal, J.; Vakharia, V. Fault severity classification of ball bearing using SinGAN and deep convolutional neural network. *Proc. Inst. Mech. Eng. Part C J. Mech. Eng. Sci.* **2022**, *236*, 3864–3877. [CrossRef]
84. Mao, W.; He, J.; Sun, B.; Wang, L. Prediction of Bearings Remaining Useful Life Across Working Conditions Based on Transfer Learning and Time Series Clustering. *IEEE Access* **2021**, *9*, 135285–135303. [CrossRef]
85. Fu, B.; Yuan, W.; Cui, X.; Yu, T.; Zhao, X.; Li, C. Correlation analysis and augmentation of samples for a bidirectional gate recurrent unit network for the remaining useful life prediction of bearings. *IEEE Sensors J.* **2020**, *21*, 7989–8001. [CrossRef]
86. Li, X.; Zhang, W.; Ma, H.; Luo, Z.; Li, X. Data alignments in machinery remaining useful life prediction using deep adversarial neural networks. *Knowl.-Based Syst.* **2020**, *197*, 105843. [CrossRef]
87. Yu, J.; Guo, Z. Remaining useful life prediction of planet bearings based on conditional deep recurrent generative adversarial network and action discovery. *J. Mech. Sci. Technol.* **2021**, *35*, 21–30. [CrossRef]
88. Karpatne, A.; Watkins, W.; Read, J.; Kumar, V. Physics-guided neural networks (pgnn): An application in lake temperature modeling. *arXiv* **2017**, arXiv:1710.11431.
89. Krupp, L.; Hennig, A.; Wiede, C.; Grabmaier, A. A Hybrid Framework for Bearing Fault Diagnosis using Physics-guided Neural Networks. In Proceedings of the 2020 27th IEEE International Conference on Electronics, Circuits and Systems (ICECS), Scotland, UK, 23–25 November 2020; pp. 1–2.

lubricants

Technical Note

Ensemble Deep Learning for Wear Particle Image Analysis

Ronit Shah [1], Naveen Venkatesh Sridharan [1], Tapan K. Mahanta [1], Amarnath Muniyappa [2], Sugumaran Vaithiyanathan [1,*], Sangharatna M. Ramteke [3,*] and Max Marian [3,*]

1 School of Mechanical Engineering, Vellore Institute of Technology, Chennai 600127, India
2 Tribology and Machine Dynamics Laboratory, Department of Mechanical Engineering, Indian Institute of Information Technology Design and Manufacturing Jabalpur, Jabalpur 482005, India
3 Department of Mechanical and Metallurgical Engineering, School of Engineering, Pontificia Universidad Católica de Chile, Macul 6904411, Chile
* Correspondence: sugumaran.v@vit.ac.in (S.V.); sangharatna.ramteke@uc.cl (S.M.R.); max.marian@uc.cl (M.M.)

Abstract: This technical note focuses on the application of deep learning techniques in the area of lubrication technology and tribology. This paper introduces a novel approach by employing deep learning methodologies to extract features from scanning electron microscopy (SEM) images, which depict wear particles obtained through the extraction and filtration of lubricating oil from a 4-stroke petrol internal combustion engine following varied travel distances. Specifically, this work postulates that the amalgamation of ensemble deep learning, involving the combination of multiple deep learning models, leads to greater accuracy compared to individually trained techniques. To substantiate this hypothesis, a fusion of deep learning methods is implemented, featuring deep convolutional neural network (CNN) architectures including Xception, Inception V3, and MobileNet V2. Through individualized training of each model, accuracies reached 85.93% for MobileNet V2 and 93.75% for Inception V3 and Xception. The major finding of this study is the hybrid ensemble deep learning model, which displayed a superior accuracy of 98.75%. This outcome not only surpasses the performance of the singularly trained models, but also substantiates the viability of the proposed hypothesis. This technical note highlights the effectiveness of utilizing ensemble deep learning methods for extracting wear particle features from SEM images. The demonstrated achievements of the hybrid model strongly support its adoption to improve predictive analytics and gain insights into intricate wear mechanisms across various engineering applications.

Keywords: tribology; lubrication; wear particle; ensemble deep learning; convolution neural network

Citation: Shah, R.; Sridharan, N.V.; Mahanta, T.K.; Muniyappa, A.; Vaithiyanathan, S.; Ramteke, S.M.; Marian, M. Ensemble Deep Learning for Wear Particle Image Analysis. *Lubricants* **2023**, *11*, 461. https://doi.org/10.3390/lubricants11110461

Received: 22 August 2023
Revised: 2 October 2023
Accepted: 20 October 2023
Published: 29 October 2023

1. Introduction

The integration of machine learning (ML) techniques offers the potential to revolutionize lubricant oil or wear particle image analysis, thus potentially contributing to lubrication interval decisions and enhancing equipment longevity and operational efficiency [1]. ML, a subset of artificial intelligence (AI), equips systems with the ability to autonomously learn from data and improve their performance over time [2]. In the area of tribology and lubrication technology, ML holds the promise of analysing intricate datasets derived from real-world operating conditions to derive more accurate and contextually relevant lubrication interval strategies [2,3]. This deviation from rule-based and static approaches to adaptive and data-driven decision making has the potential to mitigate the adverse effects of under- or over-lubrication, resulting in reduced friction, wear, and maintenance costs.

Deep learning (DL), a subset of ML, involves the use of artificial neural networks (ANN) to model and solve intricate problems. Its ability to handle large datasets and capture intricate patterns has led to remarkable advancements in diverse domains. In tribology, DL techniques offer the promise of enhanced predictive capabilities, quicker analysis of complex data, and novel insights into the underlying mechanisms governing friction, wear, and lubrication. Thereby, it is estimated that patterns and relationships between the (micro-) wear particles and the health of, for example, engines, as well as the prediction of the distance travelled by a vehicle can be identified. This might facilitate a more precise detection of wear particles and contaminants, potentially leading to engine damage and the prediction of the remaining useful life (RUL), as well as maintenance scheduling. As such, Hu et al. [4] employed ML to predict the mileage of a vehicle based on the wear particles present in the engine oil. Thereby, the researchers used a support vector machine (SVM) to classify the wear level and then used a linear regression model to predict the mileage with an accuracy of around 90%. Moreover, Sun et al. [5] employed deep learning methods for detecting and classifying wear of tungsten-carbide-copper matrix composites with high accuracy, whereby the algorithms learned from scanning electron microscopy (SEM) images.

Ensemble deep learning involves combining multiple DL models to improve accuracy and reduce overfitting by reducing the variance or errors that may be present in any one model; this has already been successfully employed in other disciplines [6,7]. In ensemble DL, the individual models are typically neural networks that are trained on different subsets of the data or with different configurations. Once the models are trained, the predictions made are combined in various ways to produce the final output. This can be performed using a simple average or weighted average of the individual model predictions, or by using more complex methods such as stacking or boosting. Ensemble DL are increasingly attracting attention, especially in competitions such as the ImageNet Large Scale Visual Recognition Challenge (ILSVRC). Winning models in such competitions often incorporate ensemble techniques due to their ability to improve the generalization ability of models, particularly when training data are limited or noisy. By leveraging the ability of convolution neural networks (CNNs) to extract features from images and classify them accurately, several studies have demonstrated the importance of utilizing this tool to detect relevant features [8–10]. Generally, CNNs are useful for image classification problems due to their capability to learn and extract meaningful features from input images automatically [11]. CNNs process images through multiple convolutional layers that enable them to learn different levels of features from input images in a hierarchical manner. Low-level features, such as, edges and corners and high-level features, such as shapes and objects, can be extracted from CNNs more effectively than traditional ML algorithms. Additionally, CNNs can handle the spatial dependencies between pixels in an image that are crucial for recognizing objects and patterns accurately. Overall, the powerful capabilities of CNNs make them an effective tool for image classification, contributing to their widespread use in various applications, such as computer vision, self-driving cars, medical image analysis and many others.

To summarize, ML methods are increasingly being employed in the context of tribology and have the potential to revolutionize wear particle image analysis to correlate features with the components' health. In this context, this contribution is based on the hypothesis that ensemble deep learning methods can identify relevant features from SEM images of wear particles with higher accuracy than individually trained ML and DL methods, thus representing a prospective tool for identifying patterns and relationships between the wear particles and the components' health, predicting the RUL and improving maintenance practices. To this end, we employed a SEM image dataset from the wear particles present in the lubricating oil at different conditions of a 4-stroke petrol engine, artificially increased the size of the image collection by data augmentation, and trained an ensemble DL model made up of Inception V3, Xception, and MobileNet V2, as well as trained the three mentioned methods individually and compared their prediction accuracies.

2. Materials and Methods

2.1. Experimental Procedure, Data Acquisition and Augmentation

The experimental data were obtained using a newly bought scooter's air-cooled and BS IV compliant single-cylinder 4-stroke petrol engine (TVS Motors, Chennai, Tamil Nadu,

India) with overhead cam, 109.7 cm^3, a max. power of 5.88 kW, a max. torque of 8.4 Nm, and a force of 1755 N. The scooter was regularly operated in the field at speeds of 700–900 min^{-1} and the distance travelled by the vehicle was tracked through global positioning system (GPS) and odometer readings. For engine lubrication, new and fully formulated SAE 10W-30 lubricating oil was utilized. A 10 ml syringe with a 110 mm-long, 3 mm-diameter tube was put into the lubricating oil tank to collect the lubricant samples (Figure 1a). Oil samples were collected from the engine at regular intervals of 300 km, 600 km, 900 km, and 1200 km (Figure 1b) and wear particle studies were carried out. To this end, oleic acid, acting as a dispersant, was mixed with extracted oil in a ratio of 1:10, ultrasonicated for 30 min to ensure a steady dispersion of wear particles, and then filtered using the filtergram technique (Figure 1c). The employed filtering flask had a 10 mm outlet conduit, a capacity of 250 ml, and a rubber tubing connecting it to the vacuum pump (VE-115N, Value, Zabrze, Poland). The flask's entrance was sealed with a laboratory rubber stopper with a hole that could be filled with a Buchner funnel containing PTFE filter paper (Nupore, Ghaziabad, Uttar Pradesh, India) with a diameter of 47 mm and a pore size of 2 μm. Following the filtering procedure, the filter paper was removed from the Buchner funnel and dried for an hour in a warm oven (WIST, Palghar, Maharashtra, India) at 35 °C. The wear particles were first removed from the filter paper using conductive carbon adhesive tape and subsequently analyzed using SEM imaging (Supra 55, Carl Zeiss, Oberkochen, Germany). The SEM images, as shown in Figure 1d, were collected using an electron current of 100 nA, an accelerating voltage of 0.02–30 kV, and a working distance of 8.5 mm. The images were then categorized/labelled and stored as *.jpg to create a uniform dataset at a scale of 10 μm. Subsequently, the dataset was transformed into binary images using Mathworks Matlab to enhance interpretability and expanded artificially by data augmentation [9] to yield a total of 400 images (100 per class) through various image transformation techniques, including rotation, shifting, flipping, adding noise, warping, blurring, zooming, etc., using AI [10] to obtain sufficient data for training. The resulting augmented dataset, which is made available under https://github.com/Sangharatna786/SEM-Images.git (accessed on 22 August 2023), was further split into 80% for training and 20% for testing the CNNs (Figure 1e), whereby the objective of the CNN was to correctly classify the wear particles to the engine condition.

2.2. Deep Learning

The employed DL CNNs were composed of artificial neurons in multiple convolution, pooling, as well as fully linked layers and utilized convolution to scale down the SEM images into a more manageable size without losing information. Thereby, the input pictures were run through a number of convolutional layers, each of which applies a different set of filters to the input image to extract key features. These filters were learned during the training process to typically capture simple features, such as edges and corners in the lower layers, and more complex features, like shapes and patterns in the higher layers. Generally, more complex features can be recognized with the growing number of layers. The spatial size of the convolved features could be decreased by the pooling layer, lowering the dimensions allowed to decrease the computational costs of data processing. After the convolutional layer, the output was passed through one or more fully connected layers to perform the classification task [12]. The final output was a probability distribution over the possible classes. Within the scope of this contribution, we employed three different CNN models, namely Inception V3, Xception, and MobileNetV2. These models, which are described in more detail in the following, reflect different advantages in terms of extraction capability, computational efficiency, and model size; these choices align with the specific needs of wear particle feature extraction from SEM images, where diverse particle sizes and complex patterns demand a range of architectural strengths while considering practical deployment and computational demands.

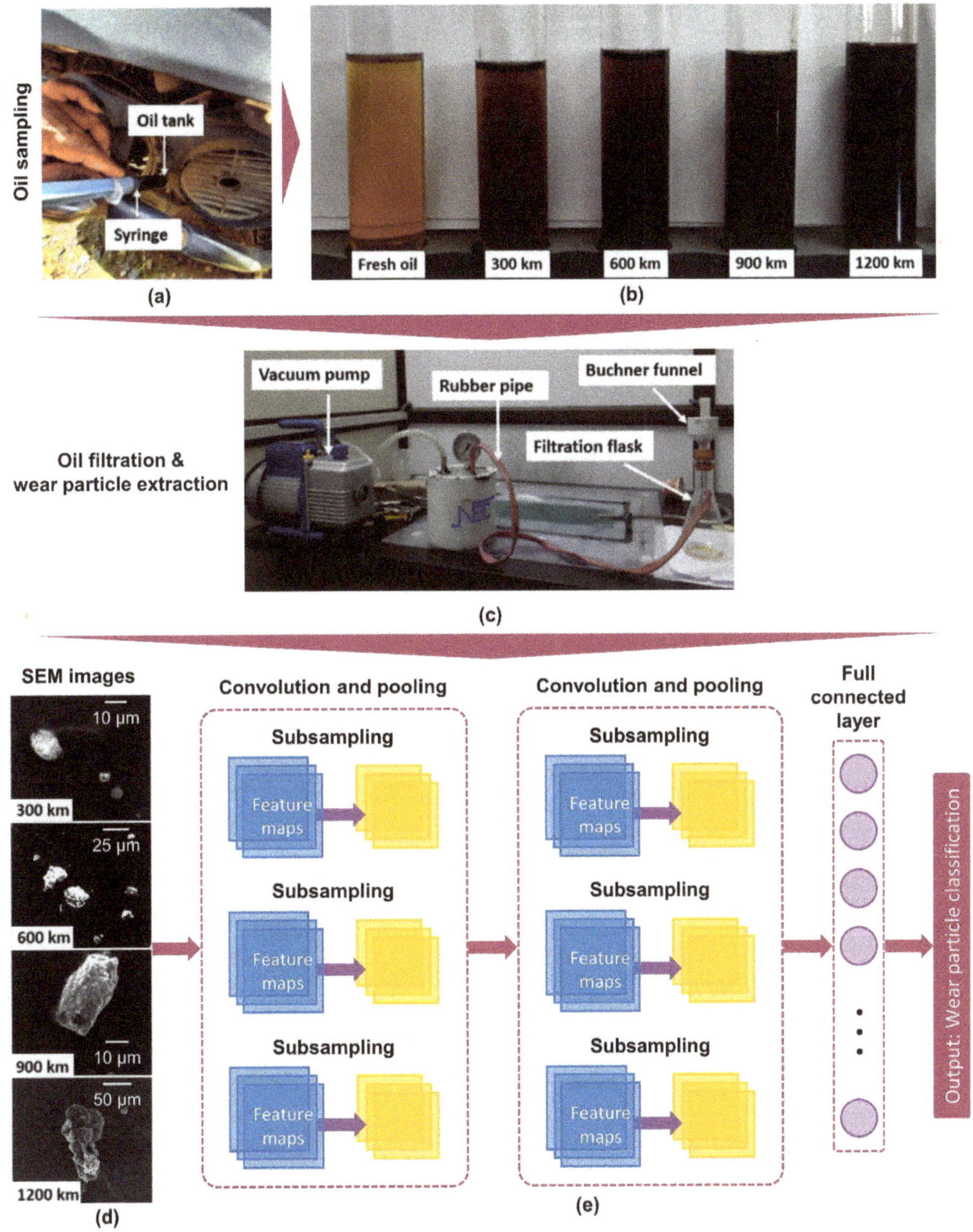

Figure 1. (**a**) Sampling lubricant from the engine, (**b**) lubricating oil samples after various intervals, (**c**) lubricant sample filtration setup, (**d**) representative SEM images of wear particles after various intervals, and (**e**) schematic of an image-processing CNN.

2.2.1. Inception V3

The deep neural network architecture Inception was introduced by Google in 2015 and is intended for tasks requiring picture recognition [13]. Inception V3 (GoogleNet V3) is based on a combination of convolutional layers of different sizes and pooling operations that extract features from the input image at different scales. At the onset of the network, the architecture employs a "stem" module, which comprises a series of convolutional and pooling layers that work together to decrease the spatial dimensions of the input image and increase the number of channels in the feature maps. InceptionV3 also uses a series of "Inception" modules that include multiple parallel convolutional and pooling operations of different sizes and aspect ratios. These operations are concatenated together along the channel dimension, allowing the network to capture features at different scales and resolutions. In addition, Inception V3 uses batch normalization and regularization techniques such as dropout and weight decay to improve the training stability and prevent overfitting. Thus, it is effective at capturing both fine-grained and global features in images due to its multi-scale approach and balance between model size and performance. Inception V3 has attained leading-edge results on various image identification benchmarks. Additionally, the architecture has been utilized as a feature extractor for different vision tasks, such as object detection and segmentation, and has been incorporated into well-known DL frameworks like TensorFlow 2.14.0 and PyTorch 2.1.0 + vu118.

2.2.2. Xception

Xception is a deep neural network architecture proposed by Google in 2016, extending the Inception architecture to use depth-wise separable convolutions in place of standard convolutions [14]. This means a factorization of standard convolutions that split the convolution into two separate operations: a depth-wise convolution, where one filter is applied to each input channel, followed by a point-wise convolution, where the output of the depth-wise convolution is subjected to a linear combination of 1×1 filters. This keeps the convolution's accuracy high while reducing the number of parameters and calculations. The Xception architecture replaces each Inception module with a series of depth-wise separable convolution blocks. Each block comprises a depth-wise convolution layer, followed by a batch normalization layer, a rectified linear unit (ReLU) activation layer, a pointwise convolution layer, another batch normalization layer, another ReLU activation layer, and a skip connection that adds the input to the output of the convolution. These blocks can be stacked to form a deep network that can learn intricate feature representations using fewer parameters and computations than traditional convolutional networks, providing strong feature extraction capabilities, especially when dealing with complex patterns in images.

2.2.3. MobileNetV2

MobileNet is a deep neural network architecture designed by Google in 2018 for mobile and embedded vision applications that require low latency and low power consumption [15]. MobileNetV2 uses a combination of depth-wise separable convolutions and linear bottleneck blocks to reduce the number of parameters and computations required for inference, while increasing the nonlinearity and preserving the information flow, thus maintaining high accuracy on image classification tasks. MobileNetV2 also introduces a new inverted residual structure that improves the accuracy and efficiency of the network. The inverted residual block consists of a linear bottleneck layer, followed by a depth-wise separable convolution and another linear bottleneck layer. The input and output of the block are connected by a shortcut connection that skips the depth-wise separable convolution, similar to the ResNet architecture. MobileNet V2 is significantly smaller and faster compared to models like Inception V3 and Xception. Also, it is a feature extractor that has been pre-trained on the Image Net dataset and may be adjusted for a range of vision tasks, including facial recognition, semantic segmentation, and object detection. MobileNet V2 has been implemented in popular DL frameworks, such as TensorFlow and PyTorch, and has achieved state-of-the-art results on mobile and embedded platforms with limited computational resources.

2.2.4. Transfer Learning and Fine-Tuning

Transfer learning is a technique that involves utilizing pre-trained models (Sections 2.2.1–2.2.3) as the starting point for a new model on a different task [16]. The rationale behind this approach is that the pre-trained model has already learned informative features from a vast dataset and these features can serve as a foundation for learning new features in a related task with less data and computational resources. Fine-tuning is a specific type of transfer learning that entails further training of the pre-trained model on the new task by adjusting the weights of some or all of its layers, whereby the degree of fine-tuning is dependent on the similarity between the initial and new tasks. After transferring pre-trained weights for Inception V3, Xception, and MobileNet V2, the model architectures were adjusted in accordance with the collected dataset. Generic image features were used in the initial layers of the pre-trained models, while domain-specific features were used for training in the following levels. Thereby, a minimum learning rate was applied for the pre-trained models to extract picture characteristics in the first few layers and encourage slow learning in the following ones. According to the chosen test circumstances, fully linked layers of pre-trained networks with 1000 neurons were changed and fixed to six neurons. A detailed specification of the pre-trained CNNs that were finally employed is summarized in Table 1.

Table 1. Detailed specification of pre-trained networks employed in this study.

Deep Learning Model	Number of Parameters	Depth
Inception V3	23.8 Million	159
Xception V2	22.9 Million	71
MobileNet V2	3.4 Million	53

2.2.5. Ensemble Learning

In order to enhance the overall performance, ensemble learning was utilized by combining the outputs of three pre-trained DL models Inception V3, Xception, and MobileNet V2 in accordance with [17]. As depicted in Figure 2, the features obtained from these models were concatenated and passed through a dropout layer with a 0.5 dropout rate, followed by a classification layer. The dropout layer helped to prevent overfitting while reducing computational time.

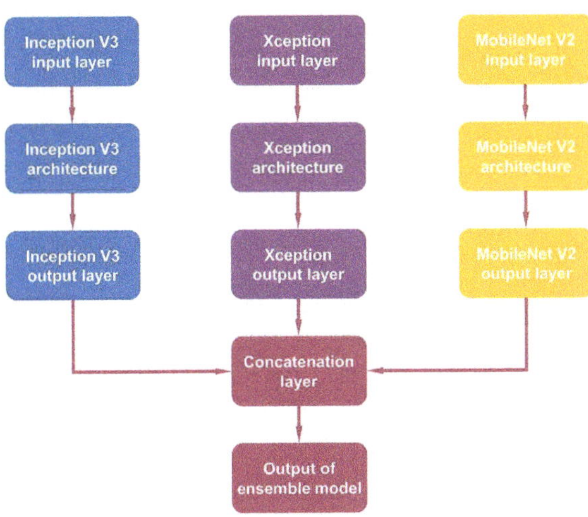

Figure 2. Workflow of the adopted ensemble deep learning approach.

3. Results and Discussions

The overall test accuracies of Inception V3, Xception, and MobileNet V2 when trained individually were 93.75%, 93.75%, and 85.93%, respectively. Thus, these models already feature superior accuracy compared to other ML approaches, such as SVM, when employed in a comparable scenario [4] (however, it should be noted that the underlying data were different and a direct comparison is not fair). The training (blue) and validation (orange) accuracies, as well as losses over training epochs for the three pre-trained models, are depicted in Figure 3a–f, whereby smooth curves could generally be observed. Furthermore, confusion matrices comparing the predicted and actual classes (i.e., travelled distances) of the testing data in its rows and columns as illustrated in Figure 4a–c were employed to assess the level of prediction of each model. Despite featuring good overall accuracy, the MobileNet V2 featured more than double or even triple the number of misclassifications (12), which indicates a lack of confidence throughout the classification in all four categories (300, 600, 900, and 1200 km), in comparison with Inception V3 (5) and Xception (4).

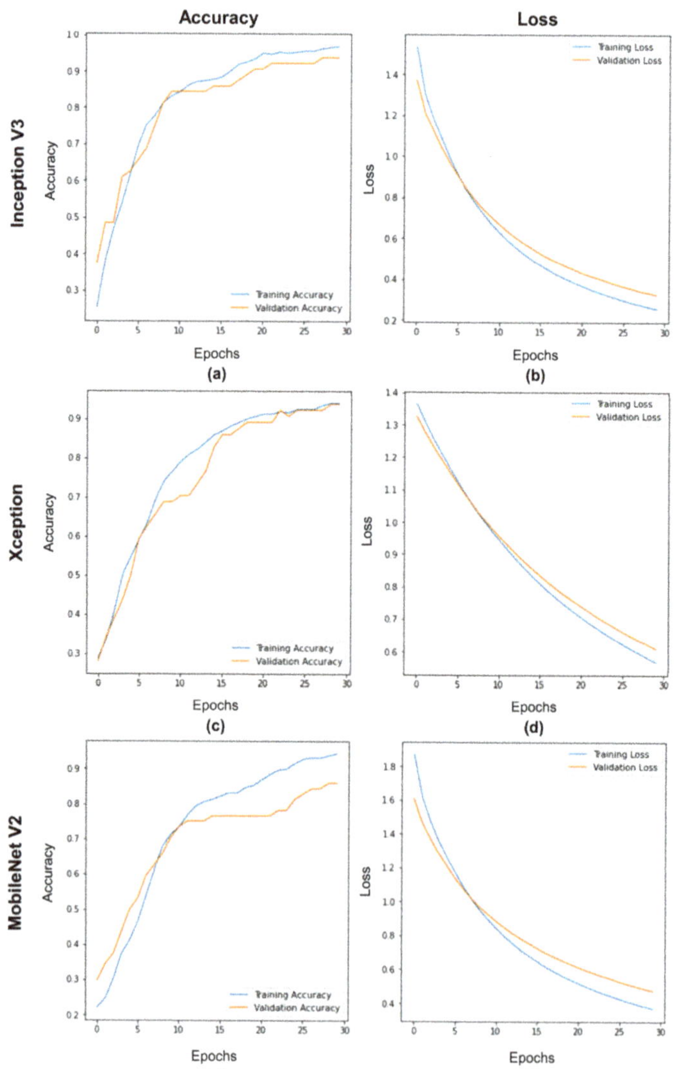

Figure 3. Training and validation (**a**,**c**,**e**) accuracies and (**b**,**d**,**f**) losses for the individually trained (**a**,**b**) Inception V3, (**c**,**d**) Xception, and (**e**,**f**) MobileNet V2 deep learning approaches.

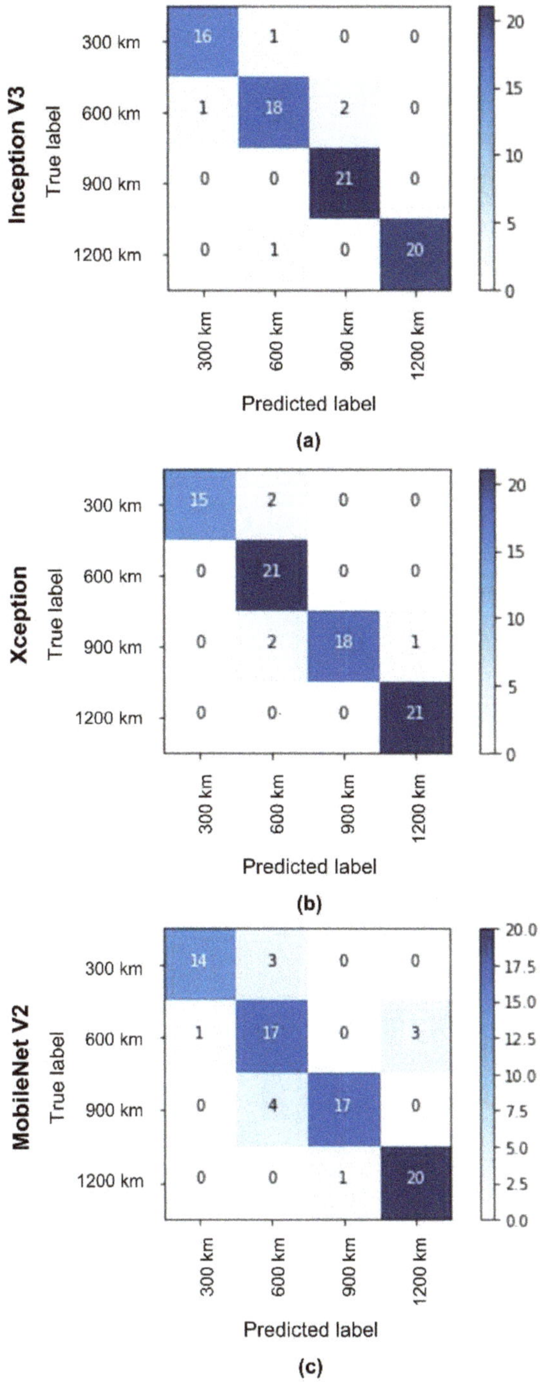

Figure 4. Confusion matrices for the testing data using the individually trained (**a**) Inception V3, (**b**) Xception, and (**c**) MobileNet V2 deep learning approaches.

In comparison to the individually trained DL approaches, the ensemble methods combining the three pre-trained deep neural networks featured a superior accuracy of 98.75%, which points towards a higher generalizability of the technique. This can also be seen in the initially already very high and fast converging training (blue) and validation (orange) accuracies, as well as losses over training epochs as shown in Figure 5a,b. As can be seen from the confusion matrix in Figure 6, the ensemble method only featured one misclassification that occurred in one of the classes (where the vehicle had travelled 600 km) and achieved perfect classification in all other classes. These findings suggest that the image features of these classes were well learned during training. The superiorness can be attributed to the ensemble's ability to capture a broader range of patterns and relationships within the data. Additionally, the model diversity mitigates the risk of overfitting by preventing it from memorizing the training data. The proposed model employed depth-wise separable convolution layers, which implemented the factorization concept resulting in reduced design dimensions and computational costs. These findings indicate that the proposed model may outperform each model regarding classification accuracy.

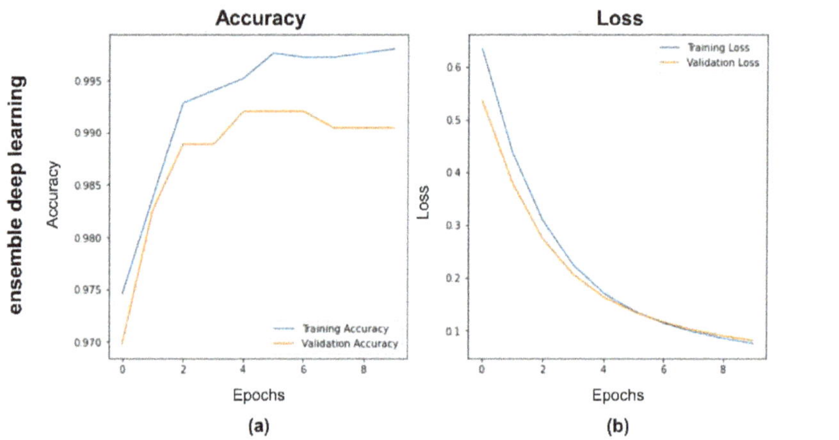

Figure 5. Training and validation (**a**) accuracies and (**b**) losses for employed ensemble deep learning approach.

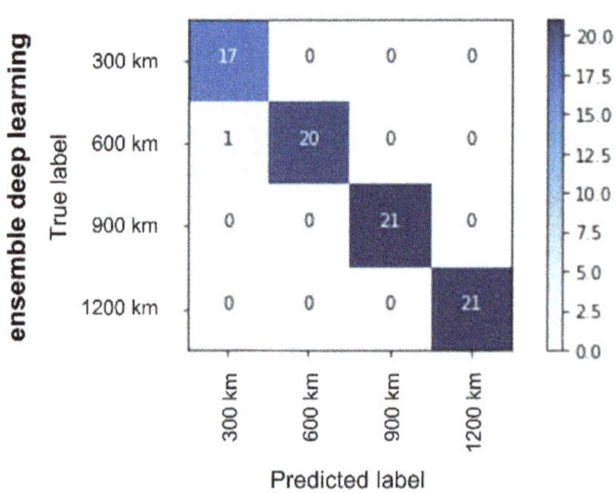

Figure 6. Confusion matrix for testing data using the ensemble deep learning approach.

4. Conclusions

The increasing integration of ML methodologies within the area of tribology shows great potential, for example, in reshaping decisions pertaining to lubrication intervals. This advancement carries the capacity to significantly augment equipment longevity and amplify operational efficacy. A promising avenue for future research involves scrutinizing wear images to discover meaningful correlations between wear particles, contaminants, and overall component health. In accordance with our investigation, predicated upon the hypothesis that ensemble DL can yield more precise prognostications of pertinent parameters in contrast to individually trained DL convolutional neural networks (CNNs), this technical note aimed to contribute to this trajectory. Leveraging SEM images depicting wear particles sourced from a diverse array of distances covered by an IC engine, our methodology encompassed the utilization of various pre-trained and fine-tuned CNN architectures, namely Inception V3, Xception, and MobileNet V2. These individual models yielded commendable classification accuracies for distance estimation of 93.75%, 93.75%, and 85.93%, respectively. In contrast, the collaborative framework of ensemble learning, harnessing the collective outputs of these three pre-trained DL models, resulted in a remarkable predictive accuracy of 98.75%. Notably, this ensemble model exhibited a substantial reduction of up to 91% in misclassifications, attributable to its inherent capacity to encapsulate a wider spectrum of patterns within the data, all while mitigating overfitting concerns and preserving a commendable level of generalizability. Thus, we postulate that the application of ensemble DL strategies emerges as a sanguine avenue for assessing the condition of lubricating oils by analysing wear particles. This, in turn, has significant implications for prognosticating, for example, the RUL of equipment, as well as refining the landscape of maintenance practices. From a research and understanding point of view, one of the primary drawbacks of ML approaches as used within this study is the lack of interpretability in "black-box" models. They generate results based on complex mathematical operations and patterns that are often difficult to decipher, making it challenging to gain insights into the underlying mechanisms. These models do not incorporate prior domain knowledge or physical principles explicitly, which can result in a disconnect between the extracted features and the actual phenomena being observed. This limitation can hinder the model's ability to provide accurate explanations or insights. Future research should, therefore, focus on making the models more transparent and interpretable. Yet, the presented approaches already can perform image feature extraction at high speed and scale. It should be emphasized that this technical note sought to demonstrate the applicability of one exemplary use case scenario. However, potential applications are not limited to analyzing wear particles from SEM images, but can be extended to extract features from any sort of images from tribo-technical systems, e.g., for predicting the wear mechanisms or surface conditions from SEM [6] or even optical microscopy images, etc., where we also assume that the presented ensemble deep learning technique features superior accuracy compared to other approaches. To fully exploit the (commercial) potential, the approach should be integrated into actual predictive maintenance systems automotive, aerospace, manufacturing, and energy sectors. Additionally, future work can focus on real-time analysis, user-friendly interfaces, cloud-based solutions, and data integration for a holistic view of equipment health.

Author Contributions: Conceptualization, N.V.S., A.M., S.V. and S.M.R.; methodology, R.S., N.V.S. and S.V.; software, R.S., N.V.S. and T.K.M.; formal analysis, R.S. and T.K.M.; investigation, R.S., T.K.M., N.V.S., A.M., S.V. and S.M.R.; resources, S.V., S.M.R. and M.M.; data curation, N.V.S., S.V. and T.K.M.; writing—original draft preparation, R.S., N.V.S., S.M.R. and M.M.; writing—review and editing, M.M.; visualization, R.S., T.K.M., S.M.R. and M.M.; supervision, S.V. and M.M.; funding acquisition, S.M.R. and M.M. All authors have read and agreed to the published version of the manuscript.

Funding: Sangharatna Ramteke and Max Marian kindly acknowledge the financial support given by ANID-Chile within the project Fondecyt de Postdoctorado N° 3230027.

Data Availability Statement: The experimental data underlying the training of the ML methods are available at https://github.com/Sangharatna786/SEM-Images.git (accessed on 22 August 2023). Further data or information can be obtained from the corresponding authors upon request.

Acknowledgments: Max Marian greatly acknowledges the support from the Vicerrectoría Académica (VRA) of the Pontificia Universidad Católica de Chile within the Programa de Inserción Académica (PIA).

Conflicts of Interest: The authors declare no conflict of interest.

References

1. Ramteke, S.M.; Chelladurai, H.; Amarnath, M. Diagnosis and Classification of Diesel Engine Components Faults Using Time–Frequency and Machine Learning Approach. *J. Vib. Eng. Technol.* **2022**, *10*, 175–192. [CrossRef]
2. Marian, M.; Tremmel, S. Current Trends and Applications of Machine Learning in Tribology—A Review. *Lubricants* **2021**, *9*, 86. [CrossRef]
3. Paturi, U.M.R.; Palakurthy, S.T.; Reddy, N.S. The Role of Machine Learning in Tribology: A Systematic Review. *Arch. Comput. Methods Eng.* **2023**, *30*, 1345–1397. [CrossRef]
4. Hu, J.; Weng, L.; Du, Y.; Gao, Z. Mileage Prediction of Electric Vehicle Based on Multi Model Fusion. *J. Transp. Syst. Eng. Inf. Technol.* **2020**, *20*, 100–106. [CrossRef]
5. Sun, W.; Gao, H.; Tan, S.; Wang, Z.; Duan, L. Wear detection of WC-Cu based impregnated diamond bit matrix based on SEM image and deep learning. *Int. J. Refract. Met. Hard Mater.* **2021**, *98*, 105530. [CrossRef]
6. Hakak, S.; Alazab, M.; Khan, S.; Gadekallu, T.R.; Maddikunta, P.K.R.; Khan, W.Z. An ensemble machine learning approach through effective feature extraction to classify fake news. *Future Gener. Comput. Syst.* **2021**, *117*, 47–58. [CrossRef]
7. Cao, Y.; Geddes, T.A.; Yang, J.Y.H.; Yang, P. Ensemble deep learning in bioinformatics. *Nat. Mach. Intell.* **2020**, *2*, 500–508. [CrossRef]
8. Shorten, C.; Khoshgoftaar, T.M. A survey on Image Data Augmentation for Deep Learning. *J. Big Data* **2019**, *6*, 60. [CrossRef]
9. Ganaie, M.A.; Hu, M.; Malik, A.K.; Tanveer, M.; Suganthan, P.N. Ensemble deep learning: A review. *Eng. Appl. Artif. Intell.* **2022**, *115*, 105151. [CrossRef]
10. Aggarwal, S.L.P. Data augmentation in dermatology image recognition using machine learning. *Skin Res. Technol.* **2019**, *25*, 815–820. [CrossRef] [PubMed]
11. LeCun, Y.; Kavukcuoglu, K.; Farabet, C. Convolutional networks and applications in vision. In Proceedings of the 2010 IEEE International Symposium on Circuits and Systems—ISCAS 2010, Paris, France, 29 May–1 June 2010; IEEE: Piscataway, NJ, USA, 2010; pp. 253–256, ISBN 978-1-4244-5308-5.
12. Naveen Venkatesh, S.; Chakrapani, G.; Senapti, S.B.; Annamalai, K.; Elangovan, M.; Indira, V.; Sugumaran, V.; Mahamuni, V.S. Misfire Detection in Spark Ignition Engine Using Transfer Learning. *Comput. Intell. Neurosci.* **2022**, *2022*, 7606896. [CrossRef] [PubMed]
13. Aghayari, S.; Hadavand, A.; Mohamadnezhad Niazi, S.; Omidalizarandi, M. Building Detection from Aerial Imagery Using Inception Resnet Unet and Unet Architectures. *ISPRS Ann. Photogramm. Remote Sens. Spat. Inf. Sci.* **2023**, *10*, 9–17. [CrossRef]
14. Chollet, F. Xception: Deep Learning with Depthwise Separable Convolutions. In Proceedings of the 2017 IEEE Conference on Computer Vision and Pattern Recognition (CVPR), Honolulu, HI, USA, 21–26 July 2017; pp. 1800–1807, ISBN 1063-6919.
15. Sridharan, N.V.; Sugumaran, V. Visual fault detection in photovoltaic modules using decision tree algorithms with deep learning features. *Energy Sources Part A Recovery Util. Environ. Eff.* **2021**, 1–17. [CrossRef]
16. Tan, C.; Sun, F.; Kong, T.; Zhang, W.; Yang, C.; Liu, C. A Survey on Deep Transfer Learning. In *Artificial Neural Networks and Machine Learning—ICANN 2018*; Kůrková, V., Manolopoulos, Y., Hammer, B., Iliadis, L., Maglogiannis, I., Eds.; Springer International Publishing: Cham, Switzerland, 2018; pp. 270–279, ISBN 978-3-030-01423-0.
17. Sridharan, N.V.; Sugumaran, V. Deep learning-based ensemble model for classification of photovoltaic module visual faults. *Energy Sources Part A Recovery Util. Environ. Eff.* **2022**, *44*, 5287–5302. [CrossRef]

lubricants

MDPI

Article

A Generalised Method for Friction Optimisation of Surface Textured Seals by Machine Learning

Markus Brase *,†, Jonathan Binder †, Mirco Jonkeren and Matthias Wangenheim

Institute of Dynamics and Vibration Research, Leibniz Universität Hannover, 30823 Garbsen, Germany; jonkeren@ids.uni-hannover.de (M.J.); wangenheim@ids.uni-hannover.de (M.W.)
* Correspondence: brase@ids.uni-hannover.de
† These authors contributed equally to this work.

Abstract: Friction behaviour is an important characteristic of dynamic seals. Surface texturing is an effective method to control the friction level without the need to change materials or lubricants. However, it is difficult to put the manual prediction of optimal friction reducing textures as a function of operating conditions into practice. Therefore, in this paper, we use machine learning techniques for the prediction of optimal texture parameters for friction optimisation. The application of pneumatic piston seals serves as an illustrative example to demonstrate the machine learning method and results. The analyses of this work are based on experimentally determined data of surface texture parameters, defined by the dimple diameter, distance, and depth. Furthermore friction data between the seal and the pneumatic cylinder are measured in different friction regimes from boundary over mixed up to hydrodynamic lubrication. A particular innovation of this work is the definition of a generalised method that guides the entire machine learning process from raw data acquisition to model prediction, without committing to only a few learning algorithms. A large number of 26 regression learning algorithms are used to build machine learning models through supervised learning to evaluate the suitability of different models in the specific application context. In order to select the best model, mathematical metrics and tribological relationships, like Stribeck curves, are applied and compared with each other. The resulting model is utilised in the subsequent friction optimisation step, in which optimal surface texture parameter combinations with the lowest friction coefficients are predicted over a defined interval of relative velocities. Finally, the friction behaviour is evaluated in the context of the model and optimal value combinations of the surface texture parameters are identified for different lubrication conditions.

Keywords: supervised learning; regression techniques; surface texturing; dynamic seals

Citation: Brase, M.; Binder, J.; Jonkeren, M.; Wangenheim, M. A Generalised Method for Friction Optimisation of Surface Textured Seals by Machine Learning. *Lubricants* **2024**, *12*, 20. https://doi.org/10.3390/lubricants12010020

Received: 2 November 2023
Revised: 14 December 2023
Accepted: 26 December 2023
Published: 9 January 2024

Correction Statement: This article has been republished with a minor change. The change does not affect the scientific content of the article and further details are available within the backmatter of the website version of this article.

1. Introduction

Friction is defined as the force of resistance acting between the contact surfaces of bodies in relative motion [1]. In total, about 20% of global energy losses are due to overcoming friction [2]. Therefore, low friction is targeted in many technical systems such as seals or bearings. In order to reduce friction in tribological systems, it is necessary to understand the individual factors that influence friction and to develop appropriate strategies to minimise the friction [1].

On the one hand, material properties, such as the crystal structure [3], hardness [4,5], elastic and shear modulus [6,7], grain size [8,9], and surface energy [10,11] of the contacting materials affect the frictional behaviour. On the other hand, the operational conditions, such as the normal loads [12,13], sliding velocities [14,15], environmental conditions [16,17], temperatures [18], and lubricants [19], have a major influence on the tribological behaviour. Of particular relevance are surface coatings or modifications of the surface topographies [20], which both can contribute significantly to the friction behaviour. Surface modifications involve techniques that artificially alter the structures of the solid surfaces through defined properties. This involves texturing the surface either by adding material to create protrusions or by removing material, displacing material and using self-moulding techniques to create dimples [21,22]. Surface modifications can be achieved by changing the surface roughnesses [23], textures [24], or a combination of both [25].

In order to reduce the friction of a tribological system, the performance of the lubricant can be improved [26], the lubricant feeding conditions can be adjusted and optimised [27], special materials and coatings can be used [28], operational conditions can be modified [29], geometries can be optimised [30], and the contact surfaces can be modified [24]. In addition, it is possible to combine different processes, such as surface texturing and surface coating [31,32]. Within this work, surface textured seals are analysed as an example application. The textures, applied to the seal surfaces, are defined by the dimple diameter, dimple distance, and dimple depth. This is why, surface modifications, specifically surface texturing, are of particular interest. Surface textures have been demonstrated to positively influence friction and wear under both dry friction conditions [33] as well as boundary [34], mixed [35], elastohydrodyamic [36], and full-film hydrodynamic lubrication conditions [37]. Surface textures exhibit different beneficial effects on friction, depending on the lubrication regime. The textures can reduce the real area of contact [38], trap wear particles [39], accelerate the formation of tribolayers [40], store lubricant [41], draw additional lubricant into the contact area [25], build-up additional hydrodynamic pressure [42], and locally increase the fluid film height [43]. However, the mechanisms through which the textured surface parameters affect the friction performance, such as the texture density or depth, are still not fully understood and require further investigation [44]. Surface texturing is highly application dependent and must be evaluated for each tribological system and lubrication regime. Furthermore, the possible number of parameters for the surface texture design is immense [45].

In this respect, machine learning is a powerful tool to predict application-dependent optimum texture parameters and to overcome or reduce time consuming and expensive trial-and-error approaches [46]. The advantages and potentials of machine learning techniques lie in the handling of high-dimensional problems and the ability to adapt models to changing conditions with reasonable effort [47], even if the physics behind the tribological system is not fully understood [48]. According to the systematic reviews by Marian et al. [47] and Paturi et al. [48], the number of tribology papers successfully investigating and applying machine learning techniques is increasing exponentially. For example, about 46% of the 330 papers evaluated were written between 2018 and 2022 [48]. In addition, around 76% of the 127 articles quantitatively analysed by Marian et al. were based on experimentally collected data. The learning algorithm used in about three quarters of the papers was a neural network category algorithm, making them overrepresented in the tribological context [47]. So, it is noticeable that authors tend to focus on a few learning algorithms at an early stage. However, other algorithms can also show comparable or even better results, as presented in Section 5.2, so these should not be excluded from the analysis from the outset. Based on the "no free lunch" theorem, it is only possible to know exactly which model is the most suitable for the present application and data, if it has been trained and tested [49]. As shown in Section 5.2, it is possible that not only one algorithm shows good results, but that there are several suitable algorithms. Also, the numerous application examples cited by Marian et al. [47] show that there is no universally applicable learning algorithm for tribological problems. According to the "no free lunch" theorem, the selection of a suitable learning algorithm must, therefore, always be made individually for the prevailing application and is a challenge in the development of models in machine learning [50]. It is, therefore, necessary to optimise the selection of texturing parameters based on data from experiments or simulations, using several machine learning (ML) algorithms.

Within this paper, the MATLAB Statistics and Machine Learning Toolbox® is utilised to build regression models. The toolbox contains 26 learning algorithms from seven categories. As an innovation, all of them are taken into account during the study to make an informed model selection based on trained and tested models with various evaluation metrics. The aim is to be able to quickly identify the best of the 26 models, without having to perform complex processes such as hyperparameter optimisation of single models, and thus enable users who are not experts in machine learning to apply the ML methods. For this purpose, a generalised method is explained on the basis of the selected aspects of the tribological example application of surface textured pneumatic piston seals. This application serves as an illustrative example. However, the methodology can also be applied to other surface textured systems, such as metallic components or rubber parts.

2. Generalised Method for Machine Learning Model Generation and Application

The procedure acquired in this paper for developing machine learning models in the context of tribological applications is shown in the flowchart below, see Figure 1. Although this is not a universal and valid method in general, the flowchart shows the most common methods, provided in the literature, that can be used.

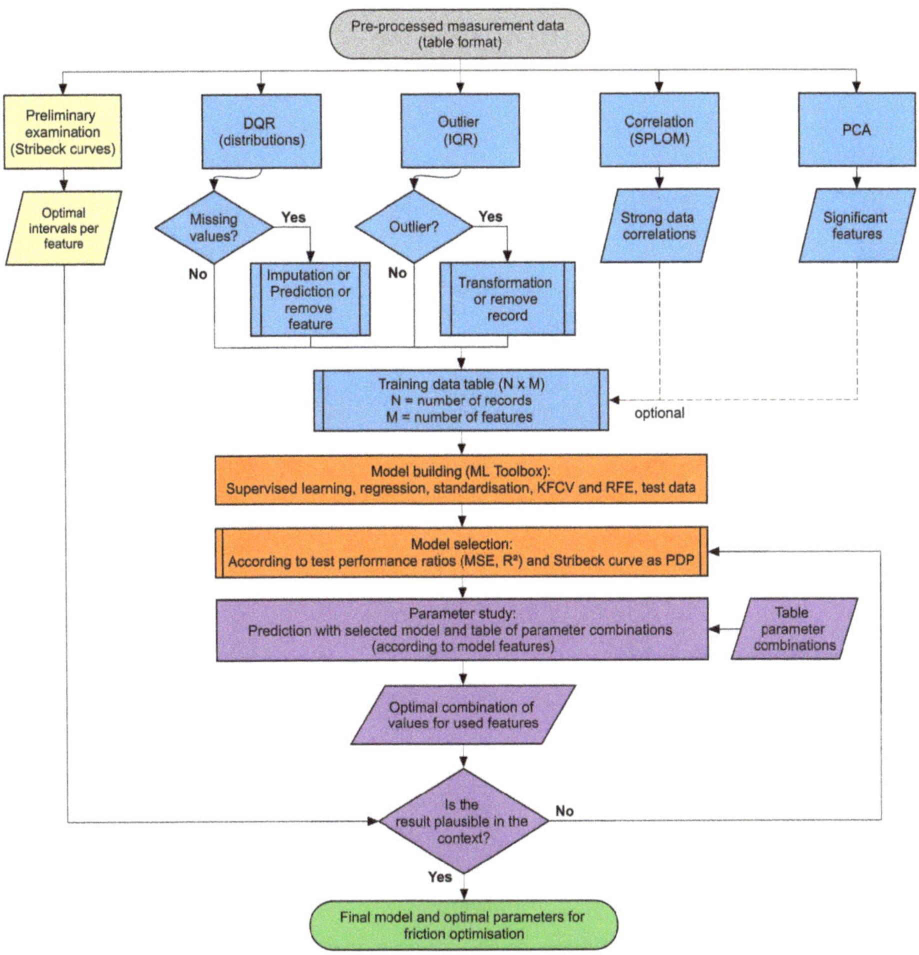

Figure 1. Flowchart of a generalised machine learning development method using standard machine learning techniques. The illustration shows an overview of the individual steps, which are briefly explained in the following chapters.

The quality of the recorded data is of great importance, so careful data acquisition must first be ensured, indicated by the grey step of the flowchart. This step is explained more in detail in Section 3 for the example application of surface textured pneumatic piston seals. Based on this, data analysis and data preparation are necessary, indicated by the blue steps of the flowchart. Using the prepared data, model building is performed, see the first orange step of the flowchart. Data analysis, data preparation, and model building are described in Section 4.

The best model is selected using mathematical performance ratios and partial dependence plots in the context of tribological relationships, like Stribeck curves, indicated by the second orange step of the flowchart, which is explained more in detail in Section 5. In addition, this chapter contains the application of the generated model for the selection of the optimal surface texture parameters, according to the purple steps of the flowchart, and the preliminary examination of the friction results, depicted by the yellow steps. Based on the green step of the flowchart, the tribological context of the friction results are discussed in Section 6.

3. Friction and Seal Surface Texture Data Acquisition

Data acquisition is a process that leads to pre-processed measurement data, highlighted as the grey step in the flowchart, shown in Figure 1. Within this step, data on friction values and surface texture parameters of dynamic pneumatic piston seals are measured, which together form the basis of the machine learning model.

3.1. Friction Measurements

The objective of the experimental testing procedure is to measure the friction forces between surface textured pneumatic piston seals and a static pneumatic cylinder tube by utilising a linear test rig, explained in Figure 2. The tribotechnical system considered to record the friction forces consists of a pneumatic piston seal and a pneumatic cylinder tube, see Figure 2b.

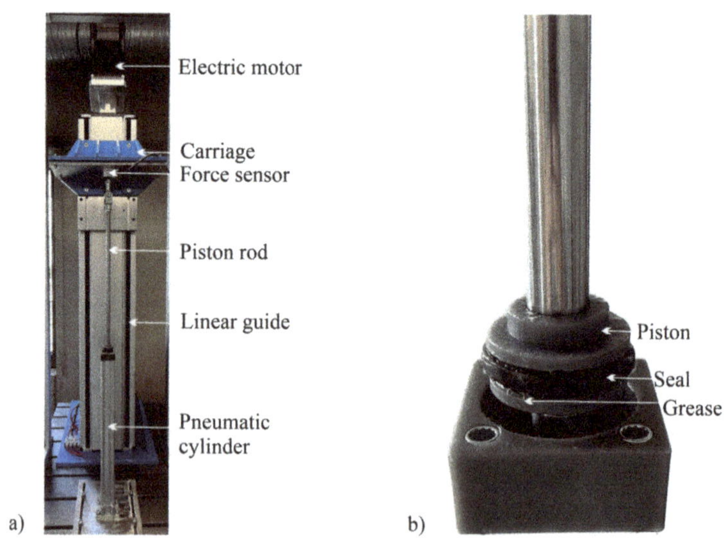

Figure 2. (**a**) Universal linear test rig. The carriage, connected to the linear guide, is driven by an electric motor. The piston rod is connected to the carriage by a force sensor. The piston rod is linked to the piston, in which the seal is installed. The seal can thus be moved relative to the static pneumatic cylinder tube at the set velocity. (**b**) Tribotechnical System. A detailed view of the piston and seal reveals the tribotechnical system. It consists of the base body (pneumatic piston seal) and the counter body (pneumatic cylinder tube). The intermediate medium is a lubricating grease.

The nominal external diameter of the seal and internal diameter of the pneumatic cylinder are equal to 40 mm. The material of the seals, which are manufactured by texturing during moulding (TDM) [51], is a fluoroelastomer with a Shore hardness of 80 A (FKM80A). The pneumatic cylinder tube is made of anodised aluminium. Throughout the test procedure, the same grease is used as in the interfacial medium [52]. An important challenge during the measurements is the creation of a lubricating film that ensures reproducible friction force measurements. Care must be taken to ensure that a consistent amount of lubricant is present inside the pneumatic cylinder for each seal that is measured. This was achieved by applying a constant mass of lubricant to the seals, cleaning the cylinder between the measurements of two seals, and applying a constant mass of lubricant to the cylinder itself. It has also been found that it is beneficial to measure friction on one seal, starting at the highest velocity and decreasing towards the lowest velocity. The reason for this is that the lubricant film is thickest at the highest velocities and is more easily dissolved than built up during the test procedure, which means that conditioning runs between velocities can be reduced. All of the experiments were performed at an ambient temperature of 20 °C. In contrast to the real technical application of a pneumatically driven actuator, the relative movement between the seal and the pneumatic cylinder was applied by the linear guide of the test rig. Within the entire test procedure, the pneumatic cylinder was depressurised. Each seal was tested at 19 test speeds ranging from 1 mm/s to 500 mm/s, moving at a predefined trapezoidal speed profile, see Figure 3a. The distance to be driven was selected between 15 mm and 450 mm depending on the velocity of the seal. For each measurement, the piston was moved in two directions from the start position to the end position and back again. This corresponded to one test cycle. A total of 12 cycles were performed for each test speed and seal. This resulted in 228 friction measurement cycles per seal, consisting of 228 downstrokes and 228 upstrokes.

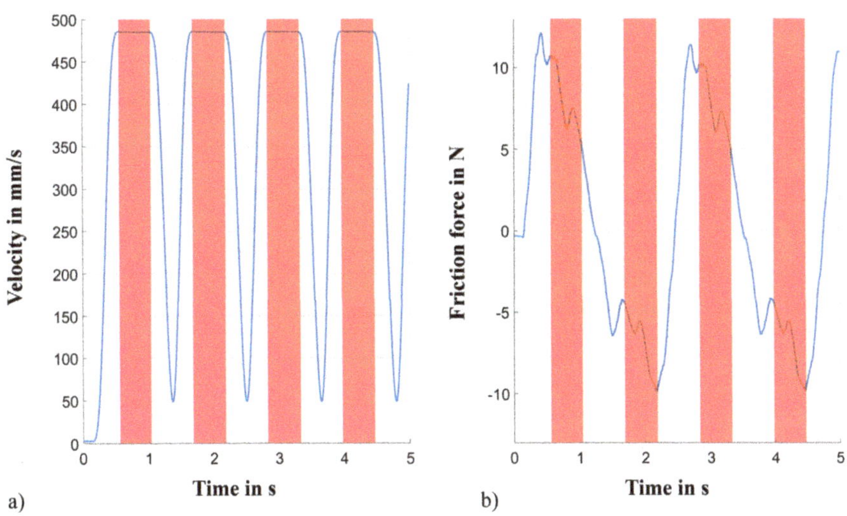

Figure 3. (**a**) Trapezoidal velocity profile of the test rig. Within the stationary area of constant velocity, marked in red, the mean value of the friction force is determined, which is used for further analyses. (**b**) Friction force as a function of time. Positive values of friction indicate the downstroke of the test rig, while negative values indicate the upstroke. The friction force signal shows a bumpy curve, as the friction distance at the higher speeds is not sufficient to generate a constant value in the friction force signal.

A conditioning run is carried out before the 12 measured cycles in order to adjust the tribological system and, in particular, the lubricant film height to the current velocity. The temperature in the contact between the seal and cylinder was not recorded. However, as the measurement results were within the 6 sigma interval and no systematic increase or decrease in the friction characteristics could be identified over the number of cycles, a temperature change in the contact was regarded to be negligible for friction evaluation. The quasi-stationary friction force signal was evaluated from the friction signal at time intervals where the velocity was stationary, see Figure 3b. To calculate the corresponding friction coefficients associated with the friction forces, which were further processed in the machine learning (ML) model, the required normal forces were provided by static FE simulations of the contact pressure distribution between seal and cylinder. The FE model, which uses a hyperelastic material model, was not the focus of this work, and was, therefore, not described in detail.

3.2. Seal Surface Texture Measurements

In contrast with other studies, the machine learning model does not use nominal texture parameters, but rather the real texture parameters of the pneumatic piston seals that are measured. The surface analyses of the seals were based on 3D microscope measurements, which were recorded using the method of focus variation. The collected data were exported as cartesian xyz data points, see Figure 4a. Based on these data points, the surface was visualised in two different areas—the textured (blue/green) and untextured (yellow/orange) areas, see Figure 4b.

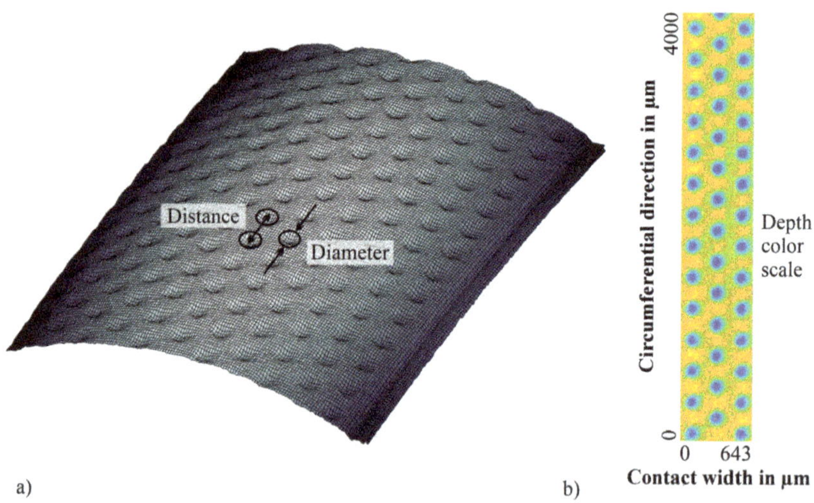

Figure 4. (**a**) Microscope surface scan of a textured seal (4 × 4 mm). The circles enclose the triangular arranged circular dimples. The texture parameters of dimple diameter and dimple distance are defined by the arrows. (**b**) Dimple depth information of the microscope scan. The yellow and orange areas represent the untextured area of the seal surface, while the green and blue areas represent the dimples with their depth.

On basis of the FE simulations already mentioned in Section 3.1, the contact width between the seal and pneumatic cylinder tube is calculated, from which the effective contact area is determined. The texture density, shown in Table 1, therefore only corresponds to the density in effective contact between the seal and cylinder and neglects the textured seal area that is not in contact, as it is tribologically irrelevant. The surface textures are defined by the dimple diameter, distance, and depth. They have a basic circular shape and are arranged in a triangular pattern. The diameter of the dimples is determined using circular approximations of the green/blue data points. The dimple depth is the mean value of all data points within these approximated circles. In addition, the distance between the dimples is calculated using the centre points of the approximated circles. This is not only done for one individual dimple, but for the entire number of dimples, which are positioned within the measuring area of 4 × 4 mm and the simulated contact width of $l = 643$ µm of the seal. The corresponding number of evaluated dimples is specified in Table 1. The related dimple parameters are determined as the mean values of the given number of individual measurements. The real texturing density is the ratio of the textured to untextured area of the analysed seal surface. Because of process tolerances that occur during the laser and vulcanisation process of the seals, odd values can be seen in Table 1.

The general seal dimensions, more precisely the inner and outer diameters of the seals and their deviation from nominal values, were not measured and considered, which is discussed in Section 6 and mentioned in the outlook.

Table 1. Real seal surface texture parameters. The values are the mean value of the specified number of analysed dimples within the simulated contact width of the seal.

Seal No.	Diameter in µm	Distance in µm	Depth in µm	Texture Density in %	Number of Analysed Dimples
1	-	-	-	-	-
2	97	195	7.9	20.4	58
3	96	243	8.6	15.3	46
4	100	244	11.4	15.8	43
5	149	244	14.5	34.3	46
6	143	292	12.9	25.2	37
7	147	294	19.3	26.8	37
8	149	294	24.8	27.2	37
9	196	293	23.6	45.0	37
10	199	390	23.3	21.9	28
11	147	293	19.1	26.3	37

4. Data Analysis, Data Preparation, and Model Development

The machine learning algorithms discussed in this paper are classified as supervised learning. Within this category, regression algorithms were used, which deal with numerical continuous output values. The training of the algorithms was conducted with a known set of input data and known responses, which are the data collected in Section 3.1. The data analysis, data preparation, and model development were guided by the method shown in Figure 1. For the development of machine learning models, high data quality is an essential requirement. This started with a comprehensive analysis of the available data to assess their quality, which included an examination of its structural characteristics and properties (data analysis). This understanding subsequently facilitated the preparation of the available data with the aim of improving its quality, as well as its transformation into the desired format (data preparation) [50,53]. Afterwards, model development began.

In order to be able to evaluate the data quality as part of the data analysis, a data quality report (DQR) according to [50] was generated in this study for the available measurement data. The report took the total number of numerical values, data completeness, cardinality, minimum and maximum values, first and third quartiles, median and arithmetic mean, and the standard deviation into account, which is shown in Table A1 of the Appendix A. It was found that there were no missing values in the features studied, that the data had a uniform character, and that there were no irregularities in the other indicators of the data quality report. Interquartile ranges (IQRs) were calculated to identify individual data points that represent mathematical outliers [53,54], which require a more detailed and individual examination. Based on this examination, it becomes apparent that the identified outliers were only default values for the test series or the texture parameters, specifically friction values, that can be evaluated as being feasible with the help of the physical relationship of the Stribeck curve. The sliding velocity can be used as an illustrative example. In Figure 5 it is visible that the sliding velocity was not sampled uniformly. The lower velocities were sampled more finely than the higher velocities. As a result, the higher values were marked as mathematical outliers, although they were not physically outliers. Therefore, all these data points were used for further analysis. The IQR calculated for the features can be found in Figure A1 of the Appendix A.

In the context of data preparation, dimensionality is particularly important. Data sets with high dimensionality lead to increased complexity, computational effort, and the risk of overfitting [55,56]. For this reason, it is more useful to have many data points for each feature, but not a larger number of features, because only features with a high significance add value to the model and its accuracy [56].

In order to reduce the number of features while preserving the most relevant information, principal component analysis (PCA) is used as a feature extraction method to assess dimensionality. As this is a requirement for PCA, the existing data sets are standardised so that the values of each feature are within the interval $[-1; 1]$ and have an arithmetic mean of zero as well as a standard deviation of one. According to the first principal component, the feature of sliding velocity has the greatest influence on the coefficient of friction, since it explains 85.34% of the variance. According to the second principal component, the texture parameters dimple diameter, dimple distance and dimple depth have the greatest influence, because they explain 13.40% of the variance. As a result, the analysis shows that the first two principal components already explain 98.74% of the variance. The features that have been examined are shown in Table 2. Based on the principal component analysis, features with little influence on the variance of the original data can be removed.

Table 2. Coefficients of the first two principal components (PC 1 and PC 2) for the examined features. The remaining features after PCA are highlighted in bold letters. Cycle and direction of motion are removed.

Feature	Sliding Velocity	Cycle	Direction of Motion	Dimple Diameter	Dimple Distance	Dimple Depth	Texture Density
PC 1 (85.34%)	1.00	0.00	0.00	0.00	0.00	0.00	0.00
PC 2 (13.40%)	0.00	0.00	0.00	0.55	0.83	0.09	0.00

The model is built using recursive feature elimination (RFE) according to the cross-validation (CV) procedure. For this purpose, the training data is divided into a training set and a test set. In this model development, 10% of the training data was used as test data. To ensure that the quality of the generated model does not depend on the division of the training data into training and validation sets, a variant of cross-validation called k-fold cross-validation (KFCV) is used [53,56]. For model development, the training data is divided into k subsets. In this study the cross-validation procedure is performed with $k = 5$ or $k = 10$ subsets, depending on the number of records.

During the RFE process, features are eliminated iterative as part of model development, which is outlined in Section 5.2. The minor influence of the cycle and the direction of motion, as already observed in the PCA (see Table 2), can be confirmed. Hence, these features are eliminated. The friction coefficients per seal and velocity are averaged over the measured up and down cycles, which is possible due to the symmetry of the seal, so that the up and downstroke force is nearly identical. The feature texture density is not removed from the training data, since it is directly related to the dimple distance and diameter. This reduces the number of data sets to $N = 190$ (10 surface textured seals and 19 tested sliding velocities), leaving $M = 5$ features.

A total of 26 learning algorithms, shown in Table 3, are used for model development using MATLAB Statistics and Machine Learning Toolbox®.

Table 3. Summary of all algorithms studied by category: Supervised Learning—Regression. Algorithm 16, which is chosen to build the best model within the prevailing application, is highlighted by underlining, compare also Table 4. The four colored algorithms correspond to the curves of Figure 6.

Category	No.	Algorithm
Linear Regression Models	1	Linear
	2	Interactions Linear
	3	Robust Linear
	4	Stepwise Linear
Regression Trees	5	Fine Tree
	6	Medium Tree
	7	Coarse Tree
Support Vector Machines	8	Linear SVM
	9	Quadratic SVM
	10	Cubic SVM
	11	Fine Gaussian SVM
	12	Medium Gaussian SVM
	13	Coarse Gaussian SVM
Gaussian Process Regression Models	14	Squared Exponential
	15	Matern 5/2
	16	Exponential
	17	Rational Quadric
Kernel Approximation Models	18	SVM Kernel
	19	Least Squares Kernel Regression
Ensembles of Trees	20	Boosted Trees
	21	Bagged Trees
Neural Networks	22	Narrow Neural Network
	23	Medium Neural Network
	24	Wide Neural Network
	25	Bilayered Neural Network
	26	Trilayered Neural Network

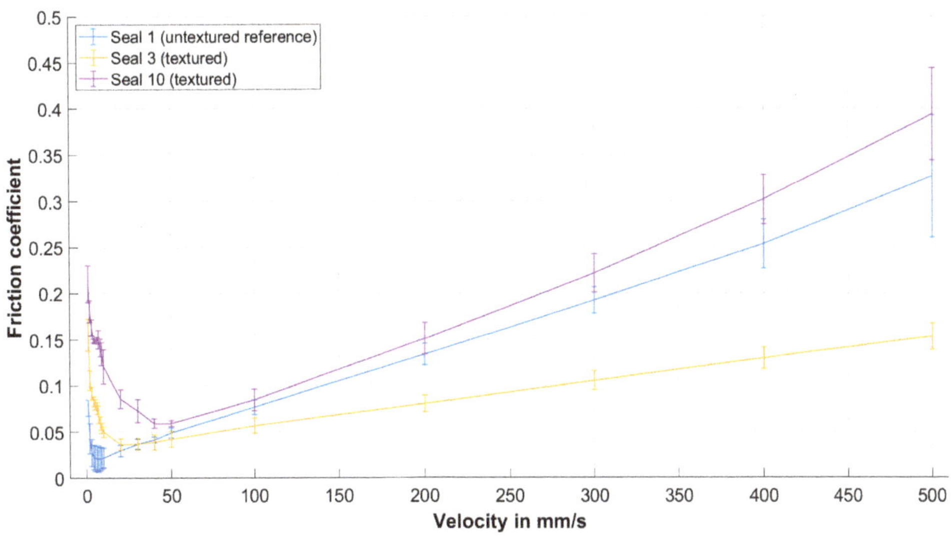

Figure 5. Stribeck curves from two different surface texture parameters, compared with the untextured reference. The single data points were provided with error bars.

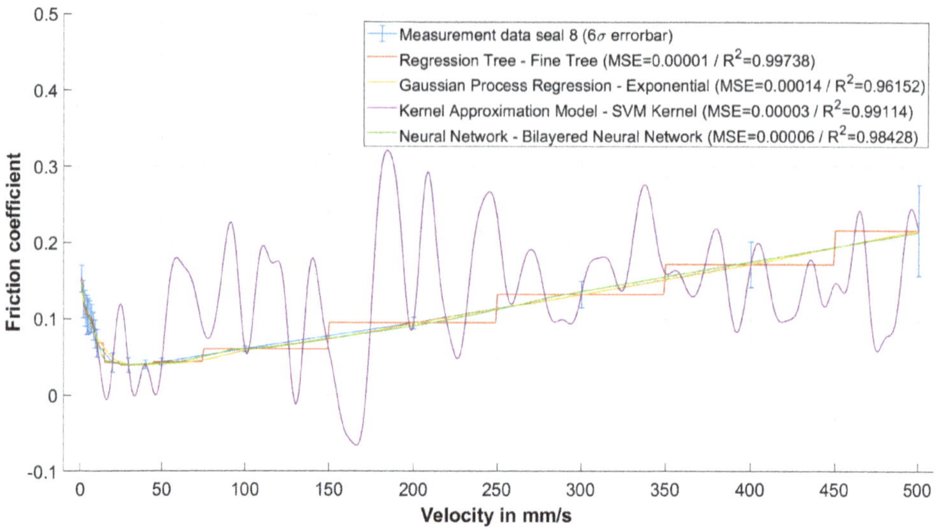

Figure 6. Pre-Comparison of the predicted Stribeck curves, based on the best algorithm each within the four best model categories according to the performance indices using seal sample 8 as an example. The curves are only based on the first model epoch, considering the maximum dataset before RFE. The individual data of the first epoch are specified in Table A2. The results of the last epoch after RFE can be found in Table 4.

Table 4. Comparison of the last epoch of the four algorithms, shown in Figure 6. The features cycle and direction of movement were already removed after recursive feature elimination (RFE), resulting in an implementation with averaged friction coefficients as described above. Only these four algorithms were evaluated up to the last epoch. The other algorithms from Table 3 were discarded due to poor metrics in the first epoch.

No.	Algorithm (Category)	*MSE*	R^2
5	Fine Tree (Regression Trees)	0.00050	0.78870
16	**Exponential (Gaussian Progress Regression Models)**	**0.00009**	**0.96077**
18	SVM Kernel (Kernel Approximation Models)	0.00320	−0.4098
25	Bilayered Neural Network (Neural Networks)	0.00070	0.68924

5. Friction Measurement Results and Machine Learning Model

5.1. Preliminary Examination of the Friction Results

From the friction coefficients per seal and sliding velocities described in Section 3.1, Stribeck curves were generated for the qualitative preliminary investigation. According to the principal component analysis (PCA) of Section 4, the features direction of the motion and cycle were not significant. Therefore, the friction coefficients per seal and velocity were averaged. Figure 5 shows the exemplary Stribeck curves of textured seals 3 and 10 compared with the untextured reference seal 1, compared with Table 1.

The curves represent the average friction coefficients per seal, while the error bars mark the 6 sigma interval at the test velocity, to obtain 99.7% as the confidence interval. It can be seen that the textured seals had higher friction coefficients in the boundary and mixed friction regime than the untextured reference. In addition, the area of mixed friction was more pronounced and extended over a larger velocity interval depending on the texture parameters. On the other hand, the Stribeck curve of seal 3 shows that texturing could lead to lower friction coefficients in the hydrodynamic friction region. The texture parameters of seal 3 were proven to be advantageous in a direct comparison with the textures of the 9 other textured seals, as the associated Stribeck curve has the lowest coefficients of friction in the hydrodynamic friction regime due to a lower slope. In contrast, the texture parameters of seal 10 resulted in high friction coefficients in all of the friction regimes. This shows that surface texturing did not automatically lead to an improved friction behaviour. In addition, the qualitative preliminary examination of the collected measurement data indicated the need for a model to identify optimal and discrete texture parameters and to fathom tribological relationships in a factual context.

5.2. Machine Learning Model

The model was built using the MATLAB Statistics and Machine Learning Toolbox® by recursive feature elimination according to the k-fold cross-validation method, see Section 4. The training data were based on the measured friction and surface data of 10 textured pneumatic piston seals, whose texture was characterised by dimples with a circular basic shape in a triangular pattern, compared with Section 3.2. The maximum dataset consisted of $N = 4560$ records (228 + 228 piston strokes multiplied with 10 textured seals) and $M = 7$ features according to Table 2, each with an associated friction coefficient output. In the first model generation step, one model was generated for each of the 26 learning algorithms specified in Table 3 using the MATLAB toolbox. In this process, $k = 10$ folds were used in the k-fold cross validation (KFCV). Furthermore, 10% of the records were separated as the test dataset, from which the mathematical evaluation metrics MSE and R^2 [53,57] were calculated. As an example, Figure 6 shows different Stribeck curves, predicted by four different algorithms based on seal 8, in comparison with the measured friction values of seal 8. The related performance indices of the four models are summarised in the legend. The figure clarifies that the evaluation of the models using solely mathematical evaluation metrics was inadequate. The Regression Tree and Kernel Approximation models showed the best performance indices, but were unable to reproduce the known dependency between the velocity and friction coefficient as a Stribeck curve with the available experimental data. In particular, the Kernel Approximation Model showed a clear overfitting. Only at discrete test velocities could the friction coefficients be accurately predicted. In addition to the performance indices, it was advisable to evaluate the models on the basis of known partial dependencies, which represented the dependence between the target response, friction value, and at least one feature. The Gaussian Process Regression and Neural Network models not only showed good performance indices, but also reproduced the partial dependence between velocity and friction coefficient as a Stribeck curve according to Figure 6. Especially in case of non-parametric models, a priori selection of learning algorithms was not advisable, as it was difficult to estimate how they reacted to the training data. As in Figure 6, the overfitted kernel approximation model showed similar correlations in other partial dependencies, which could be visualised using partial dependence plots (PDP). For example, the correlation between the dimple diameter and friction coefficient was unknown, but a strongly fluctuating correlation with many deflections was not expected from a tribological point of view.

The recursive feature elimination within the first model generation showed that the features direction of movement and cycle were not significant. The method thus confirmed the results of the principal component analysis of Section 4. The final model generation was, therefore, performed with the averaged friction coefficients, as described in Section 4. The training data consisted of $N = 190$ records and $M = 5$ features. For model building, $k = 5$ folds were used. Further, 10% of the records were separated as the test dataset in order to be able to determine the performance indices on the basis of unknown data. The comparison of the last epoch of the performance indices of the four algorithms, shown in Figure 6, is provided in Table 4. It is clear that the neural network, which had both good mathematical metrics and a good representation of tribology in the form of the Stribeck curve, had poor MSE and R^2 values in the last epoch. Therefore, throughout the course of model development and evaluation, an exponential Gaussian Process Regression model, underlined in Table 3, with the following features emerged as the best model in the applications context: velocity, dimple diameter, dimple distance, dimple depth, and real texture density inside the contact area.

The GPR model reveals the following performance indices: $MSE = 0.00009$ and $R^2 = 0.96077$. The performance indices of the model with a reduced number of features of $M = 5$ were almost identical to those of the maximum dataset of the first model generation. The MSE even improved slightly. The friction coefficients could thus be predicted with a high degree of certainty for discrete texture parameters at different sliding velocities. As shown previously, the model was able to reproduce the known dependency friction velocity as a Stribeck curve. Further evaluation of the model performance, e.g., using a residual plot, showed no anomalies that would indicate a poor model fit [57].

To identify optimal texturing parameters, the selected machine learning model was used to predict friction coefficients for discrete combinations of feature values within the intervals of the training data. Thus, there was no prediction by extrapolation. The generated Stribeck curves could be used to evaluate the tribological behaviour of the textured seals within different friction regimes. Figure 7 shows the prediction interval (grey area), in which the model could create predictions for the feature intervals specified in Table 5, compared with the mean friction curve of the untextured reference seal 1. It can be seen that there was no texture parameter combination, based on the model, that led to lower friction coefficients in the boundary and mixed friction regime.

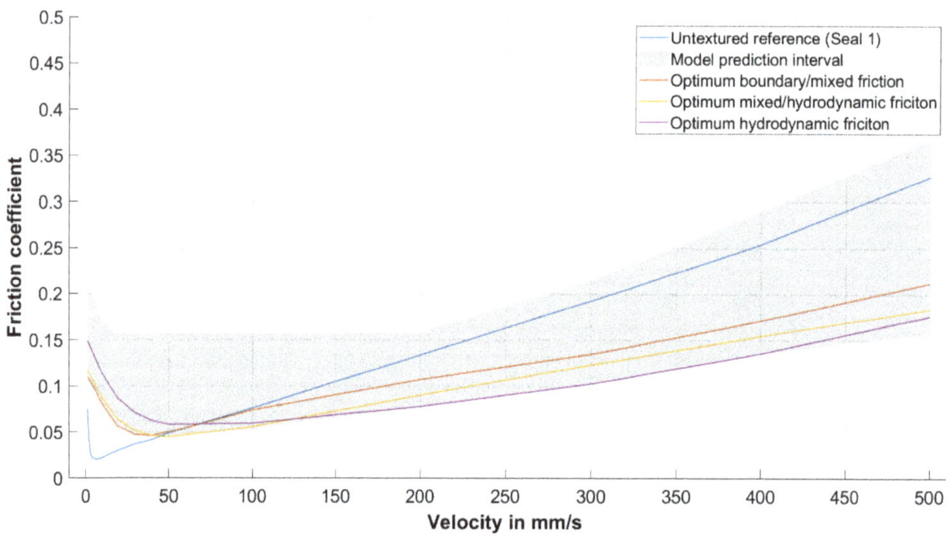

Figure 7. Model evaluation: comparison of the untextured reference seal 1 to the model prediction interval (grey area) and Stribeck curves, predicted by the model, of the identified optimum texture parameters for each friction regime.

Table 5. Feature intervals used for the machine learning model evaluation.

Feature	Unit	Min.	Max.	Resolution
Velocity	mm/s	1	500	19 values
Diameter	µm	95	205	5
Distance	µm	190	400	10
Depth	µm	5	30	1
Texture density	%	15	45	1

No texture emerged as a global optimum from the parameter study carried out. Three predicted Stribeck curves could be identified, each of which could be considered as an optimum in one of the friction states. The corresponding texture parameters are summarised in Table 6. From the model prediction, it can be seen that all three texture parameters led to lower friction coefficients in the hydrodynamic friction region compared with the untextured reference. However, low friction coefficients in the hydrodynamic friction regime were at the expense of increased friction coefficients in the boundary and mixed friction regions. In addition, the transition to hydrodynamic friction was shifted to higher speeds, so that boundary and mixed friction were more pronounced over a larger velocity interval. These results were consistent with the preliminary examination, as explained in Section 5.1.

Table 6. Optimal texture parameters for each friction regime according to the machine learning model prediction. Friction increase and friction reduction are related to the untextured reference seal 1. The data are based on the values from the graphs of Figure 7, taking into account the discrete values, as specified in Table 5.

No.	Optimum	Diameter in µm	Distance in µm	Depth in µm	Texture Density in %	Max. Friction Increase (Velocity)	Max. Friction Reduction (Velocity)
1	Boundary/Mixed friction	145	290	13	25	362% (5 mm/s)	35% (500 mm/s)
2	Mixed/Hydrodyn. friction	95	240	9	16	390% (5 mm/s)	44% (500 mm/s)
3	Hydrodyn. friction	145	290	19	26	526% (5 mm/s)	47 % (500 mm/s)

The optimum texture in the transition zone from mixed friction to hydrodynamic friction No. 2 from Table 6 can be identified as a compromise between Nos. 1 and 3. At the maximum velocity of 500 mm/s, the maximum predicted friction coefficient reduction was approximately 44%. In addition, the predicted friction coefficient of $\mu = 0.044$ at a velocity of 50 mm/s for these texture parameters was the lowest within the entire predictions of the machine learning model.

As mentioned above, the chosen textures on which the model was based were associated with a significant increase in the friction coefficients in the boundary and mixed friction regions. As can be seen in Table 6, there was a maximum increase in the friction coefficient of 526% at a low speed of 5 mm/s. So, it became clear that the improvement in the frictional behaviour strongly depended on the surface texture and especially on the operational conditions such as the sliding velocity of the seal.

6. Tribological Discussion of the Machine Learning Model Results

The Gaussian process regression (GPR) model, which was the most suitable model for the present application and the existing data in a tribological context, reproduced the property of a Stribeck curve, as explained in Section 5. Based on this property and the mathematically metric values specified in Section 5.2, the validity of the model could be assumed. As described, suitable surface textures significantly reduced friction in a hydrodynamic lubrication regime. However, an increase in friction was observed in the boundary and mixed friction regions.

There are several possible reasons for this behaviour, which is contrary to most of the literature, where the dimples showed a reduction in friction over nearly the entire range of operational conditions, e.g., due to their improved micro-hydrodynamic pressure build up or their lubricant storage effect [21,37,58–62]. In fact, however, the dimples could not only serve as a source of lubricant and thus support hydrodynamic film formation, but also as a sink in the event of mixed friction or insufficient lubrication. At low sliding velocities or during idle periods, lubricant collected in and around the dimples, causing the roughnesses of the untextured areas of the seal to be in contact with the roughnesses of the pneumatic cylinder tube surface for a longer period of time, which increased the frictional force in the mixed friction area. This effect increased with the viscosity of the lubricant, which was consistent with the present friction measurements, as a grease with a higher viscosity was used. Untextured seals, therefore, have the advantage that less lubricant was required to separate the contact surfaces in this lubrication condition [63].

Another possibility for increasing friction at low speeds, where the surfaces were not completely separated from each other as in the hydrodynamic lubrication state, was based on the texturing during moulding (TDM) manufacturing method of the seals. During production, the negative of the desired dimple texture was applied to the metallic mould by laser ablation. During the vulcanisation process, the texture is directly transferred from the mould to the rubber surface, so that protrusions in the mould became dimples in the seal. The resulting removal of material from the metal mould increased the outer diameter of the seals by an amount equal to the depth of the dimples [51]. This increased the contact pressure between the seal and the pneumatic cylinder tube and thus the friction.

A third possibility for friction increase in the boundary and mixed lubrication regimes was the texture-parameter-dependent wiping effect of the dimples, where at low sliding speeds and, therefore, low film heights, the lubricant was wiped off and the edges of the dimples interlocked with the cylinder surface. The negative contribution of the three effects to friction described above has not yet been studied and quantified in detail, and will be the subject of future research. For this purpose, the tests were to be repeated in a glass cylinder. By recording the dynamic contact between the seal and the glass cylinder with the help of a high-speed camera, lubricant sinks and wiper effects could be detected. In addition, the tests were repeated with different pistons, in which the fit between the inner diameter of the seals and the outer diameter of the pistons was varied. This changed the contact pressure between the seal surface and the internal cylinder surface, allowing manufacturing tolerances to be simulated with defined dimensions. Consequently, their effect on friction could be analysed.

On the other hand, the positive friction-reducing effects of the dimple textures in the hydrodynamic lubrication regime were evident for nearly all of the textured seals analysed in this paper (see the grey area of Figure 7). Textures with extremely large diameters, especially distances above 200 µm and 395 µm, were an exception, due to their inappropriate aspect ratio [64]. This indicates that a positive effect of the dimples was present, but that there was a limit to the positive properties of the dimples with their effect of increasing hydrodynamic pressure build-up for the prevailing specific application of a pneumatic piston seal [25].

It can also be seen that the slope of the Stribeck curve of the textured seals, and therefore the friction coefficient, was generally lower in the hydrodynamic lubrication regime, even though the lubricant was exactly the same in all of the tests. This behaviour could be explained by the advantageous choice of the texturing parameters, which reduced the shear stress inside the lubricant film by increasing the lubrication film thickness due to the increased dimple-induced micro-hydrodynamic pressure build up, which further separated the contacting surfaces [19,65]. In addition, this behaviour was supported by the lubricant storage effect of the dimples, so that sufficient lubricant was provided to separate the surfaces [34,64].

The modelling was exclusively based on the measured values recorded in Section 3. An extension to include measured values such as seal diameter tolerances, temperature, or other disturbing influences could change the results of the methodology in such a way that another of the 26 algorithms analysed was classified as the most suitable for the prevailing application, which will be analysed in future work.

7. Conclusions

The generalised method, presented in this paper, represents essential steps for building regression models through supervised learning, using experimental measured friction and surface texture data as an illustrative example of pneumatic piston seals. In particular, the parallel use of a large number of 26 different machine learning algorithms in the context of an exhaustive search led to good results, even when fundamental correlations in the prevailing data were unknown. The individual steps can be automated, so that the method is even suitable for identifying trends in ongoing experiments or production processes and for intervening at an early stage if targets are possibly missed. For example, the ML model can be used to identify the operating conditions within the limits of velocity parameters tested, where the surface textures reveal the maximum friction reduction compared with the untextured reference. This is even possible for a texture parameter combination that has not been physically tested.

The approach of this work reveals that machine learning models should be checked as much as possible using different evaluation metrics and should be classified in the specific applications context. Machine learning techniques are particularly treacherous for inexperienced users, as they usually produce good results according to the mathematical performance indicators R^2 and MSE, but may fail to represent the underlying physics, as represented by Stribeck curves. The strength of the generalised method, presented in this paper, lies in its ability to reduce factual relationships to the essential influencing parameters in order to reveal even fundamental physical relationships.

Because it can be fully automated, the method can provide early insights, particularly in tribological testing, that can be directly incorporated into testing procedures and specimen optimisations for the targeted optimisation parameter, such as friction. As Marian et al. pointed out in their systematic review that the automation of data collection and processing could additionally be applied to existing data and completed projects to extend or test relationships and conclusions through machine learning [47]. The generalised method presented in this paper, which is based on common standard machine learning procedures and a large variety of learning algorithms, is a novel and strong tool for the realisation of this approach. The main subjects and findings of the paper are listed below:

1. A novel machine learning methodology is developed to build several ML models and select the most suitable model that reliably predicts optimal surface texture parameters for different operating conditions such as lubrication regimes;
2. Both mathematical metrics and tribological relationships in the form of the Stribeck curve are taken into account to determine the most suitable ML model;
3. Surface textured pneumatic piston seals are used as an example application in this study;
4. Friction measurements of the seals and surface texture measurements of the real parts serve as the basis of data for ML modeling
5. For the example application and the underlying data, a Gaussian process regression (GPR) model has proven to be the best model in terms of mathematical metrics and the tribological representation of the Stribeck curves;
6. Depending on the prevailing friction regimes and surface textures, friction reductions of up to 47%, and friction increases of up to 526% could be identified for the surface textures, compared with an untextured reference surface;
7. The advantage of the method is that a large number of 26 ML models can be compared and the best one selected without having to perform complex processes such as hyperparameter optimisation of individual models, so that a large number of users can use the method without being ML experts.

Author Contributions: conceptualization, M.B. and M.J.; methodology, M.B., J.B. and M.J.; software, J.B.; validation, M.B. and J.B.; formal analysis, M.B. and J.B.; investigation, M.B. and J.B.; resources, M.W.; data curation, M.B. and J.B.; writing—original draft preparation, M.B., J.B. and M.J.; writing—review and editing, M.B., J.B., M.J. and M.W.; visualization, M.B., J.B. and M.J.; supervision, M.B. and M.W.; project administration, M.B. and M.W.; funding acquisition, M.W. All authors have read and agreed to the published version of the manuscript.

Funding: This project has received funding from the European Union's Horizon 2020 research and innovation programme under grant agreement No. 862100.

Institutional Review Board Statement: Not applicable.

Informed Consent Statement: Not applicable.

Data Availability Statement: Dataset available on request from the authors.

Conflicts of Interest: The authors declare no conflict of interest.

Abbreviations

The following abbreviations are used in this manuscript:

CV	Cross validation
DQR	Data quality report
GPR	Gaussian process regression
IQR	Interquartile range
KFCV	K-Fold cross-validation
ML	Machine learning
PCA	Principal component analysis
PDP	Partial dependence plot
RFE	Recursive feature elimination
SPLOM	Scatterplot matrix
SVM	Support vector machine
TDM	Texturing during moulding

Appendix A

Table A1. Data Quality Report.

Feature	Number of Values	Missing Values	Cardinality	Minimum	1st Quartil
Sliding velocity in mm/s	3990	0	19	1.00	5.00
Cycle	3990	0	11	1.00	4.00
Direction of motion	3990	0	2	−1.00	−1.00
Dimple diameter in µm	3990	0	10	96.51	100.35
Dimple distance in µm	3990	0	10	195.76	243.54
Dimple depth in µm	3990	0	10	7.91	11.41
Real texture density	3990	0	10	0.15	0.20
Friction coefficient	3990	0	3990	0.03	0.08
Feature	**Mean**	**Median**	**3rd Quartil**	**Maximum**	**Standard deviation**
Sliding velocity in mm/s	89.21	10.00	100.00	500.00	145.94
Cycle	6.24	6.00	9.00	11.00	3.04
Direction of motion	−0.50	−1.00	1.00	1.00	1.00
Dimple diameter in µm	142.57	147.23	149.23	199.10	34.74
Dimple distance in µm	278.20	292.59	293.60	389.86	48.99
Dimple depth in µm	16.55	16.79	23.32	24.78	5.99
Real texture density	0.26	0.26	0.27	0.45	0.08
Friction coefficient	0.12	0.11	0.16	0.42	0.06

Table A2. Summary of all algorithms and their performance indices within the first model epoch with the full dataset according to Section 5.2 studied by category: Supervised Learning — Regression.

Category	No.	Algorithm	*MSE*	R^2
Linear Regression Models	1	Linear	0.00259	0.27744
	2	Interactions Linear	0.00214	0.40334
	3	Robust Linear	0.00259	0.27707
	4	Stepwise Linear	0.00212	0.40908

Table A2. *Cont.*

Category	No.	Algorithm	*MSE*	R^2
Regression Trees	5	Fine Tree	0.00001	0.99738
	6	Medium Tree	0.00002	0.99337
	7	Coarse Tree	0.00011	0.96963
Support Vector Machines	8	Linear SVM	0.00262	0.26978
	9	Quadratic SVM	0.00133	0.62854
	10	Cubic SVM	0.00146	0.59357
	11	Fine Gaussian SVM	0.00057	0.84057
	12	Medium Gaussian SVM	0.00104	0.71004
	13	Coarse Gaussian SVM	0.00224	0.37614
Gaussian Process Regression Models	14	Squared Exponential	0.00046	0.87305
	15	Matern 5/2	0.00035	0.90343
	16	Exponential	0.00014	0.96152
	17	Rational Quadric	0.00036	0.90069
Kernel Approximation Models	18	SVM Kernel	0.00003	0.99114
	19	Least Squares Kernel Regression	0.00013	0.96501
Ensembles of Trees	20	Boosted Trees	0.00013	0.96412
	21	Bagged Trees	0.00018	0.95081
Neural Networks	22	Narrow Neural Network	0.00015	0.95800
	23	Medium Neural Network	0.00008	0.97772
	24	Wide Neural Network	0.00005	0.98352
	25	Bilayered Neural Network	0.00006	0.98428
	26	Trilayered Neural Network	0.00007	0.98161

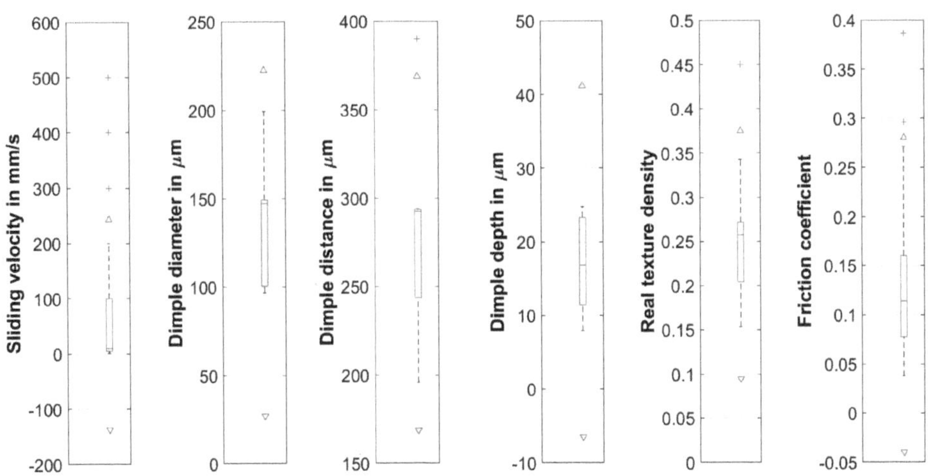

Figure A1. Boxplots of relevant features for visualising IQR. The lower and upper quartile (blue), lower and upper whisker (black), median (red line), lower limit (red downward-pointing triangle) and upper limit of the IQR (red upward-pointing triangles), and outliers (red crosses) are shown. The features cycle and direction of motion are not considered, as they only represent a numerator and the numerical representation of the direction.

References

1. Menezes, P.L.; Nosonovsky, M.; Ingole, S.P.; Kailas, S.V.; Lovell, M.R. (Eds.) *Tribology for Scientists and Engineers*; Springer: Berlin/Heidelberg, Germany, 2013.
2. Holmberg, K.; Erdemir, A. Influence of tribology on global energy consumption, costs and emissions. *Friction* **2017**, *5*, 263–284. [CrossRef]

3. Chandross, M.; Argibay, N. Friction of metals: A review of microstructural evolution and nanoscale phenomena in shearing contacts. *Tribol. Lett.* **2021**, *69*, 119. [CrossRef]
4. Bhagwat, P.; Sista, B.; Vemaganti, K. A computational study of the effects of strain hardening in micro-asperity friction models. *Tribol. Lett.* **2017**, *65*, 154. [CrossRef]
5. Reid, J.V.; Schey, J. A. The effect of surface hardness on friction. *Wear* **1987**, *118*, 113–125. [CrossRef]
6. Rejhon, M.; Lavini, F.; Khosravi, A.; Shestopalov, M.; Kunc, J.; Tosatti, E.; Riedo, E. Relation between interfacial shear and friction force in 2D materials. *Nat. Nanotechnol.* **2022**, *17*, 1280–1287. [CrossRef] [PubMed]
7. Zhou, X.; Liu, Y.; Hu, X.; Fang, L.; Song, Y.; Liu, D.; Luo, J. Influence of elastic property on the friction between atomic force microscope tips and 2D materials. *Nanotechnology* **2020**, *31*, 28. [CrossRef] [PubMed]
8. Farhat, Z.N.; Ding, Y.; Northwood, D.O.; Alpas, A.T. Effect of grain size on friction and wear of nanocrystalline aluminum. *Mater. Sci. Eng.* **1996**, *206*, 302–313. [CrossRef]
9. Penfornis, C.; Jourani, A.; Mazeran, P.-E. Effect of Grain Sizes on the Friction and Wear Behavior of Dual-Phase Microstructures with Similar Macrohardness and Composition. *Coatings* **2023**, *13*, 533. [CrossRef]
10. Kalin, M.; Polajnar, M. The effect of wetting and surface energy on the friction and slip in oil-lubricated contacts. *Tribol. Lett.* **2013**, *52*, 185–194. [CrossRef]
11. Rabinowicz, E. Influence of surface energy on friction and wear phenomena. *J. Appl. Phys.* **1961**, *32*, 1440–1444. [CrossRef]
12. Al-Bender, F.; De Moerlooze, K. On the relationship between normal load and friction force in pre-sliding frictional contacts. Part 1: Theoretical analysis. *Wear* **2010**, *269*, 174–182. [CrossRef]
13. De Moerlooze, K.; Al-Bender, F. On the relationship between normal load and friction force in pre-sliding frictional contacts. Part 2: Experimental investigation. *Wear* **2010**, *269*, 183–189. [CrossRef]
14. Horovistiz, A.; Laranjeira, S.; Davim, J.P. Influence of sliding velocity on the tribological behavior of PA66GF30 and PA66+ MoS2: An analysis of morphology of sliding surface by digital image processing. *Polym. Bull.* **2018**, *75*, 5113–5131. [CrossRef]
15. Tile, P.S.; Thomas, B. Effect of Load, Sliding Velocity, and Reinforcements on Wear Characteristics of Al7075-Based Composite and Nanocomposites Fabricated by Ultrasonic-Assisted Stir-Casting Technique. *Int. J. Met.* **2023**, 1–16. . [CrossRef]
16. Kou, B.; Li, Z.; Li, R.; Wang, Z.; Zhao, X. Influence of external environment parameters on friction coefficient between hoisting-rope and its pads. *AIP Adv.* **2023**, *13*, 6. [CrossRef]
17. Li, P.; Wang, B.; Ji, L.; Li, H.; Chen, L.; Liu, X.; Zhou, H.; Chen, J. Environmental molecular effect on the macroscale friction behaviors of graphene. *Front. Chem.* **2021**, *9*, 679417. [CrossRef] [PubMed]
18. Harsha, A.P.; Wäsche, R. Influence of temperature on friction and wear characteristics of polyaryletherketones and their composites under reciprocating sliding condition. *J. Mater. Eng. Perform.* **2018**, *27*, 5438–5449. [CrossRef]
19. Gropper, D.; Wang, L.; Harvey, T.J. Hydrodynamic lubrication of textured surfaces: A review of modeling techniques and key findings. *Tribol. Int.* **2016**, *94*, 509–529. [CrossRef]
20. Müser, M.H.; Nicola, L. Modeling the surface topography dependence of friction, adhesion, and contact compliance. *MRS Bull.* **2022**, *47*, 1221–1228. [CrossRef]
21. Lu, P.; Wood, R.J. Tribological performance of surface texturing in mechanical applications—A review. *Surf. Topogr. Metrol. Prop.* **2020**, *8*, 043001. [CrossRef]
22. Vishnoi, M.; Kumar, P.; Murtaza, Q. Surface texturing techniques to enhance tribological performance: A review. *Surf. Interfaces* **2021**, *27*, 101463. [CrossRef]
23. Bergseth, E.; Zhu, Y.; Söderberg, A. Study of surface roughness on friction in rolling/sliding contacts: Ball-on-disc versus twin-disc. *Tribol. Lett.* **2020**, *68*, 69. [CrossRef]
24. Ivanović, L.; Vencl, A.; Stojanović, B.; Marković, B. Biomimetics Design for Tribological Applications. *Tribol. Ind.* **2018**, *40*, 448–456. [CrossRef]
25. Grützmacher, P.G.; Profito, F.J.; Rosenkranz, A. Multi-Scale Surface Texturing in Tribology—Current Knowledge and Future Perspectives. *Lubricants* **2019**, *7*, 95. [CrossRef]
26. Martini, A.; Zhu, D.; Wang, Q. Friction reduction in mixed lubrication. *Tribol. Lett.* **2007**, *28*, 139–147. [CrossRef]
27. Brito, F.P.; Miranda, A.S.; Claro, J.C.P.; Teixeira, J.C.; Costa, L.; Fillon, M. The role of lubricant feeding conditions on the performance improvement and friction reduction of journal bearings. *Tribol. Int.* **2014**, *72*, 65–82. [CrossRef]
28. Bobach, L.; Bartel, D.; Beilicke, R.; Mayer, J.; Michaelis, K.; Stahl, K.; Bachmann, S.; Schnagl, J.; Ziegele, H. Reduction in EHL Friction by a DLC Coating. *Tribol. Lett.* **2015**, *60*, 17. [CrossRef]
29. Fukata, M.; Sotani, T.; Motozawa, M. Leakage and friction characteristics at sliding surface of tip seal in scroll compressors. *Int. J. Refrig.* **2021**, *125*, 104–112. [CrossRef]
30. Charitopoulos, A.; Visser, R.; Eling, R.; Papadopoulos, C.I. Design Optimization of an Automotive Turbocharger Thrust Bearing Using a CFD-Based THD Computational Approach. *Lubricants* **2018**, *6*, 21. [CrossRef]
31. Rosenkranz, A.; Costa, H.L.; Baykara, M.Z.; Martini, A. Synergetic effects of surface texturing and solid lubricants to tailor friction and wear—A review. *Tribol. Int.* **2021**, *155*, 106792. [CrossRef]
32. Zhang, K.; Deng, J.; Guo, X.; Sun, L.; Lei, S. Study on the adhesion and tribological behavior of PVD TiAlN coatings with a multi-scale textured substrate surface. *Int. J. Refract. Met. Hard Mater.* **2018**, *72*, 292–305. [CrossRef]
33. Gachot, C.; Rosenkranz, A.; Reinert, L.; Ramos-Moore, E.; Souza, N.; Müser, M.H.; Mücklich, F. Dry Friction Between Laser-Patterned Surfaces: Role of Alignment, Structural Wavelength and Surface Chemistry, *Tribol. Lett.* **2013**, *49*, 193–202. [CrossRef]

34. Erdemir, A. Review of engineered tribological interfaces for improved boundary lubrication. *Tribol. Int.* **2005**, *38*, 249–256. [CrossRef]
35. Schneider, J.; Braun, D.; Greiner, C. Laser Textured Surfaces for Mixed Lubrication: Influence of Aspect Ratio, Textured Area and Dimple Arrangement. *Lubricants* **2017**, *5*, 32. [CrossRef]
36. Marian, M.; Grützmacher, P.; Tremmel, S.; Mücklich, F.; Wartzack, S. Designing surface textures for EHL point-contacts—Transient 3D-simulations, meta-modeling and experimental validation. *Tribol. Int.* **2019**, *137*, 152–163. [CrossRef]
37. Costa, H.L.; Hutchings, I.M. Hydrodynamic lubrication of textured steel surfaces under reciprocating sliding conditions. *Tribol. Int.* **2007**, *40*, 1227–1238. [CrossRef]
38. Prodanov, N.; Gachot, C.; Rosenkranz, A.; Mücklich, F.; Müser, M.H. Contact Mechanics of Laser-Textured Surfaces. *Tribol. Lett.* **2013**, *50*, 41–48. [CrossRef]
39. Rosenkranz, A.; Heib, T.; Gachot, c.; Mücklich, F. Oil film lifetime and wear particle analysis of laser-patterned stainless steel surfaces. *Wear* **2015**, *334–335*, 1–12. [CrossRef]
40. Hsu, C.-J.; Stratmann, A.; Rosenkranz, A.; Gachot, C. Enhanced Growth of ZDDP-Based Tribofilms on Laser-Interference Patterned Cylinder Roller Bearings. *Lubricants* **2017**, *5*, 39. [CrossRef]
41. Kovalchenko, A.; Ajayi, O.; Erdemir, A.; Fenske, G.; Etsion, I. The effect of laser surface texturing on transitions in lubrication regimes during unidirectional sliding contact. *Tribol. Int.* **2005**, *38*, 219–225. [CrossRef]
42. Etsion, I. Modeling of surface texturing in hydrodynamic lubrication. *Friction* **2013**, *1*, 195–209. [CrossRef]
43. Dumont, M.-L.; Lugt, P.M.; Tripp, J.H. Surface feature effects in starved circular EHL contacts. *J. Tribol.* **2002**, *124*, 358–366. [CrossRef]
44. Wang, Z.; Ye, R.; Xiang, J. The performance of textured surface in friction reducing: A review. *Tribol. Int.* **2023**, *177*, 108010. [CrossRef]
45. Gachot, C.; Rosenkranz, A.; Hsu, S.M.; Costa, H.L. A critical assessment of surface texturing for friction and wear improvement. *Wear* **2017**, *372–373*, 21–41. [CrossRef]
46. Chen, K.; Yang, X.; Zhang, Y.; Yang, H.; Lv, G.; Gao, Y. Research progress of improving surface friction properties by surface texture technology. *Int. J. Adv. Manuf. Technol.* **2021**, *116*, 2797–2821. [CrossRef]
47. Marian, M.; Tremmel, S. Current trends and applications of machine learning in tribology—A review. *Lubricants* **2021**, *9*, 86. [CrossRef]
48. Paturi, U.M.R.; Palakurthy, S.T.; Reddy, N.S. The Role of Machine Learning in Tribology: A Systematic Review. *Arch. Comput. Methods Eng.* **2022**, *30*, 1345–1397. [CrossRef]
49. Sterkenburg, F.S.; Grünwald, P.D. The no-free-lunch theorems of supervised learning. *Synthese* **2021**, *199*, 9979–10015. [CrossRef]
50. Kelleher, J.D.; Macnamee, B.; D'Arcy, A. *Fundamentals of Machine Learning for Predictive Data Analytics: Algorithms, Worked Examples, and Case Studies*; The MIT Press: Cambridge, MA, USA; London, UK, 2015.
51. Zambrano, V.; Brase, M.; Hernandez-Gascon, B.; Wangenheim, M.; Gracia, L.; Viejo, I.; Izquierdo, S.; Valdes, J. A Digital Twin for Friction Prediction in Dynamic Rubber Applications with Surface Textures. *Lubricants* **2021**, *9*, 57. [CrossRef]
52. Klüber Lubrication München GmbH. Centoplex 2 EP—Product Information. Available online: https://www.klueber.com/de/de/produkte-service/produkte/centoplex-2-ep/9971/ (accessed on 26 October 2023).
53. Albon, C. *Machine Learning Kochbuch: Praktische Lösungen mit Python: Von der Vorverarbeitung der Daten bis zum Deep Learning*, 1st ed.; O'Reilly: Heidelberg, Germany, 2019.
54. Tukey, J.W. *Exploratory Data Analysis*; Book Addison-Wesley: Reading, PA, USA, 1977.
55. Falk, M.; Marohn, F.; Becker, R. *Angewandte Statistik MIT SAS: Eine Einführung*; Springer: Berlin/Heidelberg, Germany, 1995.
56. Wilmott, P. *Grundkurs Machine Learning*; Rheinwerk Computing: Bonn, Germany, 2020.
57. James, G.; Witten, D.; Hastie, T.; Tibshirani, R.; Taylor, J. *An Introduction to Statistical Learning*; Springer: Berlin/Heidelberg, Germany, 2023.
58. Fowell, M.; Olver, A.V.; Gosman, A.D.; Spikes, H.A.; Pegg, I. Entrainment and inlet suction: Two mechanisms of hydrodynamic lubrication in textured bearings. *J. Tribol.* **2007**, *2*, 336–347. [CrossRef]
59. Hamilton, D.B.; Walowit, J.A.; Allen, C.M. A theory of lubrication by microirregularities. *J. Basic Eng.* **1966**, *88*, 177–185. [CrossRef]
60. Hsu, S.M.; Jing, Y.; Zhao, F. Self-adaptive surface texture design for friction reduction across the lubrication regimes. *Surf. Topogr. Metrol. Prop.* **2016**, *4*, 014004. [CrossRef]
61. Olver, A.V.; Fowell, M.T.; Spikes, H.A.; Pegg, I.G. 'Inlet suction', a load support mechanism in non-convergent, pocketed, hydrodynamic bearings. *Proc. Inst. Mech. Eng. Part J Eng. Tribol.* **2006**, *220*, 105–108. [CrossRef]
62. Tang, Z.; Liu, X.; Liu, K.; Pegg, I.G. Effect of surface texture on the frictional properties of grease lubricated spherical plain bearings under reciprocating swing conditions. *Proc. Inst. Mech. Eng. Part J Eng. Tribol.* **2017**, *231*, 125–135. [CrossRef]
63. Ahmed, A.; Masjuki, H.H.; Varman, M.; Kalam, M.A.; Habibullah, M.; AL Mahmud, K.A. An overview of geometrical parameters of surface texturing for piston/cylinder assembly and mechanical seals. *Meccanica* **2016**, *51*, 9–23. [CrossRef]
64. Vencl, A.; Ivanović, L.; Stojanović, B.; Zadorozhnaya, E.; Miladinović, S.; Svoboda, P. Surface texturing for tribological applications: A review. *Proc. Eng. Sci.* **2019**, *1*, 227–239.
65. Wu, Z.; Bao, H.; Xing, Y.; Liu, L. Tribological characteristics and advanced processing methods of textured surfaces: A review. *Int. J. Adv. Manuf. Technol.* **2021**, *114*, 1241–1277. [CrossRef]

lubricants

MDPI

Article

Machine Learning for Film Thickness Prediction in Elastohydrodynamic Lubricated Elliptical Contacts

Joe Issa [1], Alain El Hajj [1], Philippe Vergne [2] and Wassim Habchi [1,*]

[1] Department of Industrial and Mechanical Engineering, Lebanese American University, Byblos 36, Lebanon; joe.issa@lau.edu (J.I.); alain.elhajj@lau.edu (A.E.H.)
[2] LaMCoS, INSA Lyon, CNRS, Univ Lyon, UMR5259, 69621 Villeurbanne, France; philippe.vergne@insa-lyon.fr
* Correspondence: wassim.habchi@lau.edu.lb

Abstract: This study extends the use of Machine Learning (ML) approaches for lubricant film thickness predictions to the general case of elliptical elastohydrodynamic (EHD) contacts, by considering wide and narrow contacts over a wide range of ellipticity and operating conditions. Finite element (FEM) simulations are used to generate substantial training and testing datasets that are used within the proposed ML framework. The complete dataset entails 915 samples; split into an 823-sample training dataset and a 92-sample testing dataset, corresponding to 90% and 10% of the combined dataset samples, respectively. The proposed ML model consists of a pre-processing stage in which conventional EHD dimensionless groups are used to minimize the number of inputs into the model, reducing them to only three. The core of the model is based on Gaussian Process Regression (GPR), a powerful ML regression tool, well-suited for small-sized datasets, producing output central and minimum film thicknesses, also in dimensionless form. The last stage is a post-processing one, in which the output film thicknesses are retrieved in dimensional from. The results reveal the capabilities and potential of the proposed ML framework, producing quasi-instantaneous predictions that are far more accurate than conventional film thickness analytical formulae. In fact, the produced central and minimum film thickness predictions are on average within 0.3% and 1.0% of the FEM results, respectively.

Keywords: machine learning; Gaussian Process Regression; elastohydrodynamic lubrication; elliptical contacts; finite elements; film thickness prediction

Citation: Issa, J.; El Hajj, A.; Vergne, P.; Habchi, W. Machine Learning for Film Thickness Prediction in Elastohydrodynamic Lubricated Elliptical Contacts. *Lubricants* **2023**, *11*, 497. https://doi.org/10.3390/lubricants11120497

Received: 18 September 2023
Revised: 20 November 2023
Accepted: 20 November 2023
Published: 22 November 2023

1. Introduction

The development of new mechanical systems focuses on maximizing their efficiency and service life. Accordingly, ensuring that machine elements operate in safe conditions is imperative to build reliable systems with low maintenance requirements. In practice, machine elements that undergo high loads, namely gears and bearings, are often lubricated to reduce component wear and optimize system efficiency. For these elements, accurately quantifying the thickness of lubricant films at the contact level is critical to maximizing the system uptime and the lifetime of its components while preventing excessive wear and reduced efficiency. Moreover, suboptimal efficiency is linked to increased energy consumption, resulting in higher operating costs, and greater greenhouse gas emissions [1].

Many lubricated mechanical components (e.g., gears, bearings, etc.) generally operate in a regime in which contact surfaces are fully separated by the lubricant, referred to as the full film separation regime. In many cases, the pressure endured by these machine elements at the contact level can exceed several gigapascals. In such cases, the solids in contact can experience elastic deformation in the contact region, a regime known as "elastohydrodynamic lubrication" (EHL).

In the general elastohydrodynamic (EHD) contact, the solids are approximated (at the contact level) by ellipsoids, each of different principal radii of curvature, R_x and R_y, in the x-

and y- space directions, respectively. Under unloaded and dry contact conditions, these solids share one point of contact, therefore with the name "point contacts". When loaded, the contact patch is elliptical (or circular, when $R_x = R_y$), hence with the name "elliptical contacts" (or "circular contacts", respectively). Two types of elliptical contacts can be recognized: wide and narrow (or slender). For wide contacts, the lubricant entrainment direction is perpendicular to the major semi-axis of the contact ellipse. Conversely, for narrow contacts, the lubricant entrainment direction is perpendicular to the minor semi-axis.

With the advent of cost-effective and accessible computational power, modeling and numerically solving the EHL problem finally became plausible and a more convenient alternative to laboratory experiments. The EHL problem is multi-physical by nature, given the multiple physics equations that need to be coupled for accurate simulation models. Different frameworks were developed to tackle this problem. One well-established approach is to use the Reynolds equation for hydrodynamic physics, which applies a thin film assumption to the Navier–Stokes equations, along with the linear elasticity equations for deformation and Newton's second law for load balance. This approach is known as the Reynolds-based approach, and equations can be solved using, for example, Finite Difference with multigrid [2] or Finite Element Methods (FEM) [3–5]. The latter method is becoming increasingly popular, due to its inherent strengths, including the adaptability of Finite Element software into meshes of custom structure and size, the accessibility of such software packages [6,7], the availability of Model Order Reduction (MOR) techniques, and most importantly full coupling (i.e., simultaneous resolution of all coupled equations), which accelerates convergence. Habchi [5] covers a fully coupled FEM model for EHD elliptical contacts, along with stabilizing formulations to accommodate for high loads. There exists another approach to modeling the EHL problem, based on Computational Fluid Dynamics (CFD) [8–10]. It involves solving the Navier–Stokes equations for the hydrodynamic part. The drawback of this approach is the extremely high computational overhead.

The complex and multi-physical nature of EHL results in a high number of parameters involved in this problem, covering the geometry and material properties of the contacting solids, their kinematics, the applied load, and the behavior of the lubricant. To simplify the analysis of EHD contacts, several sets of dimensionless groups have been proposed, representing combinations of variables. Hamrock and Dowson [11] introduced three dimensionless groups, namely, U, G and W, representing the lubricant speed, material properties, and load parameters of the problem, respectively. Moes [12] further combined the Hamrock and Dowson set into two dimensionless groups M, representing the load, and L, representing the material properties. Note that one additional dimensionless number is necessary in the study of elliptical contacts; that is, the ellipticity ratio of the contact ellipse.

Point-contact EHL simulations are known to be time-consuming, often requiring several minutes or sometimes hours to converge. This can impede the design process for engineers developing new mechanical systems. Over the years, various methods were implemented to reduce the computing time of simulations. One notable method is MOR, which reduces the size of the arising algebraic system of equations, either through projection into a reduced solution space [13,14] or through static condensation [15]. Despite the substantial decrease in computational effort and time, simulations still require considerable time to converge, particularly for high ellipticity and highly loaded cases.

As an alternative, dimensionless groups have been used to derive simple film thickness analytical equations, based on numerical and/or experimental data. Each set of such equations was optimized for a particular range of operating values and performed more poorly in extrapolation. Hamrock and Dowson [16] introduced the first set of analytical formulas, as a function of the Hamrock and Dowson dimensionless groups and the radii of curvature of the ellipsoids (R_x and R_y). A more recent set of analytical formulas with greater range was presented by Nijenbanning et al. [17]. Both sets of equations are restricted to circular and wide elliptical contacts only. Chittenden et al. [18] were the first to extend their equations to both wide and narrow elliptical contacts, but the range of application of these equations is rather limited. An extensive review of analytical formulas developed for isothermal point contacts and their respective range of interest is provided by Wheeler et al. [19].

When predicting film thicknesses in EHD contacts, a trade-off has traditionally existed between accuracy and speed. This limitation can be overcome by adopting specific data-driven approaches, such as Machine Learning (ML). Although developing sizeable datasets for ML requires significant time and effort, the resulting predictive models offer the potential for accurate and real-time film thickness predictions. To fully replace time-consuming—but accurate—physics simulations, generated models should be carefully optimized to minimize prediction errors.

ML methods are becoming increasingly popular in physical and engineering sciences. In recent years, these methods have been introduced into tribology for both classification and regression problems [20]. One popular classification example is bearing fault detection and identification, which has been tackled using a variety of ML models. For example, Kankar et al. [21] used Support Vector Machines (SVMs) [22] and Artificial Neural Networks (ANNs) [23], while Shen et al. [24] used physics-informed convolutional neural networks (CNNs). One regression example is the prediction of the remaining useful life of rolling bearings, which was addressed using, for example, Random Forests [25], and several deep learning approaches, including Stacked Autoencoder and Recurrent Neural Networks [26] and Generative Adversarial Networks [27], among others. For regression problems similar to film thickness prediction, ML methods can benefit from both the speed of analytical formulas and the accuracy of simulations. Additionally, ML methods can be used to solve differential equations. For instance, physics-informed neural networks (PINNs), which do not necessarily require a training dataset, have been used to solve hydrodynamic lubrication problems [28,29].

In terms of EHL, Marian et al. [30] were the first to employ machine learning models to predict EHD parameters, namely central and minimum film thicknesses of line and circular contacts, based on learning datasets generated using FEM simulation results. The study looked at the performance of Support Vector Regression (SVR) [31], Gaussian Process Regression (GPR) [32–34], and ANNs. Different input parameters were also tried, namely the dimensional input variables and the two aforementioned sets of dimensionless groups (Hamrock and Dowson as well as Moes). It was concluded that GPR models can offer the smallest prediction error and that dimensionless groups fail to train accurate models. This is somehow unexpected, given their popular and relatively successful use in analytical formulas. More recently, Walker et al. [35] worked on predicting the central film thickness and viscous and boundary friction in EHD line contacts using ANNs.

While the current study similarly utilizes a dataset of simulation results to build predictive machine learning models for EHL parameters, it is the first to examine central and minimal film thicknesses for the general case of elliptical contacts.

2. Finite Element Model

The FEM model employed for dataset generation is detailed in this section. It follows a full coupling or full-system approach [5] and assumes smooth solid surfaces operating under steady-state, isothermal, and Newtonian conditions, for simplicity. Moreover, a full-film regime is assumed, with the lubricant entrainment direction being in the x-direction (i.e., solids are moving at constant surface velocities u_1 and u_2 in the x-direction). The contact is subjected to a constant external applied load F.

As previously discussed, the solids are approximated by ellipsoids at the contact level, as illustrated in Figure 1a. For simpler modeling, an equivalent reduced configuration is adopted, consisting of a contact between an elastic ellipsoid of equivalent radii of curvature and solid properties and a rigid flat plane, as shown in Figure 1b. The elastic properties (i.e., Young's modulus of elasticity E and Poisson's coefficient υ) of the equivalent ellipsoid, as a function of the properties of each solid, are given by the following:

$$E = \frac{1}{\frac{1-v_1^2}{E_1} + \frac{1-v_2^2}{E_2}}, \text{ and } v = 0 \quad (1)$$

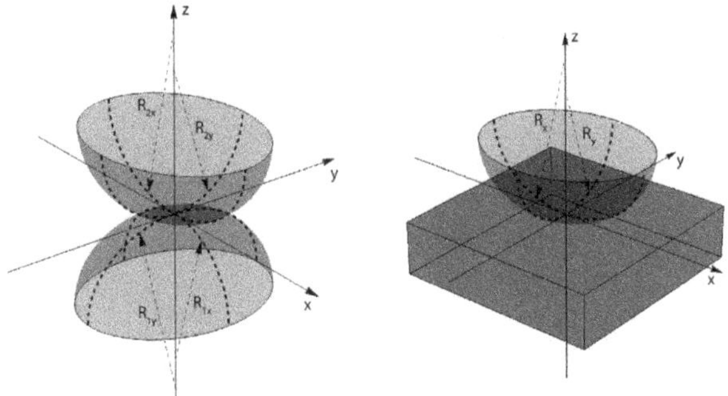

Figure 1. The contact geometry of (**a**) an actual elliptical contact and (**b**) a reduced elliptical contact.

The radii of curvature of the reduced configuration, R_x and R_y, would then be written as follows:

$$R_x = \frac{R_{x,1}\, R_{x,2}}{R_{x,1} + R_{x,2}}, \text{ and } R_y = \frac{R_{y,1}\, R_{y,2}}{R_{y,1} + R_{y,2}} \quad (2)$$

Finally, let $\overline{R}$ be the equivalent radius of curvature of the reduced geometry, written as follows:

$$\frac{1}{\overline{R}} = \frac{1}{R_x} + \frac{1}{R_y} \quad (3)$$

For a convenient and generalized study, the equations are written in dimensionless form as a function of the operating conditions and the Hertzian dry contact parameters. The latter consist of the Hertzian contact pressure p_h (i.e., the maximum pressure reached at the center of the contact) and the semi-axes of the contact ellipse in the x- and y-directions, a_x and a_y, respectively. The Hertzian contact pressure is given by the following:

$$p_h = \frac{3F}{2\,\pi\, a_x\, a_y} \quad (4)$$

The ellipticity ratio of the Hertzian dry contact patch $\theta = a_x/a_y$ can be found given the ratio of the radii $D = R_x/R_y$. For wide elliptical contacts, where $0 < D \le 1$ and $0 < \theta \le 1$, the Hertzian contact parameters are given as follows:

$$\begin{aligned} a_x &= \sqrt[3]{\frac{3F\,\overline{R}\,\theta\,\Psi_1}{\pi E}}, \text{ and } a_y = \sqrt[3]{\frac{3F\,\overline{R}\,\Psi_1}{\pi E\,\theta^2}} \\ \text{Where: } \quad \theta &\approx \frac{1}{1+\sqrt{\frac{\ln(16/D)}{2D}} - \sqrt{\ln 4} + 0.16 \ln D} \\ \Psi_1 &\approx 1 + \theta^2 \left[\frac{\pi}{2}\left(1 - \frac{\ln\theta}{4}\right) - 1\right] \end{aligned} \quad (5)$$

where Ψ_1 is the complete elliptical integral of the first kind. For narrow or slender elliptical contacts, corresponding to $D \ge 1$ and $\theta \ge 1$, the Hertzian contact parameters are written as follows:

$$\begin{aligned} a_x &= \sqrt[3]{\frac{3F\,\overline{R}\,\theta^2\,\Psi_1}{\pi\,E}}, \text{ and } a_y = \sqrt[3]{\frac{3F\,\overline{R}\,\Psi_1}{\pi\,E\,\theta}} \\ \text{Where: } \quad \theta &\approx 1 + \sqrt{\frac{D\ln(16D)}{2}} - \sqrt{\ln 4} + 0.16 \ln\left(\frac{1}{D}\right) \\ \Psi_1 &\approx 1 + \frac{1}{\theta^2}\left[\frac{\pi}{2}\left(1 + \frac{\ln\theta}{4}\right) - 1\right] \end{aligned} \quad (6)$$

The dimensionless parameters employed in the governing equations can be defined using the aforementioned variables as follows:

$$\begin{gathered} X=\frac{x}{a_x},\ Y=\frac{y}{a_y},\ Z=\frac{z}{a_x} \\ H=\frac{h\,R_x}{a_x^2},\ U=\frac{u\,R_x}{a_x^2},\ V=\frac{v\,R_x}{a_x^2},\ W=\frac{w\,R_x}{a_x^2}, \\ P=\frac{p}{p_h},\ \overline{\rho}=\frac{\rho}{\rho_0},\ \overline{\mu}=\frac{\mu}{\mu_0} \end{gathered} \tag{7}$$

where u, v, and w are the solid elastic deformations in the x-, y-, and z- directions, respectively, h and p the lubricant film thickness and pressure, respectively, and ρ_0 and μ_0 the density and viscosity of the lubricant, respectively, at ambient pressure.

2.1. Governing Equations

The computational domain Ω, shown in Figure 2, is the region over which the elastic deformation equations are applied. Its boundary includes a bottom domain, denoted as $\partial\Omega_b$, and a contact domain, denoted as Ω_c, located on the upper surface of Ω, over which the Reynolds equation is applied. Moreover, given the unidirectional lubricant flow in the $x-$direction, the problem is symmetric with respect to an $xz-$plane, denoted as $\partial\Omega_s$, passing through the center of the contact. As a result of this symmetry, the number of degrees of freedom (dofs) of the elastic and hydrodynamic parts of the problem is reduced by half. The dimensionless side length of the domain Ω was taken to be 60 [5], which is reduced to 30 in the y-direction due to symmetry ($-30 \leq X \leq 30$, $-30 \leq Y \leq 0$, and $-60 \leq Z \leq 0$). Such large dimensions are required to attain a half-space configuration. The dimensionless size of the contact domain Ω_c was taken as 6×3, accounting for symmetry ($-4.5 \leq X \leq 1.5$, $-3 \leq Y \leq 0$). Note that the origin of the domain is taken at the dry undeformed point of contact.

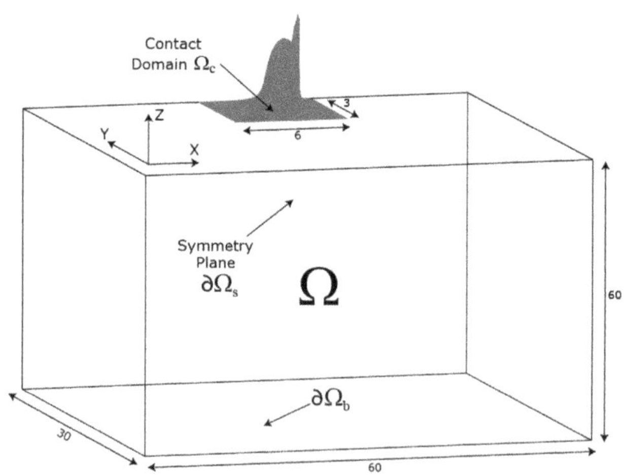

Figure 2. Computational domain of the EHD point contact.

The Reynolds equation describes the pressure distribution over the contact domain Ω_c and is written in its dimensionless form as follows:

$$\begin{gathered} \frac{\partial}{\partial X}\left(\overline{\varepsilon}\,\frac{\partial P}{\partial X}\right)+\theta^2\frac{\partial}{\partial Y}\left(\overline{\varepsilon}\,\frac{\partial P}{\partial Y}\right)=\frac{\partial(\overline{\rho}H)}{\partial X} \\ \text{Where}: \overline{\varepsilon}=\frac{\overline{\rho}H^3}{\overline{\mu}\,\lambda}\ \text{with}\ \lambda=\frac{12\,u_m\,\mu_0\,R_x^2}{a_x^3\,p_h} \end{gathered} \tag{8}$$

where $u_m = (u_1+u_2)/2$ is the mean entrainment speed. A zero-pressure boundary condition ($P=0$) is imposed on the boundary $\partial\Omega_c$ of the contact domain Ω_c, excluding the symmetry boundary $\partial\Omega_c \cap \partial\Omega_s$. For the latter, a symmetry condition is required ($\partial P/\partial Y = 0$). Lastly, the dimensionless film thickness H is defined as follows:

$$H(X,Y) = H_0 + \frac{X^2}{2} + \frac{D}{\theta^2}\frac{Y^2}{2} - W(X,Y) \tag{9}$$

The linear elasticity equations govern the dimensionless deformations U, V, and W, in the x-, y-, and z- directions, respectively. The equations are given as follows:

$$\begin{array}{l} \frac{\partial^2 U}{\partial X^2} + \theta \frac{\partial}{\partial Y}\left[\frac{1}{2}\left(\theta\frac{\partial U}{\partial Y} + \frac{\partial V}{\partial X}\right)\right] + \frac{\partial}{\partial Z}\left[\frac{1}{2}\left(\frac{\partial U}{\partial Z} + \frac{\partial W}{\partial X}\right)\right] = 0 \\ \frac{\partial}{\partial X}\left[\frac{1}{2}\left(\theta\frac{\partial U}{\partial Y} + \frac{\partial V}{\partial X}\right)\right] + \theta^2\frac{\partial^2 V}{\partial Y^2} + \frac{\partial}{\partial Z}\left[\frac{1}{2}\left(\frac{\partial V}{\partial Z} + \theta\frac{\partial W}{\partial Y}\right)\right] = 0 \\ \frac{\partial}{\partial X}\left[\frac{1}{2}\left(\frac{\partial U}{\partial Z} + \frac{\partial W}{\partial X}\right)\right] + \theta\frac{\partial}{\partial Y}\left[\frac{1}{2}\left(\frac{\partial V}{\partial Z} + \theta\frac{\partial W}{\partial Y}\right)\right] + \frac{\partial^2 W}{\partial Z^2} = 0 \end{array} \tag{10}$$

The boundary conditions of the linear elastic deformation equations are:

$$\begin{array}{ll} \frac{\partial W}{\partial Z} = -\frac{(1+D)}{2\Upsilon_1\theta}P \text{ and } \{\sigma_t\} = \{\tau_{zx}, \tau_{zy}\} = \{\varnothing\} & \text{over } \Omega_c \\ U = V = W = 0 & \text{over } \partial\Omega_b \\ V = 0 \text{ and } \{\sigma_t\} = \{\tau_{yx}, \tau_{yz}\} = \{\varnothing\} & \text{over } \partial\Omega_s \\ \sigma_n = 0 \text{ and } \{\sigma_t\} = \{\varnothing\} & \text{elsewhere} \end{array} \tag{11}$$

where σ_n and $\{\sigma_t\}$ are the normal and tangential stresses, respectively (with τ_{ij} being the latter's component in the j direction within a plane having i as normal).

Finally, for the load balance between the external load and the lubricant pressure, the equation is written as follows:

$$\int_{\Omega_c} P\, d\Omega = \frac{\pi}{3} \tag{12}$$

This equation ensures that the correct constant external load F is applied to the contact by monitoring the value of the rigid body separation term H_0. The lubricant adopted for this study is a well-characterized mineral oil, "Shell T9", for which the density-pressure response is characterized by the Murnaghan equation given by the following:

$$\rho(p) = \rho_0\left[1 + \frac{K_0' p}{K_{00}\exp(-\beta_K T_0)}\right]^{\frac{1}{K_0'}} \tag{13}$$

with the ambient temperature $T_0 = 30$ °C, $K_0' = 10.545$, $K_{00} = 9.234$ GPa, $\beta_K = 6.09 \times 10^{-3}$ K^{-1}, and $\rho_0 = 872$ kg/m^3 [36]. The viscosity–pressure dependence of this lubricant is characterized by the modified Yasutomi-WLF model given by the following:

$$\begin{array}{ll} & \mu(p) = \mu_g \exp\left[\frac{-2.303 C_1 (T_0 - T_g) F}{C_2 + (T_0 - T_g) F}\right] \\ \text{with} & T_g(p) = T_{g0} + A_1 \ln(1 + A_2 p) \\ & F(p) = (1 + b_1 p)^{b_2} \end{array} \tag{14}$$

where T_g is the glass transition temperature, with T_{g0} being its ambient-pressure value, and with parameters $A_1 = 188.95$ °C, $A_2 = 0.53$ GPa^{-1}, $b_1 = 7.37$ GPa^{-1}, $b_2 = -0.62$, $C_1 = 15.90$, $C_2 = 14.16$ °C, $T_{g0} = -68.47$ °C, and $\mu_g = 10^{12}$ Pa·s [36]. Given these values, the ambient-pressure viscosity $\mu_0 = 0.0125$ Pa·s, and the reciprocal asymptotic isoviscous pressure coefficient [37] $\alpha^* = 21.21$ GPa^{-1}.

2.2. Overall Numerical Procedure

The domain Ω was discretized into a mesh of second-order (quadratic) Lagrange finite elements of a tetrahedral shape (10 nodes). The mesh across the contact domain Ω_c was taken as the two-dimensional projection of the three-dimensional mesh, that is, a mesh of

6-node Lagrange triangular elements. To decrease the computational time, the mesh was refined the most at Ω_c and made increasingly coarser in further regions of the domain Ω. Moreover, a mesh refinement process was conducted to ensure grid-independent solutions.

All equations were discretized using FEM and solved simultaneously. Since the Reynolds equation is a convection-diffusion equation, artifact oscillations may arise in the high-load solution when using the standard Galerkin formulation [5]. Several formulations have been introduced to remedy this issue, namely the Streamline Upwind Petrov-Galerkin (SUPG), Galerkin Least-Squares (GLS), and Isotropic Diffusion (ID) [5]. In the model employed in this paper, the SUPG stabilizing term is combined with the ID term, which allows for the resolution of cases with Hertzian pressures of several gigapascals.

Furthermore, the penalty term introduced by Wu [38] was added to the Reynolds equation to comply with the cavitation boundary condition. The resulting non-linear algebraic system of equations is then solved using the damped-Newton method [39], starting from a carefully chosen initial guess. For more details on the FEM modeling of the EHL problem (convergence criteria, meshing, FEM formulations, etc.), interested readers are referred to [5].

2.3. Experimental Validation

The FEM model was previously validated against experiments for circular contacts on numerous occasions, but not for elliptical contacts. Therefore, experimental validation for wide and narrow elliptical contacts is required.

Experiments were conducted using an optical interferometry roller-on-disc apparatus in a temperature-controlled environment at 30 °C, with a steel roller ($E = 210$ GPa and $v = 0.3$) against a glass plane ($E = 72$ GPa and $v = 0.23$). For the wide contact experiments, the roller's radii of curvature at the point of contact $R_x = 13.05$ mm and $R_y = 84$ mm, and the experiments were conducted for a constant load of $F = 150$ N, corresponding to a Hertzian pressure of $p_h = 0.484$ GPa. Under these conditions, the contact ellipticity ratio $\theta \approx 0.295$. A total of 20 different values of entrainment speeds ranging from $u_m = 0.05$ m/s to $u_m \approx 6.458$ m/s were considered. For narrow contacts, the roller's radii of curvature at the point of contact $R_x = 12.7$ mm and $R_y = 4.82$ mm, and the experiments were conducted for a constant load of $F = 13$ N, corresponding to a Hertzian pressure of $p_h = 0.526$ GPa. Under these conditions, the contact ellipticity ratio $\theta = 1.89$. A total of 15 different values of entrainment speeds ranging from $u_m = 0.8214$ m/s to $u_m \approx 6.46$ m/s were considered. For both cases, the lubricant used was the Shell T9 lubricant. Note that film thickness measurements for wide contacts extend to much lower speeds compared to slender ones. This is because the film thickness range covered by the employed test rig is fixed. It spans from a few tens of nanometers up to slightly less than a micron. Given that the same lubricant is employed for both wide and slender contacts, with the same inlet temperature, a wide contact would generate thicker films than a slender one (at similar speeds), allowing for measurements at smaller entrainment speeds for the former. In addition, for wide contacts, the higher external load F enhances contact stability, allowing for measurements of even thinner films (i.e., at even lower speeds).

As can be seen in Figure 3, which shows central and minimum film thicknesses against mean entrainment speeds on a log-log scale, the FEM model generally complies with the experimental data, especially at low speeds where the deviation is minimal. For higher entrainment speeds, the discrepancy is seen to increase, which can be justified by the prevalence of thermal and shear-thinning effects that were not accounted for in the model. Coefficient of determination R^2 values (between experimental and numerical data) are reported within Figure 3 for each set.

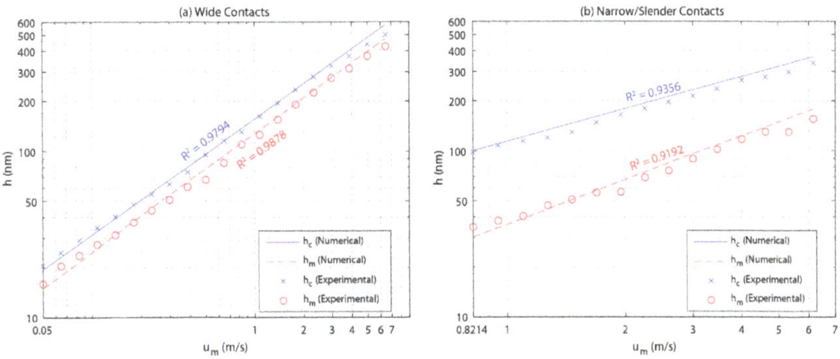

Figure 3. Comparison of numerical and experimental central and minimum film thickness variations as a function of the entrainment speed for (**a**) wide contacts and (**b**) narrow contacts.

3. Machine Learning

In general, supervised ML models for regression are algorithms that learn the relationship between variables in a dataset and predict a continuous target or output variable. The ML model is provided with input features that will be used to interpret the variation of the output variable in a learning process known as model training. Once the dataset is generated or compiled, it is split into at least two datasets: a training and a testing dataset. A variety of ML models are trained on the former set with varying parameters. Then, the performance and accuracy of these models are evaluated on the testing dataset based on appropriate evaluation metrics to empirically find the optimal model for a given application. An important procedure conducted by ML engineers is feature selection, which involves selecting the necessary input features (here, the relevant EHD parameters) to properly capture the variation in the outputs (here, the central and minimum film thicknesses). This section covers the development of the dataset, the feature selection process, the description of the employed ML model (namely GPR), its underlying choices for the kernel function, and the employed data standardization techniques and performance metrics.

3.1. Data Generation

Prior to generating the operating conditions of every sample in the dataset, a range of operating conditions should be specified. The ranges are defined such that every datapoint belongs to the full-film elastohydrodynamic lubrication regime. In other words, the lubricant film should be sufficiently thick to prevent a surface asperity contact, indicative of the mixed lubrication regime. In addition, the pressure endured by the components should not result in plastic deformation, which is avoided in practice as it would lead to failure. To achieve this, specific ranges for the Hertzian pressure p_h, mean entrainment speed u_m, and ratio of radii D were established, and the values of the remaining parameters were calculated accordingly. Moreover, cutoff values for the Moes dimensionless parameters M and L [12] were set to enforce the same full-film EHL conditions. This additional restriction ensures that all considered samples remain within the vicinity of the piezoviscous elastic lubrication regime. Further verification will be applied through the careful visual inspection of all obtained pressure and film thickness profiles to ensure that they exhibit all typical characteristics of this regime (i.e., a flat central film thickness region with a minimum thickness in the vicinity of the side lobes, a near-Hertzian pressure distribution with a zero pressure gradient on the inlet side to ensure that the chosen inlet length is sufficient to avoid numerical starvation, etc.). The Moes dimensionless groups are written as a function of the Hamrock and Dowson dimensionless groups U, G, and W [11], which are given as follows:

$$U = \frac{\mu_0 u_m}{2ER_x}, \; G = 2\alpha^* E, \text{ and } W = \frac{F}{2ER_x^2} \tag{15}$$

Then, the Moes dimensionless groups are given by the following:

$$M = W(2U)^{-3/4}, \text{ and } L = G(2U)^{1/4} \tag{16}$$

The ranges of interest of all parameters are presented in Table 1.

To provide representative sampling across the ranges of interest, Latin Hypercube Sampling (LHS) [40] was employed. LHS is a space-filling design method that maximizes the minimum distance between all datapoints. Such a method is employed as it is ideal for ML models to have spaced-out samples.

Two separate datasets, each of 625 points, were generated using LHS. One dataset was for wide elliptical contacts, and another one was for narrow contacts. Both datasets were generated for steel–steel contacts, lubricated with Shell T9. The *lhsdesign* function in MATLAB [41] generated a unit hypercube, which was then mapped linearly to match the ranges of the operating conditions in Table 1. Note that argument "Smooth" was set to "off" to obtain equally spaced steps in each dimension, and the argument "Criterion" was set to "maximin" to maximize the minimum distance between datapoints, as desired. The distribution of LHS datapoints of each dataset can be seen in Figure 4a,b for wide and narrow contacts, respectively. After applying the constraints in Table 1, the former dataset was left with 413 samples, while the latter had 503 samples. The distribution of constrained LHS datapoints of each dataset is shown in Figure 4c,d for wide and narrow contacts, respectively. Note that for both wide and narrow contacts, the low-speed high-load cases and the high-speed low-load cases were filtered out in the constrained datasets. This is because the former would yield extremely thin films (below 10 nm, roughly) where the direct contact between asperities is likely to occur (mixed lubrication), while the latter would yield extremely thick films with low pressures and little-to-no solid elastic deformations (hydrodynamic lubrication). For every simulated case, the values of parameters M, L, and θ were recorded in a table, along with target variables H_c^* and H_m^*, as detailed in Section 3.2. The datasets were finally combined into one larger dataset of 915 samples, which was split into an 823-sample training dataset and a 92-sample testing dataset, corresponding to 90% and 10% of the combined dataset samples, respectively.

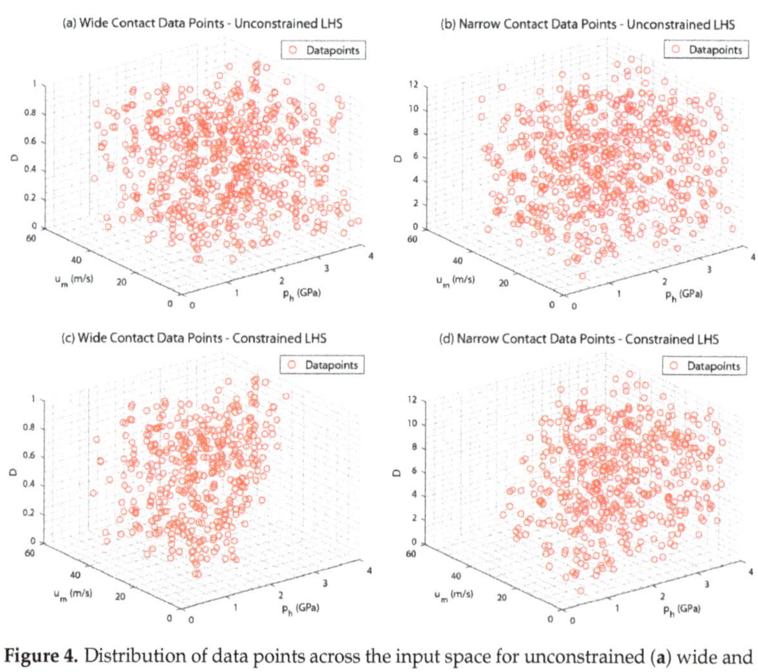

Figure 4. Distribution of data points across the input space for unconstrained (**a**) wide and (**b**) narrow contacts, as well as constrained (**c**) wide and (**d**) narrow contacts.

Table 1. Operating conditions' ranges of interest for dataset generation.

	Parameter	Lower Bound	Upper Bound	Unit
Ranges of interest	u_m	0.01	50	m/s
	p_h	0.4	4	GPa
	D	1/12	12	-
Constraints	M	10	3000	-
	L	1	20	-

3.2. Feature Selection

Feature selection is a key component in any ML process, consisting of the careful selection of the input parameters that would be needed for the prediction of specified outputs. The simplest and most obvious choice for the EHL problem would be a brute-force approach, whereby all physical inputs of the contact (i.e., kinematics, load, solid and fluid material properties, etc.) are selected as features. This would result though in a relatively high number of input parameters, some of which may be correlated. Knowledge of the EHL problem and its characteristics will turn out to be fundamental in selecting the right inputs, as will be discussed in what follows.

This study aims to develop ML models that can accurately predict the central and minimum film thicknesses of EHD elliptical contacts. A common practice in ML is to reduce the number of input features into a model and hence the popularity of methods, such as Principal Component Analysis (PCA) [42]. This practice boosts the model's simplicity and interpretability, as well as its computational efficiency, both in training and prediction. In this work, the number of input features was minimized based on the existing knowledge of the EHL problem, its underlying physics, and governing mechanisms. The Moes dimensionless groups, M and L [12], defined in Equation (16), were exclusively used as input features for GPR, as this set has the lowest number of dimensionless groups due to the optimum similarity analysis employed to derive it. This set was also proven to be relatively successful at film thickness prediction in analytical formulas [19]. Furthermore, the ellipticity of the contact θ (or the ratio of radii D) is required to convey the shape or aspect ratio of the contact. The feature selection process is summarized in Figure 5, where the numbers between brackets refer to the equation(s) defining their corresponding parameters. This figure details the pre-processing and post-processing phases applied prior to and following the ML phase. The values of parameters R_x, E, and α^* were fixed, along with the density and viscosity models and their parameters for the Murnaghan equation of state (13) and the modified Yasutomi-WLF Equation (14), respectively. Only parameters D, p_h, and u_m were varied based on LHS. The number of input features was then reduced by deriving parameter F using Equation (4) as a prerequisite to finding the Hamrock and Dowson dimensionless groups, U, G, and W using Equation (15). Then, these groups were further combined into the Moes dimensionless groups M and L using Equation (16). Furthermore, these features were transformed into their logarithmic values $\ln(M)$ and $\ln(L)$, respectively. At this stage, it is essential to introduce another commonly-adopted definition for the dimensionless film thickness:

$$H^* = \frac{h}{R_x\sqrt{2U}} \tag{17}$$

This definition is used in the ML process to conform with the Moes dimensionless groups (M and L) as input parameters. The same M, L, and θ values result in the same H^* and not H (not accounting for lubricant compressibility, and assuming an idealistic exponential viscosity-pressure response) [12]. Analytical formulas featuring this definition were previously established by Evans and Snidle [43] and Nijenbanning et al. [17]. This allows the model to be generalized to different solid material properties. A sufficiently accurate generalization would also be attained for fluids with a non-exponential viscosity–pressure behavior, by using α^* to describe their response. The output features H_c^* and H_m^*

are also transformed into their logarithmic values, $\ln(H_c^*)$ and $\ln(H_m^*)$. The logarithmic transformation is based on analytical formulas indicating a power-law relationship between the input and output features. It reduces the non-linearity of the problem and allows for easier and more accurate ML modeling, as will be discussed in Section 4. Dimensional film thicknesses were then retrieved using Equation (17), and the ML-model-performance metrics were evaluated based on the output dimensional film thicknesses, as discussed in Section 3.3. Note that fixing a subset of EHD parameters while maintaining the ability to generalize onto other values is made possible through the use of dimensionless groups.

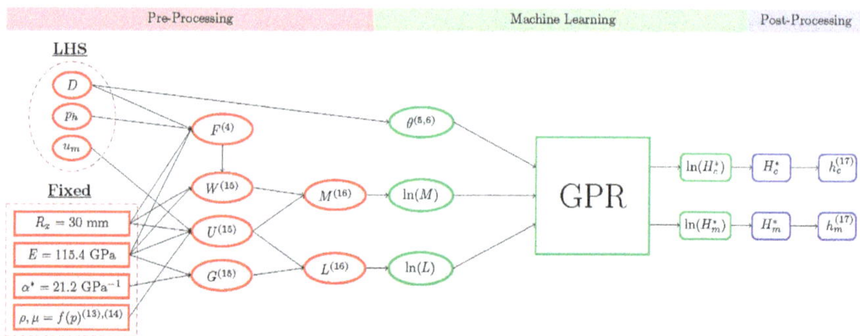

Figure 5. Diagram illustrating the dataset generation, dimensionality reduction, and feature selection for this study.

3.3. Gaussian Process Regression (GPR)

Several models may be used for regression given tabular data, each of different working principles. For the sake of brevity, only GPR [33] will be considered here, given its superior performance over SVR [31] and its better suitability for limited datasets compared to ANNs [23]. GPR is a non-parametric—the number of parameters is a function of the number of training points—probabilistic model that can account for noise and uncertainty. This model employs a kernel function to capture data nonlinearity. The kernel function should be carefully selected or designed to inform the model on the relationship and covariance between variables and therefore potentially achieve better performance. The different kernel function choices that are considered in this work are detailed in Appendix A.

ML models typically have hyperparameters that should be tuned to improve performance. Hyperparameters traditionally were fixed before training begins and not altered during the process. However, models now employ optimization algorithms to fine-tune some hyperparameters during training. This helps to avoid overfitting and improve generalization to unseen data [44]. In GPR, the hyperparameters include the kernel length-scale, which can be different for every input feature, and determines the flexibility and sensitivity of the model (range over which one datapoint can influence other datapoints) and the kernel signal variance, which controls the scale of the function and the distribution of data. The values of the kernel hyperparameters are initialized prior to training the models. Then, through an optimization process, these values are updated until convergence is reached. In the *scikit-learn* GPR implementation employed in this work, the *l-bfgs-b* algorithm [45] maximizes the log marginal likelihood [33], which reduces errors and improves model accuracy.

In GPR, the predicted function is modeled as a Gaussian process distribution. In other words, the predicted function does not have one specific parametric form, but rather it is a distribution over several functions. This allows the model to approximate uncertainty and noise in data by calculating the standard deviation of predictions. Let $\tilde{x}$ and $\hat{x}$ be the sets of values of input features from two sample datasets $\tilde{X}$ and $\hat{X}$ of dimensions $\tilde{n}$ and $\hat{n}$, respectively, with $\tilde{x}_i$ and $\hat{x}_i$ being their individual inputs ($i = 1 \ldots N_f$, where N_f corresponds to the number of input features). A multivariate Gaussian process is defined by a mean function $m(x)$ and a kernel function $k(\tilde{x}, \hat{x})$ evaluated for every possible pairwise combination $(\tilde{x}, \hat{x})$. The former is the expected value for an input x, while the latter represents the dependence and correlation between inputs $\tilde{x}$ and $\hat{x}$. A function $y(x)$ following a Gaussian distribution would then be denoted as follows:

$$y(x) \sim \mathcal{N}(m(x), k(\tilde{x}, \hat{x})) \tag{18}$$

In a Gaussian process, any subset of points belonging to the same dataset follows a joint normal, Gaussian distribution. As such, the training dataset output values y_{tr} and the testing dataset output values y_{te} are drawn from a joint multivariate Gaussian distribution, in a GPR model. Initially and before training or observing any data, a prior distribution over functions is assumed. The prior distribution can be seen as an initial assumption about the model parameters prior to observing any data. Then, by applying Bayes' rule [33] and conditioning to the training dataset, the posterior distribution is obtained. This distribution combines the prior with the likelihood function, both of which are assumed to be Gaussian. The latter is derived from training data and reflects the probability of observing the output value, given the set of input values. The posterior distribution is also a multivariate Gaussian process representing the updated belief about model parameters, characterized by a posterior mean function and a posterior kernel function. The posterior mean function represents the noise-free or average value of predictions, while the posterior kernel function can quantify noise and uncertainty in predictions. The output (and input) data will be standardized around a zero mean, as will be discussed later. Accordingly, the mean function $m(x)$ is taken as zero, allowing for notational and computational simplicity. Note that since the dataset was developed using deterministic FEM simulations, the data are assumed to be noise-free (no uncertainty). The noise variance value is negligible, and the uncertainty of predicted values is not considered, i.e., the posterior kernel function is not considered, and the predicted output is a specific value and not a range of values. Therefore, to make predictions, the posterior mean function is simply evaluated for the input values of data points to be predicted. Given the following prior distribution over functions:

$$y(x) \sim \mathcal{N}(0, k(\tilde{x}, \hat{x})) \tag{19}$$

the prior joint multivariate Gaussian distribution between all dataset points (split into training and testing subsets, denoted by the subscripts "*tr*" and "*te*", respectively) is written as follows:

$$\begin{bmatrix} y_{tr} \\ y_{te} \end{bmatrix} \sim \mathcal{N}\left(\begin{bmatrix} 0 \\ 0 \end{bmatrix}, \begin{bmatrix} k(X_{tr}, X_{tr}) & k(X_{te}, X_{tr})^T \\ k(X_{te}, X_{tr}) & k(X_{te}, X_{te}) \end{bmatrix} \right) \tag{20}$$

and the prediction function of the GPR model for testing datapoints is as follows:

$$\hat{y}_{te} = k(X_{te}, X_{tr})[k(X_{tr}, X_{tr})]^{-1} y_{tr} \tag{21}$$

The GPR prediction, Equation (21), can be seen as a linear regression or weighted sum equation [32], where every output in y_{tr} is multiplied by a given weight. The weight stems from the evaluation of the kernel function, and its value is based on the similarity between the input values of the testing and training points. For a more detailed derivation of Equation (21), interested readers are referred to [33]. Notice that only two kernel matrices need to be evaluated for noise-free predictions. The first is $k(X_{tr}, X_{tr})$ for every possible pairwise combination of training samples, resulting in a matrix of size $N_{tr} \times N_{tr}$, where N_{tr} is the number of samples in the training set. The second kernel matrix $k(X_{te}, X_{tr})$ evaluates the kernel function for every possible pairwise combination of samples, where one sample is from the testing set and the other is from the training set. The size of $k(X_{te}, X_{tr})$ would be $N_{te} \times N_{tr}$, where N_{te} is the number of samples in the testing set. The computational complexity is imposed by the inversion of matrix $k(X_{tr}, X_{tr})$, which results in an algorithm scaling with $O(N_{tr}^3)$. For a relatively small dataset of several hundred datapoints, this is not a concern. However, it may turn out to be prohibitive for larger datasets. The exact GPR implementation used in *scikit-learn* is based on Algorithm 2.1 in [33].

Finally, note that all input features fed into the ML model and the outputs should be normalized or standardized prior to training and predicting, as detailed in Appendix B. Also, listed in Appendix B are the performance metrics that are employed in evaluating the pertinence and accuracy (in comparison with FEM simulation results) of the different considered ML model configurations, namely the *adjusted R-squared*, *mean absolute percentage error* (*MAPE*), and *maximum absolute percentage error* (*MAXAPE*). The "error" terminology in these metrics simply corresponds to output deviations with respect to FEM simulation results.

4. Results and Discussion

In this section, the performance of different GPR models is evaluated in order to find an optimal configuration, for which the predictive performance is then compared to that of a conventional analytical film thickness formula for the central and minimum film thickness in elliptical EHL contacts.

First, several kernel functions and a combination of these functions were considered, in order to determine the best performing configuration. The performance of all functions based on the testing dataset is summarized in Table 2. Clearly, ARD-Matern kernels offer the best fit, while the ARD-RBF kernel seems to be the worst. Moreover, combining the ARD-Matern kernel for $\nu = 3/2$ with that for $\nu = 5/2$ further reduces prediction errors with respect to the FEM simulations. For this configuration, *MAPE* values drop to 0.31% and 1.00% only, for the central and minimum film thicknesses of testing points, respectively, with a *MAXAPE* of 3.05% and 6.97%, respectively. The corresponding kernel function is obtained by simply combining or adding the ARD-Matern kernel for $\nu = 3/2$ with that for $\nu = 5/2$.

Table 2. Performance metrics of GPR models based on the testing dataset for different kernel functions.

		h_c			h_m		
Kernel Function		*Adj.* R^2 (-)	*MAPE* (%)	*MAXAPE* (%)	*Adj.* R^2 (-)	*MAPE* (%)	*MAXAPE* (%)
ARD-RBF		0.9871	2.28%	22.02%	0.9841	6.89%	68.20%
RQ		0.9988	0.71%	5.15%	0.9979	1.88%	11.74%
ARD-Matern	$\nu = 3/2$	0.9995	0.39%	5.33%	0.9987	1.39%	7.66%
	$\nu = 5/2$	0.9990	0.53%	9.12%	0.9975	1.58%	12.86%
	$\nu = 3/2 \oplus \nu = 5/2$	0.9999	0.31%	3.05%	0.9992	1.00%	6.97%

One noticeable trend depicted in Table 2 is the larger errors for minimum film thickness predictions compared to central film thickness. Such a trend is not unusual though, and it has often been observed in analytical film thickness formulae [19]. This is because the governing physical mechanisms for the central film thickness are better understood than those for the minimum film thickness. Even after decades of trials, the quest for a reliable minimum film thickness formula remains elusive. From the earliest theoretical studies on EHL, it was identified that the central film thickness is governed by lubricant rheology in the low-pressure inlet region of the contact, though proper quantification of the inlet pressure was only recently carried out [46]. This has led to the identification of a single pressure–viscosity parameter that would govern the central film thickness response in analytical formulae. There is still no general consensus though on the definition of this parameter, and several variants have been proposed over the years [47]. Nonetheless, it is at least clear that central film thickness is governed by the inlet rheology. The minimum film thickness was also thought to be governed by the inlet rheology. Only recently did Habchi et al. [48] discover that it is also influenced by the high-pressure rheology in the central part of the contact. This implies that an additional high-pressure definition for a pressure–viscosity parameter would be needed—whether in analytical formulae or ML frameworks—to properly describe the governing physics of minimum film thickness. Such a definition is yet to be developed, and it is definitely beyond the scope of the current work. Nonetheless, once available, it would be straightforward to include as an additional input feature for analytical formulae or ML models. The lack of a proper understanding of the governing mechanisms of the minimum film thickness is probably the reason why corresponding recent analytical formulae have shifted—with somewhat more success—towards a prediction of the ratio of the central-to-minimum film thickness, rather than a direct prediction of the minimum film thickness itself [49].

The performance of the best performing model (and kernel) based on the testing dataset is illustrated in Figure 6. Figure 6a,b show the predicted central and minimum film thickness values, respectively (using the ML model), compared to their "true" values (given by the FEM simulations). Note that the closer the predictions are to the diagonal, the more accurate they are. The performance metrics for each feature are displayed within its corresponding figure. Clearly, the proposed ML framework is perfectly capable of predicting both the central and minimum film thickness for all considered testing points, with a slightly better precision for the former. Figure 6c,d show the percentage residuals/errors of the central and minimum film thickness, respectively, for all testing points, as a function of their corresponding so-called true value, obtained using the FEM model described in Section 2. The results reveal that errors are rather scattered, with no significant error bias towards low, medium, or high film-thickness valuese.

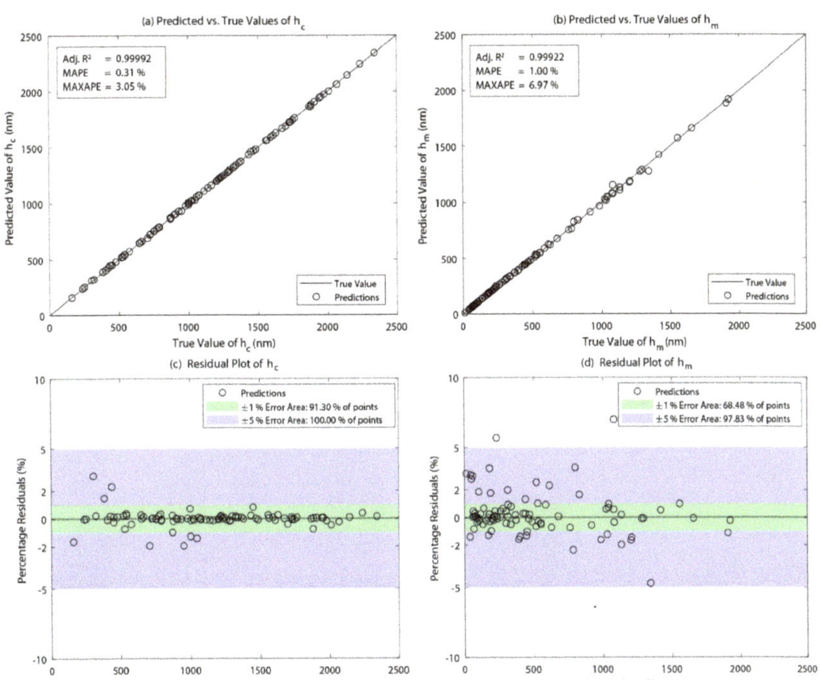

Figure 6. Predicted vs. true values and percentage residual plots of central ((**a**) and (**c**), respectively) and minimum film thicknesses ((**b**) and (**d**), respectively) for the best-performing ML model.

Next, in order to showcase the benefits of transforming the features M, L, H_c^*, and H_m^* into their logarithmic values, the models were re-trained using their untransformed/original values. The corresponding results for the testing dataset are presented in Table 3. In comparison with the results of Table 2, it can be seen that all models for all kernel functions perform worse when the scale of input features M and L and outputs H_c^* and H_m^* is linear instead of logarithmic. For the best-performing model (last row in both tables), the logarithmic transformation reduces the

MAPE from 0.52% to 0.31% and *MAXAPE* from 7.48% to 3.05% for the central film thickness. For the minimum film thickness, *MAPE* is reduced from 1.50% to 1.00% and *MAXAPE* from 7.32% to 6.97%. Reductions are more significant for other choices of kernel functions. This shows the importance of transforming these features into their logarithmic scales, to reduce the ML model non-linearity. In fact, since the earliest theoretical EHL studies, it has been known that the film thickness varies as a power function of the Moes parameters M and L (or their underlying Hamrock and Dowson parameters, G, U, and W), as evidenced by most analytical formulas using these parameters [19]. Note that the values of the *Adj. R-squared* can be misleading, as they are very close to unity despite the large percentage errors. For a more detailed observation of the performance of this model, Figure 7 shows the same results as Figure 6, but without the logarithmic transformation.

Table 3. Performance metrics of GPR models using the initial scale (instead of logarithmic) of M, L, H_c^*, and H_m^* for different kernel functions.

		h_c			h_m		
Kernel Function		*Adj.* R^2 (-)	*MAPE* (%)	*MAXAPE* (%)	*Adj.* R^2 (-)	*MAPE* (%)	*MAXAPE* (%)
ARD-RBF		0.9420	4.65%	49.15%	0.8857	36.28%	566.52%
RQ		0.9969	1.06%	8.36%	0.9927	5.38%	30.63%
ARD-Matern	$\nu = 3/2$	0.9993	0.47%	5.98%	0.9974	2.11%	11.89%
	$\nu = 5/2$	0.9968	0.87%	20.90%	0.9882	6.84%	239.95%
	$\nu = 3/2 \oplus \nu = 5/2$	0.9998	0.52%	7.48%	0.9992	1.50%	7.32%

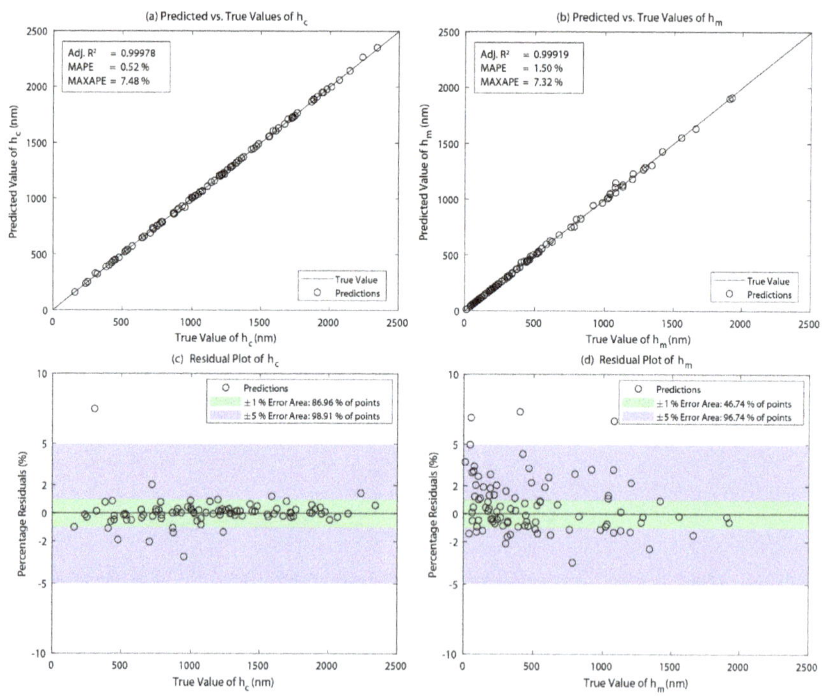

Figure 7. Predicted vs. true values and percentage residual plots of central ((**a**) and (**c**), respectively) and minimum film thicknesses ((**b**) and (**d**), respectively) for the best-performing ML model, without logarithmic scaling of M, L, H_c^*, and H_m^*.

Trained ML models can generate thousands of predictions per second, hence offering a similar speed to analytical formulas, such as, for example, the Hamrock and Dowson equations [16] (still probably the most widely used film thickness formulas to date [50], despite decades of development). It would therefore be interesting to compare the accuracy of these formulas to that of the proposed ML model, based on the same testing set, as employed here. Figure 8 shows the performance metrics of the Hamrock and Dowson formulas, in a similar fashion to Figures 6 and 7. Out of fairness, narrow elliptical contacts were left out of the testing set, since the Hamrock and Dowson formulas were originally developed for wide and circular contacts only. Not only are the analytical formulas' predictions significantly less accurate than the ML model, but they are also mostly of a non-conservative nature (i.e., they overpredict the film thickness). In addition, the usual significant loss of accuracy of analytical formulae for the minimum film thickness—discussed earlier—can be clearly seen in Figure 8b,d. The results of Figure 8 are in general agreement with those of Wheeler et al. [19] who reported relative deviations in central and minimum film thickness predictions using the Hamrock and Dowson formulae that are as high as approximately 22% and 95%, respectively, compared to EHL simulation results.

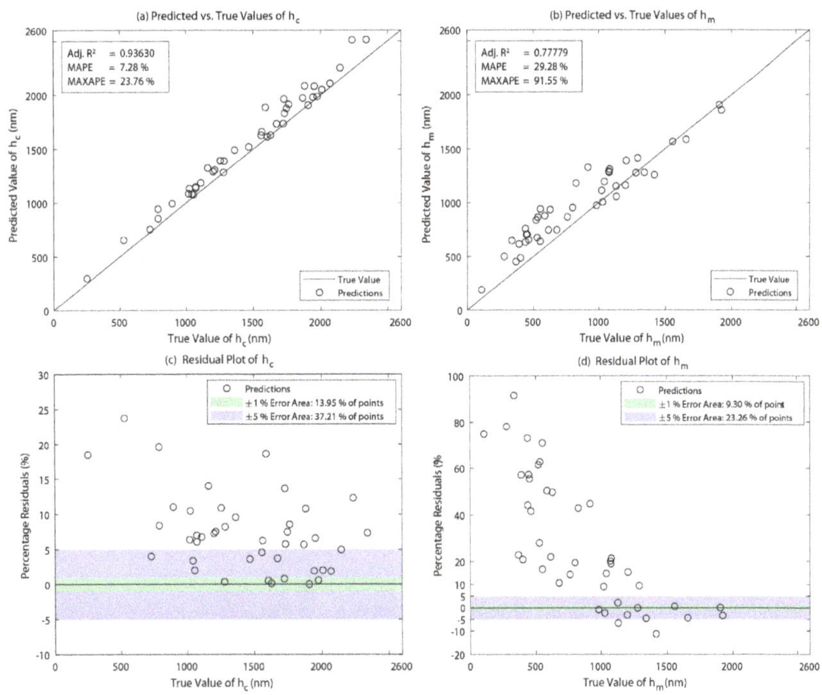

Figure 8. Predicted vs. true values and percentage residual plots of central ((**a**) and (**c**), respectively) and minimum film thicknesses ((**b**) and (**d**), respectively) for the Hamrock and Dowson analytical formulas for wide and circular cases of the testing dataset only.

The superior predictive performance of ML regression frameworks, like GPR, over analytical film thickness formulas may be attributed to the fact that the former are non-parametric (i.e., they operate in a functional space of infinite dimensions or, in other words, with an infinite number of possible regression functions), whereas the latter are parametric, and they usually rely on a pre-defined single regression function with a certain fixed number of parameters (e.g., Hamrock and Dowson [16]) or the combination of a limited number of pre-defined functions (e.g., Nijenbanning et al. [17], which employs a combination of four pre-defined functions, one for each of the known EHL regimes: rigid-isoviscous, elastic-isoviscous, rigid-piezoviscous, and elastic-piezoviscous).

Remark 1. *Out of fairness, it should be emphasized that the range of M and L that was originally covered in developing the Hamrock and Dowson formulas is significantly smaller than the one covered by the testing dataset. Actually, only a few samples of the testing dataset fall within that range. Therefore, the current comparison is not entirely fair. Nonetheless, it is quite illustrative, since it gives the reader an idea of the range of errors that are involved in using such analytical formulas outside their range of application, a common practice in the EHL literature. In addition, the Hamrock and Dowson formulas were developed using the simplistic exponential pressure–viscosity relationship, which fails to accurately capture the high-pressure rheology of typical lubricants. As such, this would yield relatively inaccurate minimum film thickness predictions, as discussed earlier. The impact on central film thickness should be minimal, however, as long as the correct pressure–viscosity coefficient is employed. This is because the central film thickness is governed by the low-pressure inlet rheology of the lubricant [46], which is well captured, even by such simplistic rheological models.*

5. Conclusions

This study extends the use of Machine Learning (ML) approaches for EHL film thickness predictions to the general case of elliptical contacts by considering wide and narrow contacts over a wide range of ellipticity and operating conditions. FEM simulations are used to generate substantial training and testing datasets that are used within the proposed ML framework. The complete dataset entails 915 samples, split into an 823-sample training dataset and a 92-sample testing dataset, corresponding to 90% and 10% of the combined dataset samples, respectively. The proposed ML model consists of a pre-processing stage in which the conventional EHD dimensionless groups are used to minimize the number of inputs to the model, reducing them to only three. The core of the model is based on GPR, a powerful ML regression tool, well-suited for small-sized datasets, producing output central and minimum film thicknesses also in a dimensionless form. The last stage is a post-processing one, in which the output film thicknesses are retrieved in a dimensional form.

First, the ML model was tuned to find the most suitable choice/combination of kernel functions, which was then used to make film thickness predictions for the testing dataset. The results revealed the power of the proposed ML approach, producing predictions that are far superior to analytical film thickness formulae in terms of accuracy, for a similar negligible computational effort. Then, the importance of transforming the input Moes dimensionless parameters and the output film thicknesses into a logarithmic scale was quantified. Such a transformation reduces the non-linearity in the ML model, leading to improved prediction accuracy, since central and minimum film thicknesses are known to vary as a power function of the Moes parameters.

To conclude, this study constitutes a first approach towards establishing a generalized ML framework for elliptical EHD contacts, through the use of conventional EHL dimensionless groups, namely the Moes parameters. It was shown that such groups can, in fact, be employed as input features and produce accurate models, contrarily to what was suggested by Marian et al. [30]. The generality of the proposed framework is not attributed to the ML model itself, but rather to the well-known central and minimum film thickness similitude of EHD contacts with similar values of the Moes parameters. Nonetheless, some improvements can still be applied to enhance the predictive accuracy. For instance, the influence of lubricant compressibility may be incorporated either a priori by employing several density–pressure responses and adding their corresponding parameters to the input features of the dataset or a posteriori by using a correction factor for the central film thickness [51,52]. This is because the influence of lubricant compressibility is known to be restricted to the central film thickness, with no noticeable effect on the minimum film thickness. Incorporating the influence of compressibility would enhance the accuracy of central film thickness predictions. As for minimum film thickness predictions, these could be enhanced through the derivation of a dedicated pressure–viscosity parameter at high-pressure, to be added as an additional input feature.

Author Contributions: Conceptualization, W.H.; methodology, J.I. and W.H.; software, J.I. and A.E.H.; validation, P.V., formal analysis, J.I. and W.H.; investigation, J.I. and W.H; resources, W.H. and P.V.; data curation, J.I. and W.H.; writing—original draft preparation, J.I.; writing—review and editing, J.I., P.V. and W.H.; visualization, J.I.; supervision, W.H.; project administration, W.H.; funding acquisition, W.H. All authors have read and agreed to the published version of the manuscript.

Funding: This research received no external funding.

Data Availability Statement: The data presented in this study are available on request from the corresponding author.

Conflicts of Interest: The authors declare no conflict of interest.

Nomenclature

α^*	Reciprocal asymptotic isoviscous pressure coefficient (Pa^{-1})
β_K	Murnaghan EoS isothermal bulk modulus temperature coefficient (K^{-1})
μ	Lubricant low-shear/Newtonian viscosity (Pa·s)
$\overline{\mu}$	Dimensionless lubricant low-shear/Newtonian viscosity
μ_g	Lubricant viscosity at glass transition temperature (Pa·s)
μ_0	Lubricant low-shear/Newtonian viscosity at ambient pressure (Pa·s)
μ_{tr}, σ_{tr}	Mean and standard deviation of features within the training dataset
v	Equivalent solid Poisson coefficient
v_1, v_2	Poisson coefficient of solids 1 and 2
Ω	Equivalent solid computational domain
Ω_c	Contact computational domain
$\partial\Omega_c$	Boundaries of Ω_c
$\partial\Omega_b$	Fixed boundary of Ω
$\partial\Omega_s$	Symmetry boundary of Ω
Ψ_1	Complete elliptic integral of the first kind
ρ	Lubricant density (kg/m^3)
$\overline{\rho}$	Lubricant dimensionless density
ρ_0	Lubricant density at ambient pressure (kg/m^3)
σ_n	Normal component of 3D stress tensor (Pa)
σ_f, l, α, ν	GPR model hyperparameters
$\{\sigma_t\}$	Vector of tangential components of 3D stress tensor (Pa)
θ	Contact ellipticity ratio
τ_{ij}	Shear stress in the j-direction within a plane having i as normal (Pa)
a_x, a_y	Hertzian elliptical contact semi-axes in the x, y-directions (m)
A_1, C_2	Modified Yasutomi-WLF viscosity model parameters (°C)
A_2, b_1	Modified Yasutomi-WLF viscosity model parameters (Pa^{-1})
b_2, C_1	Modified Yasutomi-WLF viscosity model parameters
D	Ratio of contact equivalent radii of curvature R_x and R_y
E	Equivalent solid Young's modulus of elasticity (Pa)
E_1, E_2	Young's moduli of elasticity of solids 1 and 2 (Pa)
F	Contact external applied load (N)
G, U, W	Hamrock and Dowson material, speed, and load dimensionless groups
h	Lubricant film thickness (m)
h_c	Central film thickness (m)
h_m	Minimum film thickness (m)
H_c, H_c^*	Dimensionless central film thickness
H_m, H_m^*	Dimensionless minimum film thickness
H_0	Dimensionless rigid-body separation
H, H^*	Dimensionless lubricant film thickness
K_{00}	Isothermal bulk modulus at zero absolute temperature (Pa)
K_0'	Pressure rate of change of isothermal bulk modulus at zero pressure
L, M	Moes dimensionless material properties and load parameters
m, k	Mean and kernel functions
$\widetilde{n}, \hat{n}$	Sizes of sample datasets $\widetilde{X}$, $\hat{X}$

N_f	Number of input features
N_{tr}, N_{te}	Number of samples in the training and testing datasets
p	Pressure (Pa)
p_h	Hertzian contact pressure (Pa)
P	Dimensionless pressure
R_{x1}, R_{x2}	Principal radii of curvature of solids 1 and 2 in the xz-plane (m)
R_{y1}, R_{y2}	Principal radii of curvature of solids 1 and 2 in the yz-plane (m)
R_x	Radius of curvature of equivalent elastic solid in the xz-plane (m)
R_y	Radius of curvature of equivalent elastic solid in the yz-plane (m)
$\overline{R}$	Equivalent radius of curvature of reduced contact geometry (m)
T_g	Glass transition temperature (K)
T_{g0}	Glass transition temperature at zero pressure (K)
T_0	Ambient temperature (K)
u, v, w	Equivalent solid deformation components in the x, y, z-directions (m)
u_1, u_2	Surface velocities of solids 1 and 2 in the x-direction (m/s)
u_m	Contact mean entrainment speed in the x-direction (m/s)
U, V, W	Solid dimensionless deformation components in x, y, z-directions
x, y, z	Space coordinates (m)
y, y_{tr}, y_{te}	Gaussian distribution for standard, training and testing subsets
$\hat{y}_{te}$	Prediction function of GPR model for testing samples
$\widetilde{x}, \hat{x}$	Sample input features
$\widetilde{x}_i, \hat{x}_i$	Input i of $\widetilde{x}$, $\hat{x}$
x_i	Input feature number i
$\breve{x}_i$	Normalized value of input feature number i
$y_i, \hat{y}_i, \overline{y}$	Output variable i, its predicted and mean values in the testing dataset
X, Y, Z	Dimensionless space coordinates
X_{tr}, X_{te}	Training and testing sample datasets
$\widetilde{X}, \hat{X}$	Sample datasets

Appendix A. Kernel Function Definitions

The first step in developing a GPR model is the selection of a kernel function, informing the model about the smoothness and general patterns in the data. Various kernel or covariance functions can be employed in GPR, including the Radial Basis function (RBF), also known as the Exponential Quadratic or Squared Exponential kernel function, the Rational Quadratic (RQ) function, and the Matern functions [33]. Given the multiple input features of varying scale/significance for the EHL problem, the Automatic Relevance Determination (ARD) variation of these kernels will be employed, when possible. ARD kernels feature a specific length-scale for every input feature and can hence automatically determine the relevance of every parameter via the optimization process [53]. The ARD-RBF is written as follows:

$$k(\widetilde{x}, \hat{x})_{ARD-RBF} = \sigma_f^2 \exp\left[-\frac{1}{2}\sum_{i=1}^{N_f}\left(\frac{\widetilde{x}_i - \hat{x}_i}{l_i}\right)^2\right] \quad \text{(A1)}$$

where σ_f is the signal variance, N_f is the number of input features, and l_i is a strictly positive hyperparameter, known as the length-scale of feature i $(l_i > 0)$. Version 1.0.2 of *scikit-learn* does not offer an ARD variation of the RQ kernel; therefore, the RQ kernel featuring one common length-scale l for all N_f input features will be used instead and is written as follows:

$$k(\widetilde{x}, \hat{x})_{RQ} = \sigma_f^2 \left(1 + \frac{\sum_{i=1}^{N_f} (\widetilde{x}_i - \hat{x}_i)^2}{2\alpha l^2}\right)^{-\alpha} \quad \text{(A2)}$$

where α is the scale mixture parameter, which controls the trade-off between capturing long-scale variations and short-scale fluctuations. Lastly, the general ARD-Matern kernel function is written as follows:

$$k(\tilde{x},\hat{x})_{ARD-Matern} = \frac{\sigma_f^2}{\Gamma(\nu)2^{\nu-1}}\left[\sqrt{2\nu\sum_{i=1}^{N_f}\left(\frac{\tilde{x}_i-\hat{x}_i}{l_i}\right)^2}\right]^{\nu} K_\nu\left(\sqrt{2\nu\sum_{i=1}^{N_f}\left(\frac{\tilde{x}_i-\hat{x}_i}{l_i}\right)^2}\right) \quad \text{(A3)}$$

where K_v is the modified Bessel function of the second kind [54], Γ is the gamma function, and ν is an additional strictly positive hyperparameter that controls the function smoothness $(\nu > 0)$. In fact, for $\nu = 1/2$, the function is reduced to the Absolute Exponential kernel [33]. When $\nu \to \infty$, the RBF function is obtained. For $\nu = 3/2$, the kernel is once differentiable (i.e., the function and its derivative are continuous), and for $\nu = 5/2$, the kernel is twice differentiable (i.e., the function, its first and second derivatives are continuous). Note that ν remains fixed during optimization, unlike the length-scale hyperparameters, which are optimized. More importantly, these kernel functions can be added or multiplied to model more complex behavior as seen in [34].

Appendix B. Data Standardization and Performance Metrics

The input features fed into the ML model and the outputs should be normalized or standardized prior to training and predicting. This pre-processing step is a typical practice in ML and presents two main advantages [55]. The first advantage is that it prevents the dominance of features of higher values and scale over features of lower values and scale (i.e., the dominance of M over L and D in EHL, for example). This effect is reflected by the value of the Euclidean distance employed in the kernels. The second advantage of this practice is that it can accelerate the convergence of the hyperparameter optimization algorithm. Moreover, the inversion of matrix $k(X_{tr}, X_{tr})$ becomes a more stable operation when input data are standardized. Given the zero-mean assumption of the output values introduced in Equation (19) and for computational reasons, a z-score normalization (or standardization) is preferred for GPR [56]. The resulting distribution conforms with the Gaussian distribution, which improves the performance. The scaling transformation centers the data around zero and normalizes them with respect to the standard deviation. The normalized value $\breve{x}_i$ for a given feature x_i reads as follows:

$$\breve{x}_i = \frac{x_i - \mu_{tr}(x_i)}{\sigma_{tr}(x_i)} \quad \text{(A4)}$$

where $\mu_{tr}(x_i)$ and $\sigma_{tr}(x_i)$ correspond to the feature's mean value and its standard deviation, respectively, both obtained from the training set only. This transformation is applied to each of the N_f input features and both output variables individually and to each of the $N_{te} + N_{tr}$ data points of the entire dataset. Every feature has its own mean and variance values and hence its own transformation equation. Note that when normalizing the testing dataset, it is important to use the training dataset mean and standard deviation to avoid data leakage, which indirectly informs the ML model about the testing dataset during training. For outputs, this transformation is reversed after prediction, by inverting Equation (A4) to restore the original scale of values.

The performance of trained ML models is assessed using the testing set based on several metrics. One popular metric used to evaluate the trained models is the *R-squared* metric or coefficient of determination [57]. The advantages of this metric include its generality and dimensionless property, which make comparing different models convenient. It is calculated over the testing dataset and is given as follows:

$$R^2 = \frac{\sum_{i=1}^{N_{te}} (\hat{y}_i - \overline{y})^2}{\sum_{i=1}^{N_{te}} (y_i - \overline{y})^2} \quad \text{(A5)}$$

where $\overline{y}$ is the mean value of the output variable obtained from the testing dataset, $\hat{y}$ is the predicted output, and y is the true value from the testing dataset, obtained from the FEM simulations. The upper bound of R^2 is the ideal value of $+1$, and it has no lower bound (R^2 can tend towards $-\infty$). However, this metric suffers from an inherent flaw as it tends to increase (or remain constant) with the addition of more input features, even if the added features are not relevant to the problem. The *R-squared* does not account for the number and/or relevance of the individual N_f input features. To overcome this, the *adjusted R-squared* metric will be used instead, which is given as a function of *R-squared* and the number of input features N_f, as follows:

$$Adj.\ R^2 = 1 - \frac{(1 - R^2) \times (N_{te} - 1)}{N_{te} - N_f - 1} \quad \text{(A6)}$$

Another metric that will be considered is the mean absolute percentage error (*MAPE*) [58]. Given that percentage error is commonly used in computational engineering to determine convergence, mesh independence, and model validity, it can provide familiar insight into the ML model compliance with the dataset. The *MAPE* equation is given by the following:

$$MAPE = \frac{1}{N_{te}} \sum_{i=1}^{N_{te}} \left| \frac{\hat{y}_i - y_i}{y_i} \right| \times 100 \quad \text{(A7)}$$

This metric has a lower bound of 0% (its ideal value) but has no upper bound and can go up to $+\infty$. *MAPE* is known to be biased towards underpredictions [59]. This suits the EHL problem, as underpredictions are conservative. Therefore, variants of the *MAPE*, such as the symmetric mean absolute percentage error (*SMAPE*) [60], which overcomes this "bias" will not be considered. Lastly, the maximum absolute percentage error reached, *MAXAPE*, is treated as an additional metric that represents the worst prediction error the model reaches. Its equation is written as follows:

$$MAXAPE = \max\left(\left| \frac{\hat{y}_{te} - y_{te}}{y_{te}} \right| \times 100 \right) \quad \text{(A8)}$$

Note that evaluating the metrics on the dimensionless or dimensional film thicknesses should not result in significant discrepancies. For *MAPE* and *MAXAPE*, discrepancies should even be nil. In this work, the metrics were evaluated based on dimensional film thicknesses.

References

1. Holmberg, K.; Erdemir, A. Influence of Tribology on Global Energy Consumption, Costs and Emissions. *Friction* **2017**, *5*, 263–284. [CrossRef]
2. Venner, C.H.; Lubrecht, A.A. *Multi-Level Methods in Lubrication*; Elsevier: Amsterdam, The Netherlands, 2000; ISBN 0-08-053709-X.
3. Oh, K.P.; Rohde, S.M. Numerical Solution of the Point Contact Problem Using the Finite Element Method. *Int. J. Numer. Methods Eng.* **1977**, *11*, 1507–1518. [CrossRef]
4. Ahmed, S.; Goodyer, C.E.; Jimack, P.K. An Adaptive Finite Element Procedure for Fully-Coupled Point Contact Elastohydrodynamic Lubrication Problems. *Comput. Methods Appl. Mech. Eng.* **2014**, *282*, 1–21. [CrossRef]
5. Habchi, W. *Finite Element Modeling of Elastohydrodynamic Lubrication Problems*; John Wiley & Sons: Hoboken, NJ, USA, 2018; ISBN 1-119-22512-4.
6. Lohner, T.; Ziegltrum, A.; Stemplinger, J.-P.; Stahl, K. Engineering Software Solution for Thermal Elastohydrodynamic Lubrication Using Multiphysics Software. *Adv. Tribol.* **2016**, *2016*, e6507203. [CrossRef]
7. Tan, X.; Goodyer, C.E.; Jimack, P.K.; Taylor, R.I.; Walkley, M.A. Computational Approaches for Modelling Elastohydrodynamic Lubrication Using Multiphysics Software. *Proc. Inst. Mech. Eng. Part J J. Eng. Tribol.* **2012**, *226*, 463–480. [CrossRef]

8. Hartinger, M.; Dumont, M.-L.; Ioannides, S.; Gosman, D.; Spikes, H. CFD Modeling of a Thermal and Shear-Thinning Elastohydrodynamic Line Contact. *J. Tribol.* **2008**, *130*, 041503. [CrossRef]
9. Hajishafiee, A.; Kadiric, A.; Ioannides, S.; Dini, D. A Coupled Finite-Volume CFD Solver for Two-Dimensional Elasto-Hydrodynamic Lubrication Problems with Particular Application to Rolling Element Bearings. *Tribol. Int.* **2017**, *109*, 258–273. [CrossRef]
10. Havaej, P.; Degroote, J.; Fauconnier, D. A Quantitative Analysis of Double-Sided Surface Waviness on TEHL Line Contacts. *Tribol. Int.* **2023**, *183*, 108389. [CrossRef]
11. Hamrock, B.J.; Dowson, D. Isothermal Elastohydrodynamic Lubrication of Point Contacts: Part II—Ellipticity Parameter Results. *J. Lubr. Technol.* **1976**, *98*, 375–381. [CrossRef]
12. Moes, H. Optimum Similarity Analysis with Applications to Elastohydrodynamic Lubrication. *Wear* **1992**, *159*, 57–66. [CrossRef]
13. Habchi, W. Reduced Order Finite Element Model for Elastohydrodynamic Lubrication: Circular Contacts. *Tribol. Int.* **2014**, *71*, 98–108. [CrossRef]
14. Scurria, L.; Fauconnier, D.; Jiránek, P.; Tamarozzi, T. A Galerkin/Hyper-Reduction Technique to Reduce Steady-State Elastohydrodynamic Line Contact Problems. *Comput. Methods Appl. Mech. Eng.* **2021**, *386*, 114132. [CrossRef]
15. Habchi, W.; Issa, J.S. An Exact and General Model Order Reduction Technique for the Finite Element Solution of Elastohydrodynamic Lubrication Problems. *J. Tribol.-Trans. ASME* **2017**, *139*, 051501. [CrossRef]
16. Hamrock, B.J.; Dowson, D. Isothermal Elastohydrodynamic Lubrication of Point Contacts: Part III—Fully Flooded Results. *J. Lubr. Technol.* **1977**, *99*, 264–275. [CrossRef]
17. Nijenbanning, G.; Venner, C.H.; Moes, H. Film Thickness in Elastohydrodynamically Lubricated Elliptic Contacts. *Wear* **1994**, *176*, 217–229. [CrossRef]
18. Chittenden, R.J.; Dowson, D.; Dunn, J.F.; Taylor, C.M.; Johnson, K.L. A Theoretical Analysis of the Isothermal Elastohydrodynamic Lubrication of Concentrated Contacts. I. Direction of Lubricant Entrainment Coincident with the Major Axis of the Hertzian Contact Ellipse. *Proc. R. Soc. Lond. A Math. Phys. Sci.* **1997**, *397*, 245–269. [CrossRef]
19. Wheeler, J.-D.; Vergne, P.; Fillot, N.; Philippon, D. On the Relevance of Analytical Film Thickness EHD Equations for Isothermal Point Contacts: Qualitative or Quantitative Predictions? *Friction* **2016**, *4*, 369–379. [CrossRef]
20. Sose, A.T.; Joshi, S.Y.; Kunche, L.K.; Wang, F.; Deshmukh, S.A. A Review of Recent Advances and Applications of Machine Learning in Tribology. *Phys. Chem. Chem. Phys.* **2023**, *25*, 4408–4443. [CrossRef]
21. Kankar, P.K.; Sharma, S.C.; Harsha, S.P. Fault Diagnosis of Ball Bearings Using Machine Learning Methods. *Expert Syst. Appl.* **2011**, *38*, 1876–1886. [CrossRef]
22. Cristianini, N.; Shawe-Taylor, J. *An Introduction to Support Vector Machines and Other Kernel-Based Learning Methods*; Cambridge University Press: Cambridge, UK, 2000; ISBN 978-0-521-78019-3.
23. Krogh, A. What Are Artificial Neural Networks? *Nat. Biotechnol.* **2008**, *26*, 195–197. [CrossRef]
24. Shen, S.; Lu, H.; Sadoughi, M.; Hu, C.; Nemani, V.; Thelen, A.; Webster, K.; Darr, M.; Sidon, J.; Kenny, S. A Physics-Informed Deep Learning Approach for Bearing Fault Detection. *Eng. Appl. Artif. Intell.* **2021**, *103*, 104295. [CrossRef]
25. Bienefeld, C.; Kirchner, E.; Vogt, A.; Kacmar, M. On the Importance of Temporal Information for Remaining Useful Life Prediction of Rolling Bearings Using a Random Forest Regressor. *Lubricants* **2022**, *10*, 67. [CrossRef]
26. Han, T.; Pang, J.; Tan, A.C.C. Remaining Useful Life Prediction of Bearing Based on Stacked Autoencoder and Recurrent Neural Network. *J. Manuf. Syst.* **2021**, *61*, 576–591. [CrossRef]
27. Suh, S.; Lukowicz, P.; Lee, Y.O. Generalized Multiscale Feature Extraction for Remaining Useful Life Prediction of Bearings with Generative Adversarial Networks. *Knowl.-Based Syst.* **2022**, *237*, 107866. [CrossRef]
28. Zhao, Y.; Guo, L.; Wong, P.P.L. Application of Physics-Informed Neural Network in the Analysis of Hydrodynamic Lubrication. *Friction* **2023**, *11*, 1253–1264. [CrossRef]
29. Almqvist, A. Fundamentals of Physics-Informed Neural Networks Applied to Solve the Reynolds Boundary Value Problem. *Lubricants* **2021**, *9*, 82. [CrossRef]
30. Marian, M.; Mursak, J.; Bartz, M.; Profito, F.J.; Rosenkranz, A.; Wartzack, S. Predicting EHL Film Thickness Parameters by Machine Learning Approaches. *Friction* **2022**, *11*, 1–22. [CrossRef]
31. Smola, A.J.; Schölkopf, B. A Tutorial on Support Vector Regression. *Stat. Comput.* **2004**, *14*, 199–222. [CrossRef]
32. Schulz, E.; Speekenbrink, M.; Krause, A. A Tutorial on Gaussian Process Regression: Modelling, Exploring, and Exploiting Functions. *J. Math. Psychol.* **2018**, *85*, 1–16. [CrossRef]
33. Rasmussen, C.E.; Williams, C.K.I. *Gaussian Processes for Machine Learning*; Adaptive computation and machine learning; MIT Press: Cambridge, MA, USA, 2006; ISBN 978-0-262-18253-9.
34. Duvenaud, D. Automatic Model Construction with Gaussian Processes. Ph.D. Thesis, University of Cambridge, Cambridge, UK, June 2014.
35. Walker, J.; Questa, H.; Raman, A.; Ahmed, M.; Mohammadpour, M.; Bewsher, S.R.; Offner, G. Application of Tribological Artificial Neural Networks in Machine Elements. *Tribol. Lett.* **2023**, *71*, 3. [CrossRef]
36. Wheeler, J.-D.; Molimard, J.; Devaux, N.; Philippon, D.; Fillot, N.; Vergne, P.; Morales-Espejel, G.E. A Generalized Differential Colorimetric Interferometry Method: Extension to the Film Thickness Measurement of Any Point Contact Geometry. *Tribol. Trans.* **2018**, *61*, 648–660. [CrossRef]

37. Blok, H. Inverse Problems in Hydrodynamic Lubrication and Design Directives for Lubricated Flexible Surfaces. *Proc. Intl. Symp. Lob. Wear Houst.* **1963**, *7*.
38. Wu, S.R. A Penalty Formulation and Numerical Approximation of the Reynolds-Hertz Problem of Elastohydrodynamic Lubrication. *Int. J. Eng. Sci.* **1986**, *24*, 1001–1013. [CrossRef]
39. Deuflhard, P. *Newton Methods for Nonlinear Problems: Affine Invariance and Adaptive Algorithms*; Springer Science & Business Media: Berlin/Heidelberg, Germany, 2005; ISBN 978-3-540-21099-3.
40. Viana, F.A.C. A Tutorial on Latin Hypercube Design of Experiments. *Qual. Reliab. Eng. Int.* **2016**, *32*, 1975–1985. [CrossRef]
41. The MathWorks, Inc. *MATLAB Version: 9.9.0 (R2020b)*; The MathWorks, Inc.: Natick, MA, USA; p. 2020.
42. Jolliffe, I.T.; Cadima, J. Principal Component Analysis: A Review and Recent Developments. *Philos. Trans. R. Soc. A Math. Phys. Eng. Sci.* **2016**, *374*, 20150202. [CrossRef] [PubMed]
43. Evans, H.P.; Snidle, R.W. The Isothermal Elastohydrodynamic Lubrication of Spheres. *J. Lubr. Technol.* **1981**, *103*, 547–557. [CrossRef]
44. Hopgood, A.A. *Intelligent Systems for Engineers and Scientists: A Practical Guide to Artificial Intelligence*; CRC Press: Boca Raton, FL, USA, 2021; ISBN 978-1-00-048410-6.
45. Morales, J.L.; Nocedal, J. Remark on "Algorithm 778: L-BFGS-B: Fortran Subroutines for Large-Scale Bound Constrained Optimization". *ACM Trans. Math. Softw.* **2011**, *38*, 1–4. [CrossRef]
46. Habchi, W.; Bair, S. Quantifying the Inlet Pressure and Shear Stress of Elastohydrodynamic Lubrication. *Tribol. Int.* **2023**, *182*, 108351. [CrossRef]
47. Bair, S. The Unresolved Definition of the Pressure-Viscosity Coefficient. *Sci. Rep.* **2022**, *12*, 3422. [CrossRef]
48. Habchi, W.; Sperka, P.; Bair, S. Is Elastohydrodynamic Minimum Film Thickness Truly Governed by Inlet Rheology? *Tribol. Lett.* **2023**, *71*, 96. [CrossRef]
49. Habchi, W.; Vergne, P. A Quantitative Determination of Minimum Film Thickness in Elastohydrodynamic Circular Contacts. *Tribol. Lett.* **2021**, *69*, 142. [CrossRef]
50. Lubrecht, A.A.; Venner, C.H.; Colin, F. Film Thickness Calculation in Elasto-Hydrodynamic Lubricated Line and Elliptical Contacts: The Dowson, Higginson, Hamrock Contribution. *Proc. Inst. Mech. Eng. Part J J. Eng. Tribol.* **2009**, *223*, 511–515. [CrossRef]
51. Venner, C.H.; Bos, J. Effects of Lubricant Compressibility on the Film Thickness in EHL Line and Circular Contacts. *Wear* **1994**, *173*, 151–165. [CrossRef]
52. Habchi, W.; Bair, S. Quantitative Compressibility Effects in Thermal Elastohydrodynamic Circular Contacts. *J. Tribol.* **2012**, *135*, 011502. [CrossRef]
53. Neal, R.M. *Bayesian Learning for Neural Networks*; Springer Science & Business Media: Berlin/Heidelberg, Germany, 1996; ISBN 978-1-4612-0745-0.
54. Abramowitz, M.; Stegun, I.A. *Handbook of Mathematical Functions with Formulas, Graphs, and Mathematical Tables*; U.S. Government Printing Office: Washington, DC, USA, 1948.
55. Hackeling, G. *Mastering Machine Learning with Scikit-Learn*; Packt Publishing Ltd.: Birmingham, UK, 2017; ISBN 978-1-78829-849-0.
56. Neal, R.M. Monte Carlo Implementation of Gaussian Process Models for Bayesian Regression and Classification. *arXiv* **1997**, arXiv:physics/9701026.
57. Wright, S. Correlation and Causation. *J. Agric. Res.* **1921**, *20*, 557–585.
58. de Myttenaere, A.; Golden, B.; Le Grand, B.; Rossi, F. Mean Absolute Percentage Error for Regression Models. *Neurocomputing* **2016**, *192*, 38–48. [CrossRef]
59. Armstrong, J.S.; Collopy, F. Error Measures for Generalizing about Forecasting Methods: Empirical Comparisons. *Int. J. Forecast.* **1992**, *8*, 69–80. [CrossRef]
60. Armstrong, J.S. *Long-Range Forecasting*, 2nd ed.; Wiley-Interscience: Rochester, NY, USA, 1985.

lubricants

Article

An AI-Extended Prediction of Erosion-Corrosion Degradation of API 5L X65 Steel

Ariel Espinoza-Jara [1,2], Igor Wilk [3], Javiera Aguirre [4,5] and Magdalena Walczak [1,*]

1 Department of Mechanical and Metallurgical Engineering, Escuela de Ingeniería, Pontificia Universidad Católica de Chile, Vicuña Mackenna 4860, Macul, Santiago 7820436, Chile; aoespinoza1@uc.cl
2 Department of Civil and Environmental Engineering, Politecnico di Milano, 20133 Milano, Italy
3 Department of Technical Physics, Computer Science and Applied Mathematics, Lodz University of Technology, 90-005 Lodz, Poland
4 Corrosion and Wear of Materials Unit, DICTUC, Vicuña Mackenna 4860, Santiago 7820436, Chile
5 Escuela de Construcción Civil, Facultad de Ingeniería, Pontificia Universidad Católica de Chile, Av. Vicuña Mackenna 4860, Macul, Santiago 7820436, Chile
* Correspondence: mwalczak@uc.cl; Tel.: +56-95504-4627

Abstract: The application of Artificial Neuronal Networks (ANN) offers better statistical accuracy in erosion-corrosion (E-C) predictions compared to the conventional linear regression based on Multifactorial Analysis (MFA). However, the limitations of ANN to require large training datasets and a high number of inputs pose a practical challenge in the field of E-C due to the scarcity of data. To address this challenge, a novel ANN method is proposed, structured to a small training dataset and trained with the aid of synthetic data to produce an E-C neural network (E-C NN), applied for the first time in the study of E-C wear synergy. In the process, transfer learning is applied by pre-training and fine-tuning the model. The initial dataset is created from experimental data produced in a slurry pot setup, exposing API 5L X65 steel to a turbulent copper tailing slurry. To the previously known E-C scenario for selected values of flow velocity, particle concentration, temperature, pH, and the content of the dissolved Cu^{2+}, new experimental data of stand-alone erosion and stand-alone corrosion is added. The prediction of wear loss by E-C NN considers individual parameters and their interactions. The main result is that E-C ANN provides better prediction than MFA as evaluated by a mean squared error (MSE) values of 2.5 and 3.7, respectively. The results are discussed in the context of the cross-effect between the proposed prediction model and the resulting estimation of relative contribution to E-C synergy, which is better predicted by the E-C NN. The E-C NN model is concluded to be a viable alternative to MFA, delivering similar prediction with better sensitivity to E-C synergy at shorter computation times when using the same experimental dataset.

Keywords: erosion-corrosion wear; ANN; multifactorial analysis; synthetic data; erosion-corrosion data; erosion data; corrosion data

Citation: Espinoza-Jara, A.; Wilk, I.; Aguirre, J.; Walczak, M. An AI-Extended Prediction of Erosion-Corrosion Degradation of API 5L X65 Steel. *Lubricants* **2023**, *11*, 431. https://doi.org/10.3390/lubricants11100431

Received: 28 July 2023
Revised: 13 September 2023
Accepted: 14 September 2023
Published: 5 October 2023

1. Introduction

In the mining industry, slurry pipeline systems are the most cost-effective solution for long-distance transport of large quantities of particulate solids. However, due to the nature of the carrier fluid, one of the main threats to a pipeline system, consisting of pipes, pumps, valves, etc., is the degradation by erosion and corrosion processes acting simultaneously in a synergistic phenomenon referred to as erosion-corrosion (E-C). In this context, the challenge is to keep the pipeline integrity to prevent failures that can result in leakage accidents [1] affecting nearby communities and the environment.

In metallic materials, the wear rate of E-C is defined by the weight loss of the material due to the physical damage induced by solid particles impacting over the surface and by the corrosion mechanisms that involve ionic exchange between the surface and the carrier fluid (electrolyte). In this process, the erosion mechanism is enhanced by corrosion and vice versa through a complex synergy resulting in E-C weight loss higher than the sum of losses by erosion and corrosion separately [2]. The latter is often described in terms of weight loss by Equations (1) and (2):

$$T_{wl} = E + C + S, \quad (1)$$

$$S = E_C + C_E, \quad (2)$$

where T_{wl}, E, C, and S describe the total weight loss by E-C, weight loss by sheer mechanical erosion (stand-alone erosion), weight loss by sheer electrochemical corrosion (stand-alone corrosion), and weight loss attributed to the synergy effects, respectively. The synergy S can be further decomposed into erosion-enhanced corrosion (C_E) and corrosion-enhanced erosion (E_C).

A full description of the E-C phenomenon involves a comprehensive analysis of individual mechanisms as well as their synergistic effects, which can be done in terms of key variables. These variables can be categorized as those proper of the target material and those proper of the slurry, comprising the carrier fluid and the suspended particles. Because the carrier fluid is in motion, its characteristics both as a mechanical medium and as an electrolyte must be distinguished. A summary of all the variables relevant for the modeling of erosive wear was presented by Javaheri et al. [3]. In this work, we extend the scope of relevant variables to explicitly include the synergy with corrosion by compiling E-C data from the literature [4–20], resulting in the fishbone diagram shown in Figure 1.

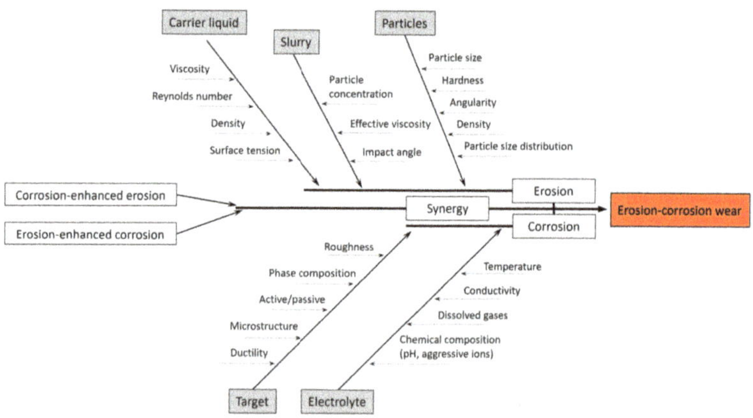

Figure 1. Summary of variables relevant to Erosion-Corrosion (E-C) wear.

Because of the apparently unpredictable nature of the wear caused by E-C, it is considered mandatory to develop a prediction tool capable of estimating the wear rate of materials exposed to slurry flow. The currently used models have not proven sufficiently accurate in estimating the service life of slurry transport systems and even revision of design standards has been advised [21]. However, description of physical problems, such as wear, commonly involves a deterministic approach, relying on the physical laws and structure–property relations. Due to the complex nature of the E-C mechanism, a fully deterministic model for accurate prediction of wear rate remains elusive. Alternative approaches include stochastic, e.g., [22–25], or statistical modeling. The latter relies on establishing correlations between key variables to create a multi-factorial predictive framework. These empirical models, although not including the physics of the phenomena, allow for gaining insight into the interdependence of factors contributing to the E-C phenomena. By analyzing the statistical relationships between principal variables, predictions and effective mitigation strategies can be developed.

The conventional linear regression for multi-factorial analysis (MFA) and response surface method (RSM) has been primarily applied to the study of E-C phenomena, but recent studies have demonstrated the potential of Artificial Intelligence (AI)-based prediction methods, particularly Artificial Neural Networks (ANN), in producing better predictions in terms of statistical accuracy [26–30]. This method has gained popularity in various research areas with applications such as financial analysis, logistics management, weather predictions, etc., showcasing successful outcomes [31] with a substantial number of variables. However, when employing these new AI-based techniques to E-C wear, two main challenges arise: the requirement for large training datasets and a high number of inputs. These restrictions present a significant challenge because scarcity of available data is owed to the substantial time and resource costs associated with its production [32,33]. This is an inherent restriction, the impact of which should be evaluated for each particular application. Whereas AI-enhanced models have been shown to be applicable in the study of friction, wear, and roughness evolution [34], which involve physical mechanisms relevant to E-C, it is estimated as worthwhile to explore and validate its effectiveness specifically for E-C phenomena.

Whereas machine learning (ML) models have mostly been employed in laboratory settings to predict abrasion wear measured by pin-on-disc set-up [32], a significant knowledge gap emerges when it comes to their direct application in comprehending E-C wear mechanisms, particularly in the complex context of turbulent environments. Furthermore, there remains an unaddressed challenge concerning the scarcity of data for neural network applications in wear problems. Additionally, existing E-C wear studies are often reliant on MFA, a tool valued for its capacity to elucidate variable relationships but found lacking in robustness in predictive interpolation. Both these gaps underscore the need for novel approaches and models within the field. In this context, the present work introduces a novel approach to predicting E-C wear by comparing the effectiveness of ANN and MFA based on linear regressions. The method involves developing an ANN-based predictive model tailored for small training datasets and exploring its applicability for extreme cases of factor combinations. In particular, it is hypothesised that ANN can predict the E-C rate with higher accuracy than the conventional MFA approach on the same E-C dataset. The previously published experimental E-C dataset [35] is now expanded by incorporating data of stand-alone erosion and stand-alone corrosion measured in a slurry pot configuration for six parameters: flow velocity, particle concentration, temperature, pH, oxygen content, and copper ion content. Both the MFA and ANN models are utilized to predict wear loss, considering the individual contribution and interaction of the parameters. To overcome the limited experimental data, a synthetic dataset is generated by interpolating factors and incorporating predictions from the MFA model. This approach is equivalent to the data augmentation strategy in ML applications used when there are missing data, unbalanced data, under-sampling, or small dataset problems [33]. This synthetic dataset is then used to pre-train the ANN, followed by fine-tuning using the experimental data. The performance of the MFA and ANN models is compared for stand-alone erosion, corrosion, and combined E-C scenarios. Finally, the relevance of the results is discussed regarding the relative contribution of synergistic factors.

2. Experiment and Methods

The present study builds on the experimental data of E-C weight loss reported by Aguirre et al. [35] along with their MFA. The experimental extension consists of including new data for the stand-alone erosion and stand-alone corrosion mechanisms for E-C wear weight loss, while maintaining the same levels of the original studies' variables which are presented in Appendix A. This extension completes an experimental dataset that establishes the relationship between the controlled variables of flow velocity (V), particle content (P), temperature (T), pH, content of dissolved oxygen (DO), and content of copper ions (Cu^{2+}), and their influence on the effective wear rate, encompassing E-C and the stand-alone corrosion and erosion. Then, through regression analysis, empirical models are derived from the dataset to interpolate the variables and generate a larger dataset. This expanded dataset is used to create and pre-train a Deep Learning Neural Network (DLNN).

Subsequently, transfer learning is applied to fine-tune the DLNN using the experimental data. The entire workflow is summarized in Figure 2.

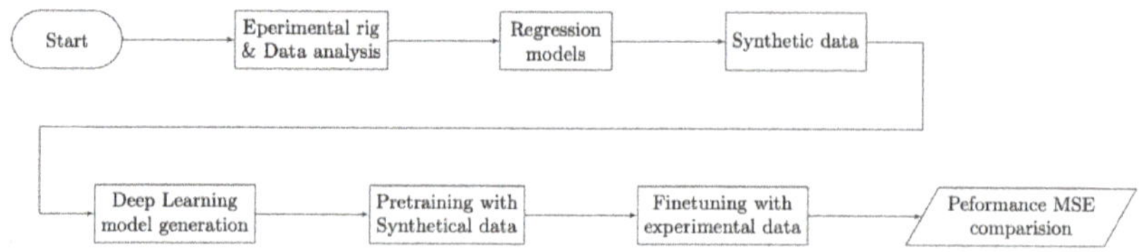

Figure 2. Summary of the work flow, from collection of experimental E-C data to performance evaluation through MSE.

2.1. Experimental Determination of E-C Wear

The experiment for measuring the weight loss of stand-alone corrosion and stand-alone erosion was carried out using the same experimental setup and materials as those reported by Aguirre et al. [35]. The target material were sample cylinders of API 5L X65 steel of 15 mm in diameter and 10 mm in height. Before conducting the experiments, the API 5L X65 cylinder samples were prepared, polishing their surface with SiC paper, starting from 320 to 1200 grit, and then thoroughly cleaning them with ethanol in an ultrasonic bath to remove any contaminants. The cleaned samples were then weighed using an analytical balance with a precision of 0.1 mg to obtain their initial weight, ensuring accurate measurement of weight loss during the test. The cylinders were mounted at the rotation axis of a rotating cylinder electrode (RCE) setup (Figure 3A), allowing for separation of erosion from corrosion via electrochemical polarization. The stand-alone erosion measurements were carried out with the corrosion suppressed by applying −1.2 V vs. Ag/AgCl (cathodic protection) to the target material. The stand-alone corrosion scenarios were implemented by not adding the solid particles.

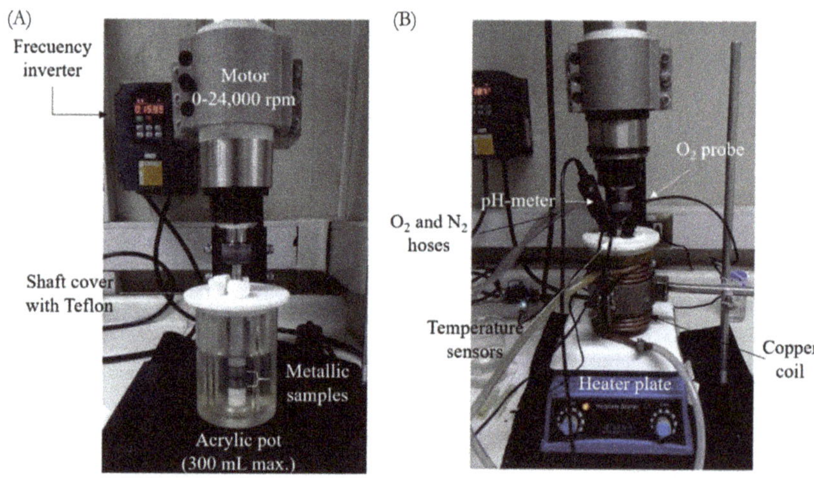

Figure 3. Experimental slurry pot: (**A**) overview of the RCE set-up with location of the test sample in a pot filled with water for visibility and (**B**) complete configuration for controlling all the studied factors. Adapted from [35].

The electrolyte used for preparing the slurry was prepared using distilled water, sulfuric acid, and sodium hydroxide in concentrations necessary to produce and maintain the desired pH during the exposure. The values of the pH, temperature, and content of Cu^{2+} ions were selected to be relevant for realistic operating environments. The duration of exposure to erosion or corrosion was 75 min. After exposure, the samples were removed and immediately cleaned to remove any residual deposits or contaminants that may have adhered during exposure. The cleaning procedure consisted of an ultrasonic bath of the samples for 180 s, first in acetone and then in an inhibiting acid solution (Clark solution consisting of 1000 mL of hydrochloric acid, 20 g of antimony trioxide (Sb_2O_3), and 50 g of stannous chloride ($SnCl_2$)) according to the ASTM standard G1-03 [36]. The cleaned samples were then carefully weighed again to determine the weight loss. The variability of the difference in weight before and after exposure was determined to approximate the statistical experimental error of E-C, stand-alone erosion, and stand-alone corrosion, which were found to be 0.078, 0.173, and 1.292 $mg{\cdot}cm^{-2}h^{-1}$, respectively.

2.1.1. Experiment Design

The test parameters used in the fractional factorial design of the experiments are summarized in Table 1, indicating the lowest, central, and highest values assigned to each parameter. The specific values were chosen to cover the range of conditions met during the practice of handling the slurry from the copper tailing. In particular, the high levels of *P* and Cu^{2+} ions correspond to the worst-case scenario of the residual copper in the tailing. The high and low levels of dissolved oxygen correspond to fully oxygenated and oxygen-free electrolytes, respectively, mimicking the oxygen consumption in a closed system handling slurry.

The design included stand-alone corrosion and erosion experiments. A total of 105 observations are resultant, with 35 observations for each experiment, including the earlier E-C experiment conducted by Aguirre et al. [35] and included in Appendix A Table A3. Each set of the 35 runs included 32 distinct factor combinations, as well as one combination representing the factors' central values. In addition, two replications of the central combination were added to assess the variability of the results and explain the intrinsic experimental error. For instance, in stand-alone corrosion, the particle concentration factor (*P*) is irrelevant and has been set to 0 since there are no particles involved. Similarly, in the case of stand-alone erosion, factors such as pH, ($P \times Cu^{2+}$) concentration, and dissolved oxygen (*DO*) are not relevant. The experimental results for E and C results are presented in Appendix A Tables A1 and A2 respectively.

Table 1. Summary of test parameters used for the fractional factorial design of experiment, indicating the lowest, central, and highest values.

Factor	Unit	Symbol	Low Level	Central Level	High Level
Velocity	m/s	*V*	3	5	7
Particles concentration	wt %	*P*	45	55	65
Temperature	°C	*T*	25	35	45
pH	pH	*pH*	5	8	11
Dissolved oxygen	ppm	*DO*	0	5	10
Copper ion concentration	ppm	Cu^{2+}	0	250	500

2.1.2. Multifactorial Analysis of Experimental Data

The experimental results from Table 1 were used to construct a response surface and develop a polynomial model for predicting wright loss in relation to stand-alone erosion and stand-alone corrosion as schematized in Figure 4 . Estimation of weight loss within the range of the input factors was obtained by fitting the experimental data to a polynomial equation as an output of the multifactorial analysis. The relationship between the input factors and the resulting erosion and corrosion effects is then quantified by the coefficients of the polynomial equation, expressing the relative importance of each factor.

The analysis was conducted using the R programming language [37] with specific libraries to handle the data and calculate the main factor contributions. The "Dplyr" and "FrF2" [38] libraries were used to process the experimental data and assess the key factors' impacts. Additionally, the " Leaps" [39] library, along with the "regsubsets" function, was used to determine the best linear regression fitting. This process involved optimizing the output polynomial by considering the most significant factors based on Mallows' Cp [40] and Schwartz's information criterion (bic) [41]. These criteria were essential for selecting the most relevant factors and obtaining an accurate regression model for the data. Finally, the fitting metrics, the multiple R-squared value, and the adjusted R-squared value have been obtained using the "lm" function contained in the R language default libraries.

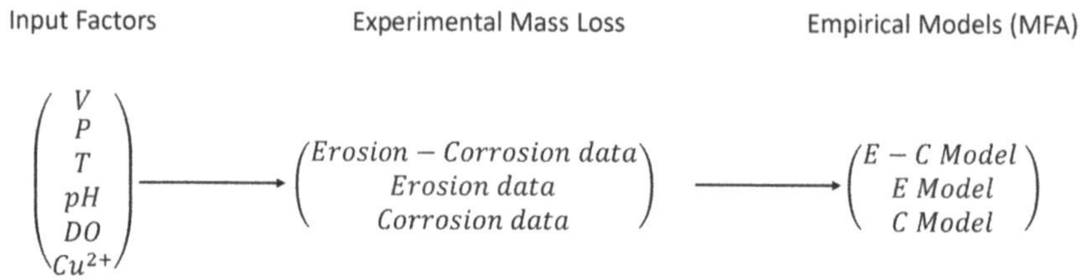

Figure 4. Schematic representation of generating the empirical model by MFA using experimental data.

2.2. E-C Neural Network

The model of E-C wear based on the neural network was obtained by the development of model architecture, followed by transfer learning consisting of stages of pre-training and fine-tuning. Since the amount of experimental data was insufficient for developing an NN model, synthetic data were generated in addition.

2.2.1. Model Architecture

In this study, the GridSearchCV tool from the Scikit Learn library [42] was used to generate and assess various combinations of hyperparameters, i.e., the main parameters used for setting up the NN. The hyperparameter grid encompassed options for the number of layers (ranging from 2 to 6), number of nodes in the first layer following the input layer (128, 64, 32, and 16), a single node in the last layer to select particular mode (erosion–corrosion, erosion, corrosion), activation functions (sigmoid, relu, tanh, and softmax), loss functions (Poisson, Hinge), and batch size (3 and 10). The optimal architecture was determined using experimental data, as schematized in Figure 5. The selection of this architecture was guided by its ability to accurately model and predict E-C wear within the particular context of this study. It represents the combination of hyperparameters that demonstrated the highest performance and predictive accuracy for the weight-loss data.

The input layer and output layers were defined by the experimental problem, with six input parameters available in this case to derive one output parameter (wear rate). In order to facilitate a more efficient convergence and improve the accuracy of the NN model, two additional input parameters were incorporated: corrosion activation and erosion activation (attaining the value of 0 or 1 for deactivate or activate, respectively), obtaining eight input parameters in total. Although these parameters do not have the physical meaning of an experimental factor, they are crucial in enhancing the model's performance by adding information on whether the combination of input factors corresponds to the experimental run of stand-alone erosion, stand-alone corrosion, or the full E-C scenario. The inclusion of corrosion and erosion activation parameters aids the learning process of the NN because these parameters act as conduits, facilitating exploration of the input space and accelerating the convergence of the model's results. Their presence helps the NN to better model complex interactions between the different wear mechanisms in the sense that faster convergence and enhanced overall predictive capability of the network is observed.

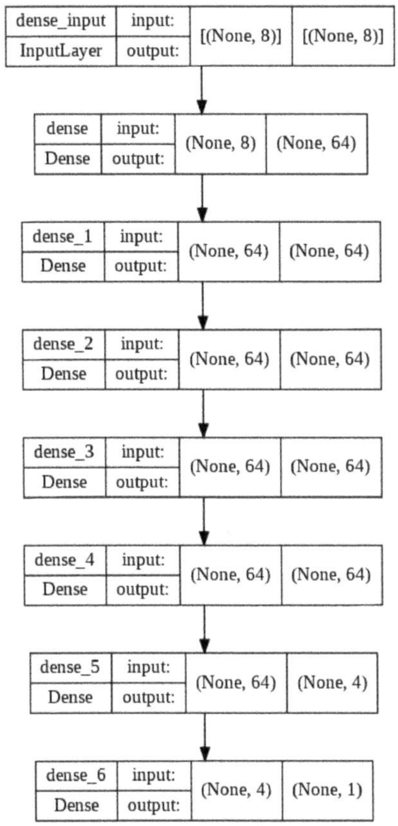

Figure 5. Schematic representation of the NN architecture after GridSearchCV optimization process.

2.2.2. Pre-Training Using Synthetic Data

The training process employed a two-step approach to optimize the NN model's performance. In the initial step, the model was pre-trained using an extended dataset that combined the original experimental data with synthetically generated ones. The pre-training phase provided the model with a broader scope of E-C patterns by exposing it to the most diverse range of scenarios represented in the synthetic data.

The original dataset consisted of 105 observations, with 35 observations per E-C, stand-alone erosion, and stand-alone corrosion, respectively. In order to enhance the dataset and capture a broader range of variations, 1395 synthetic data points were generated, resulting in a consolidated dataset of 1500 observations. The synthetic data were included to ensure a comprehensive dataset for training the ANN model. The synthetic data were generated by fitting a robust polynomial model derived from the results of stand-alone erosion and stand-alone corrosion experiments, as well as the erosion–corrosion findings. This polynomial model provided an effective means of interpolating the experimental data and extrapolating them to unexplored regions of the input space, as shown schematically in Figure 6.

The synthetic data set were used for the pre-training stage, and in this case only, was divided into subsets of Training (75%) and Validation (25%) in order to find the best epoch number to avoid an overfitted model. For the resulting architecture, the best relationship on MSE between Training and Validation was in the 135 epoch, obtaining a validation MSE of 2.5 at epoch 135 and MSE of 17.9 at epoch 136, which is indicative of a starting point of overfitting for this synthetic dataset.

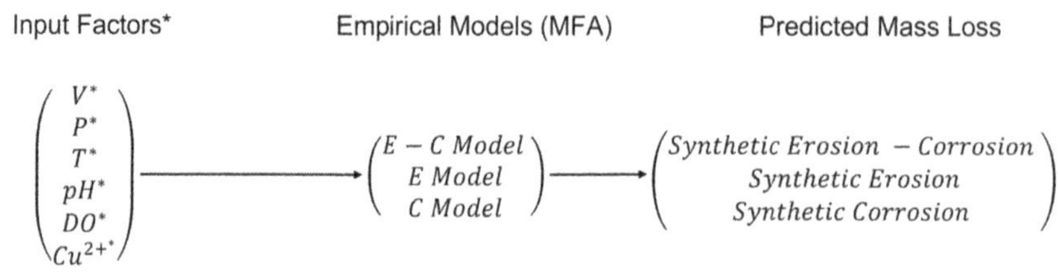

Figure 6. Schematic representation of generating synthetic data using not measured values of factors (*) in the range of experimental factors except for the very experimental values.

2.2.3. Fine-Tuning, Validation and Evaluation

After pre-training, the model underwent fine-tuning using a curated dataset derived solely from the original experimental data, without any synthetic data. The aim of fine-tuning was to improve the model's performance and enable it to capture the specific nuances and characteristics present in the experimental conditions. Progress during the fine-tuning process was monitored through a validation subset to minimize prediction errors. The E−C NN model resulting from this process was then evaluated using a separate test subset.

The experimental dataset, consisting of 105 observations, was first shuffled to ensure randomness and eliminate potential bias. The shuffled dataset was then divided into three subsets: 50% for fine-tuning training (53 observations), 25% for validation (26 observations), and 25% for testing (26 observations).

During the fine-tuning process, the model learned from the input–output patterns in the data and adjusted its internal parameters to minimize prediction errors until reaching a satisfactory level of convergence.

The validation subset served as an independent dataset to assess the model's performance during training. However, as it is incorporated into the training process, its evaluation can become biased. To provide an unbiased evaluation of the model's performance, the testing subset was used as a benchmark. This subset allowed assessment of the model's ability to accurately predict E-C weight loss using unseen data, providing insights into its reliability and robustness.

For the fine-tuning stage, the model was trained during 50 epochs, which was the optimum for this hyperparameter delivered by the GridSearchCV algorithm [42] without producing overfitting. The MSE values obtained during the validation stage and for the test sub-dataset were 3.6 and 6.9.

2.2.4. Sensibility Analysis of E-C NN

To analyse the sensitivity of the E-C NN model to each parameter, a systematic process was followed. The training and fine-tuning process for the E-C NN model was repeated 34 times, with each iteration predicting the wear rate for all experimental combinations. After obtaining the 34 predictions for the experimental wear rates, the results were averaged to obtain a single prediction for each wear rate.

Next, the average prediction was evaluated using the fractional factorial design algorithm to assess the independent effects of each factor on the E-C, erosion, and corrosion as stand-alone wear rate and to be comparable with the main effects of experimental factors. This analysis allowed us to understand the individual contribution of each factor and their significance in influencing the wear rate. It is important to remark that the most relevant factors to E-C were identified as the parameters that exhibited the most significant changes in the predicted weight loss.

3. Results

3.1. MFA Model of E-C Wear

Figure 7 summarizes the main effects of E-C obtained experimentally by Aguirre et al. [35], demonstrating that the velocity (V), temperature (T), and content of dissolved oxygen (DO) are the significant factors as determined by MFA (Equation (3)):

$$(E-C)_{rate} = -7.880 + 0.811V + 0.0726P + 0.054T + 0.032DO + 0.013Cu^{2+} + 0.037V \times DO - 0.0002P \times Cu^{2+} \quad (3)$$

The derived equation represents a parametric relationship between the E-C rate and the importance of the influencing factors. The units of the variables are summarized in Table 1. The units of the coefficients are chosen to rescale the respective variable to the units of wear rate ((mg·cm^{-2}h^{-1})). Positive coefficients indicate a positive correlation, suggesting that an increase in these factors leads to an increase in the E-C wear rate. Conversely, negative coefficients indicate a negative correlation, indicating that an increase in those factors results in a decrease in the E-C rate. The interaction terms, such as ($V \times DO$) and ($P \times Cu^{2+}$), take into account the combined effects of multiple factors on the E-C rate, which can be either synergistic or antagonistic when positive or negative, respectively. The variability of the E-C rate determined for the standard deviation of the central levels is approximately 0.078 (mg·cm^{-2}h^{-1}), which is comparable to the statistical error of the experiment, visualised in Figure 7 as a gray band. This indicates that the obtained results are reliable and within an acceptable range of variability. An extended description of the experimental procedure, data collection, data analysis, and the derivation of the MFA model is provided in the work of Aguirre et al. [35].

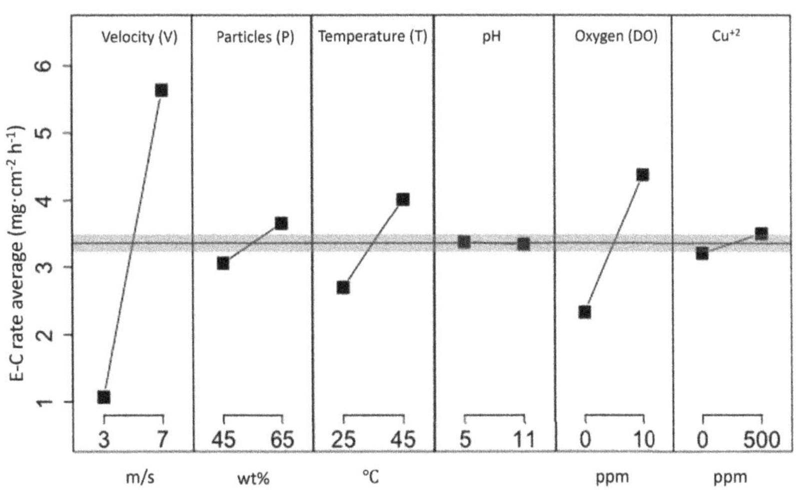

Figure 7. Main effects of experimental factors on the E-C rate. The range of experimental standard deviation is marked in gray. Adapted from [35].

3.2. MFA Model of Stand-Alone Erosion Wear

In an analogy to the E-C main effects shown in Figure 7, the results specific to stand-alone erosion are presented in Figure 8. The experimental standard deviation of the erosion data is approximately 0.173 (mg·cm^{-2}h^{-1}), which is higher than that of the E-C data. The significant main effects are those of velocity, particle content (P), and oxygen content (DO).

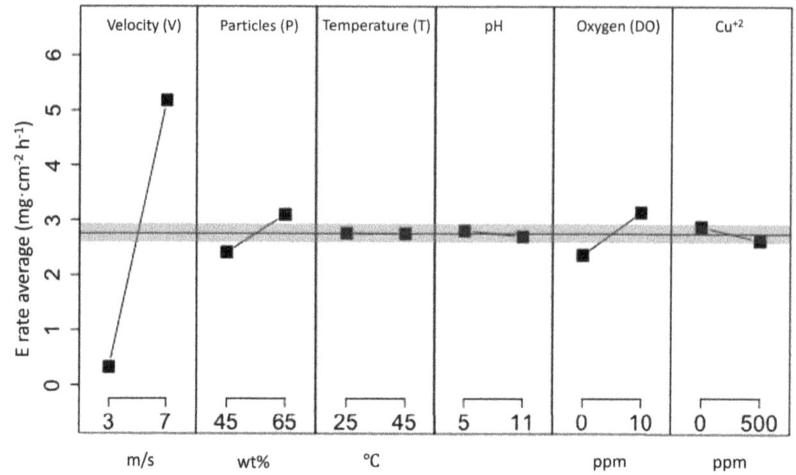

Figure 8. Main effects of experimental factors on the stand-alone erosion rate. The range of experimental standard deviation is marked as a gray stripe.

The linear model of stand-alone erosion with second-order terms obtained in an analogy to Equation (3) by MFA is given by Equation (4):

$$E_{rate} = -0.0128P - 0.0117V^2 + 0.002P \times V^2. \tag{4}$$

The residual standard error of this fitted erosion MFA model is 1.328 (mg·cm^{-2}h^{-1}), the multiple R-squared value is 0.8438, and the adjusted R-squared value, accounting for the number of predictors and degrees of freedom, was calculated to be 0.8276. The average deviation of the observed erosion rate values from the predicted values indicates that approximately 84.38% of the variability in the erosion rate can be explained by the model. The metrics suggest that the linear model with second-order terms provides a reasonable fit to the data, as evidenced by the relatively low residual standard error and high R-squared values. The experimental error visualised in Figure 8 as a gray band indicated that the effect of the temperature pH and content of Cu^{2+} is not relevant.

3.3. MFA Model of Stand-Alone Corrosion Wear

The main effects of factors in the corrosion experiment were analysed analogously to those of the E-C and stand-alone erosion scenarios. The results are summarized visually in Figure 9 and the linear model with second-order terms for the stand-alone corrosion is given by Equation (5):

$$C_{rate} = 0.0282Cu + 0.0273T - 0.0448pH + 0.02113DO + 0.0002Cu \times T - 0.0032Cu \times pH + 0.0004Cu \times DO \tag{5}$$

The experimental variability of the corrosion rate at the central levels was assessed by the standard deviation to be approximately 1.292 (mg·cm^{-2}h^{-1}), which is notably higher than that of E-C and stand−alone erosion. Considering the experimental error visualised in Figure 9 as a gray band, the significant main effects are only those of pH and the content of the dissolved (Cu^{2+}). The metrics of the fitted corrosion MFA model are a residual standard error of 2.324 (mg·cm^{-2}h^{-1}), multiple R-squared value of 0.8821, and adjusted R-squared value, accounting for the number of predictors and 28 degrees of freedom amounting to 0.8526.

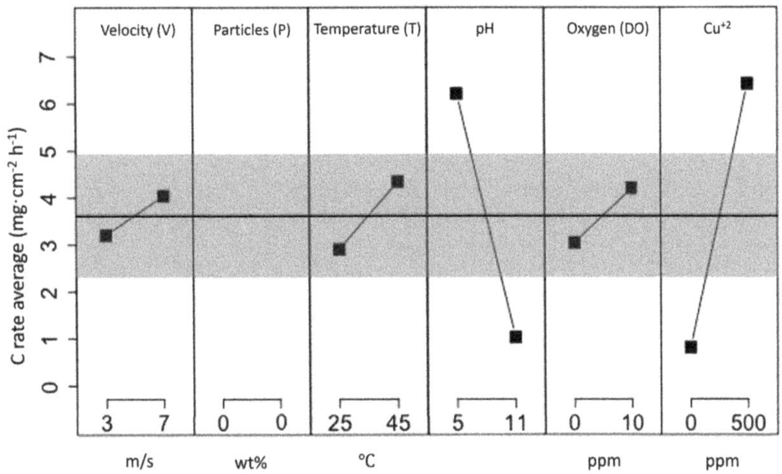

Figure 9. Main effects of experimental factors on the stand-alone corrosion rate. The range of experimental standard deviation is marked in gray.

3.4. E-C NN Model Predictions

Figure 10 summarises the results of the E-C NN model predictions in comparison with the experimental data of the E-C rate and the stand-alone erosion and corrosion rates. The effect of transfer learning is visualised by considering the direct E-C NN and the fine-tuned E-C NN. In general, the tendencies of all the main effects are correctly predicted by the E-C NN model; however, the slopes differ slightly. The slopes produced by the fine-tuned model are closer to the experimental data, which is particularly evident in the case of the E-C predictions. The effect is less pronounced in the case of stand-alone erosion, whereas for the stand-alone corrosion there is almost no effect of transfer learning except for the pH factor, in which fine-tuning provides notably better prediction.

The effect of the particles on stand-alone corrosion is included to represent the statistical error, as this is the physical meaning of corrosion weight loss in the absence of abrasive particles. In the case of the experimental data, the statistical error represents the experimental accuracy.

The predictions of the fine-tuned E-C NN model are compared with the MFA predictions and experimental data in Figure 11. In general, both models produce the same tendencies for all the factors, but the values predicted by the E-C NN are closer to the experimental ones as compared to the MFA model. This effect is most notable in the case of stand-alone erosion, in which the weight loss is systematically underestimated for all the factors.

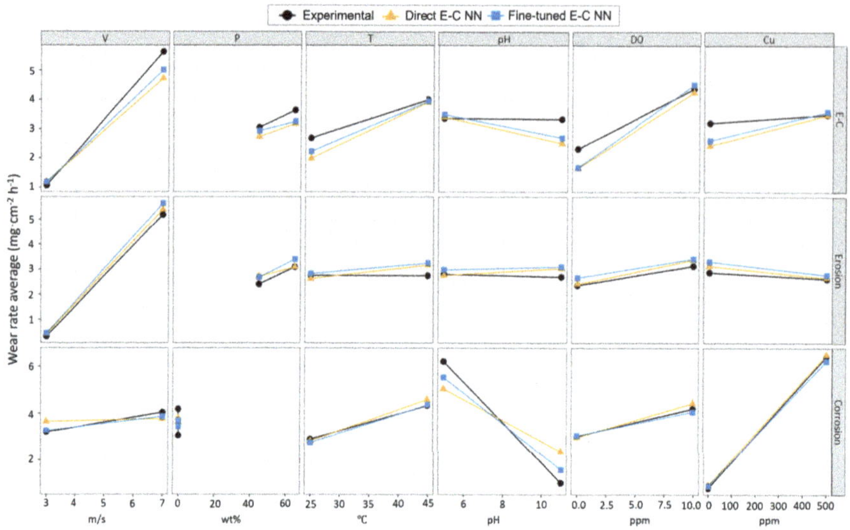

Figure 10. Effect of pre-training on the prediction of the rates of E-C (**upper row**), stand-alone erosion (**middle row**), and stand-alone corrosion (**bottom row**) for the main factors as compared to the experimental values.

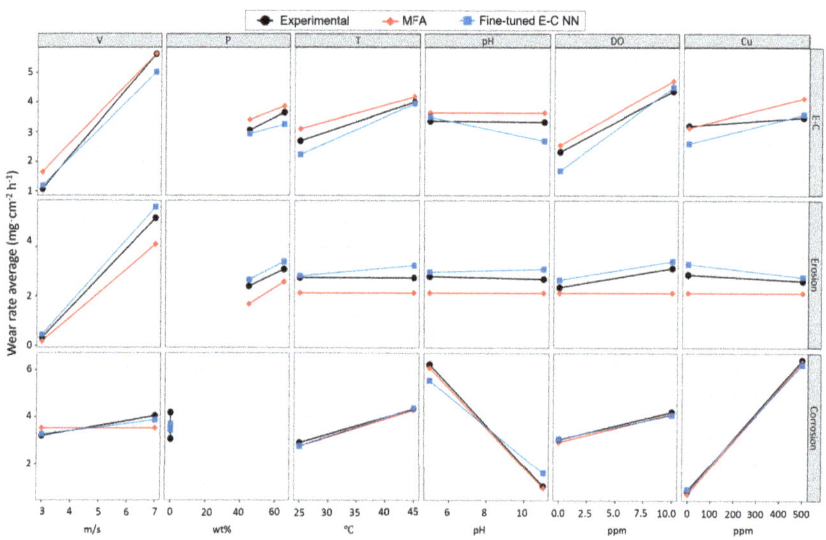

Figure 11. Comparison of main effects of experimental factors on the wear rate of E-C (**upper row**), stand-alone erosion (**middle row**), and stand-alone corrosion (**bottom row**) as determined by the experiment, MFA, and pre-trained E-C NN.

4. Discussion

4.1. Validity of Synthetic Data

Due to the multi-factor nature of the E-C wear (Figure 1) MFA models have been the method of choice to describe the E-C process. However, availability of experimental data is restricted because of the involved effort, considering that the results are specific to particular experimental set-ups [3]. In addition, to assure applicability of results to actual slurry handling systems, the design of experiments is restricted to the range of parameters of practical interest rather than relevance for the relative importance of the mechanisms of wear loss. In this context, generation of synthetic data from MFA model corresponds to the current practice in predicting wear rate. The synthetic data are then the best interpolation between the measured values.

In this work, a comprehensive MFA model of E-C wear is presented, i.e., derived from the data considering both E-C (Equation (3)) and the stand-alone erosion and corrosion scenarios (Equations (4) and (5), respectively). The inclusion of synergy-free scenarios of erosion and corrosion resulted in models not capable of predicting the very values measured experimentally (see Figure 11). The deviation is particularly notable in the case of stand-alone corrosion, where even the sign of the wear rate was changed from weight loss (experiment) to weight gain (MFA model). This observation is attributed to the limited capacity of MFA to capture a possible shift in the dominant mechanism of wear or the rise of synergy. The complexity of the involved physical, chemical, and mechanical processes, although generally acknowledged, has not been studied sufficiently to provide an analytical description of each mechanism. The lack of mechanistic descriptions hinders the evaluation of the validity of the synthetic data, as a monotonic response of the system at the intermediate values is inherently assumed. However, for the purpose of the current work, and for a lack of reason to assume otherwise, the synthetic data derived from the MFA model are considered sufficiently valid.

The accuracy of the MFA model certainly depends on the experimental error. Although all the weight loss data were measured using the same analytical balance, the statistical error of the stand-alone corrosion data is notably higher than that of erosion-corrosion and stand-alone erosion data. This error is mostly explained by the outlier data point measured for high velocity (7 m/s), low particle concentration (45 wt.%), high temperature (45 °C), low pH (5), high content of dissolved oxygen (10 ppm), and high content of dissolved Cu^{2+} (500 ppm), which correspond to aggressive combination parameters indicating a possible shift in the mechanism of E-C wear as discussed by Aguirre et. al [43]. Nonetheless, the accuracy of the MFA model (1.922 (mg·cm^{-2}h^{-1})) is higher than the statistical error of the experiment (1.292 (mg·cm^{-2}h^{-1})), justifying at least the validity of the synthetic data.

Notwithstanding the above discussed limitations, by combining the experimental data with the synthetic data, the dataset became more diverse and representative of a wide range of E-C scenarios. This augmentation of the dataset was necessary for the NN model to learn from a broader spectrum of conditions, enabling it to generalise better and make more accurate predictions.

4.2. Applicability of E-C NN

The E-C NN model was developed using experimental and synthetic data, which were separated into subsets used independently to train and evaluate the model. In addition, development of the model comprised the approach of transfer learning, consisting of pre-training followed by fine-tuning. By combining the pre-training with synthetic data and subsequent fine-tuning with a small dataset from the experiments, the NN model could leverage the advantages of both data types. In other words, the pre-training provided the model with a "fundamental understanding" of E-C behaviour, whereas the fine-tuning aligned the model's predictions with real-world scenarios of the specific case as shown in Figure 10. As result, the fine-tuned E-C NN predicts the wear rate of a determined system better than MFA, as shown in Figure 11.

Overfitting of the E-C NN model model was avoided at the stage of fine-tuning. Indeed, there is no evidence of global overfitting as a validation MSE of 3.9 and the overall prediction MSE of 2.4 were observed. It is noted that in Figure 12, certain data points appear to be closer to zero error, indicating the possibility of localised overfitting. However, the overall model does not exhibit signs of overfitting. The absence of overfitting indicates that the model was able to generalise well, without overemphasizing the noise or irrelevant patterns in the data.

Figure 12 compares the performance of the MFA and E-C NN model, both direct and fine-tuned, and Table 2 compares the statistics of the models' predictions in terms of squared error. In general, the fine-tuned E-C NN provides the best prediction as expressed by the mean squared error of 2.535 compared with 3.694 for the MFA model. Notably, the mean error of the direct E-C NN model is similar to that of the MFA, but with a significantly lower median of 0.019 as compared with 0.345 for the MFA.

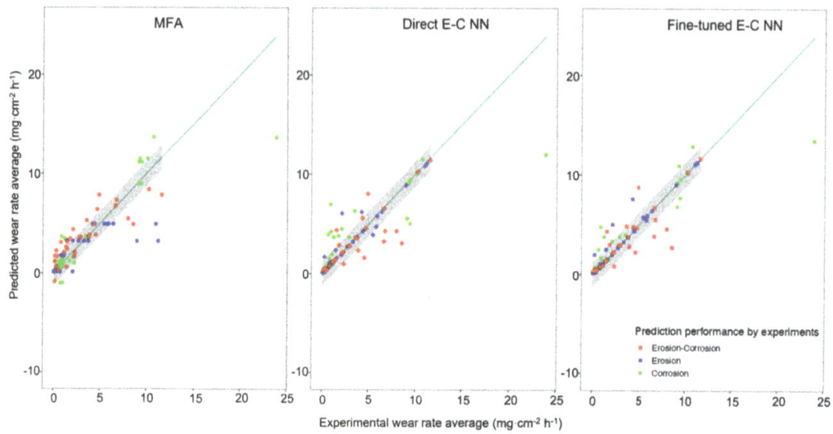

Figure 12. Comparison of accuracy in predicting experimental values of wear rate by the MFA and E-C NN models (direct and transfer learner). Each data point corresponds to an experimental run. The gray area represents standard deviation due to experimental error. The green line is added as guide for the eye to represent perfect prediction.

Table 2. Tabulation of squared errors of the MFA and E-C NN model predictions shown in Figure 12.

	MFA	Direct E-C NN	Fine-Tuned E-C NN
Min.	0.000	0.000	0.000
1st.Qu.	0.037	0.001	0.000
Median.	0.345	0.019	0.005
Mean	3.694	3.623	2.535
3rd.Qu.	2.519	0.909	0.202
Max.	104.642	141.747	108.842

The fine-tuned E-C NN model demonstrates a superior capability for predicting wear rate as compared to the MFA models applied directly to the experimental E-C data. Moreover, the E-C NN model is able to capture complex relationships and non-linearities in the dataset, which is evident in the velocity parameter (V). This parameter in the MFA was introduced using squared velocity to have better fitting, with the rationale that erosive wear is mostly caused by the transfer of kinetic energy, which is proportional to V^2 rather than V [44]). In the case of E-C NN, no such introduction was necessary because the model is capable of inferring this relation, highlighting the advantage of ANNs in modeling intricate patterns not necessarily captured by a conventional MFA model. Interestingly, even the direct E-C NN model, i.e., without transfer learning, outperforms the MFA models, suggesting that the inherent flexibility and non-linear nature of ANNs can provide an advantage in modeling the complex E-C processes.

In Figure 12, the gray area indicates the experimental error. It is observed that the prediction of stand-alone corrosion by the E-C NN tends to be overestimated as more run points are located above the equiproportional line. This behaviour could be attributed to the random shuffling of the dataset, which might have resulted in an imbalance of erosion, corrosion, and E-C cases in the training and evaluation sub-datasets. To verify the sufficiency of the size of the dataset, a separate study with an extended design for the experiment should be used. Further, it is interesting to note that the E-C predictions are more dispersed beyond the gray band compared to the other predictions. This could indicate that the model encountered fewer instances of E-C during the training process, suggesting a potential improvement in future applications. Despite these observations, the overall results obtained from the E-C NN model outperform those obtained from the MFA, demonstrating the superiority of the neural network approach in predicting wear rates for E-C phenomena.

The impact of corrosion error demonstrated in Figure 11 for the scenario of stand-alone corrosion in the absence of particles has also been predicted by the E-C NN. This prediction, interpreted as a predicted error, is certainly influenced by the experimental data error. Notably, the corrosion experiment was conducted using the same configuration as the stand-alone erosion and E-C experiment design, involving repetitions with no changes in particle concentration. As a result, variations in the measured wear rate can be attributed to the experimental reproducibility of measurements.

Further, it should be noted that the E-C NN takes into account all the factors and their effects to predict the wear rate, whereas MFA often excludes some parameters to achieve a better fit with linear regression. While this approach may be mathematically appealing, it leads to the omission of the direct and synergistic effects of the factors, limiting the model's ability to capture the full complexity of the phenomenon being studied. In contrast, the E-C NN comprehensively considers all contributing factors, resulting in more accurate and reliable predictions.

The computational cost of training the E-C NN model was remarkably low, even with a relatively small dataset. The pre-training process took only 135 s, and the fine-tuning process was completed in just 0.2 s using Google Colab CPU session and TensorFlow Keras as the framework. These efficient times underscore the feasibility of employing ANN models for E-C analysis, even in scenarios where data availability is limited or computational resources are constrained. The results demonstrate that ANN models can offer a practical and accessible approach to studying E-C phenomena, providing accurate predictions and valuable insights without the need for extensive computational resources.

4.3. E-C Synergy

The synergy of the physical, chemical, and mechanical processes involved in E-C wear provide a total E-C weight loss that differs from the sum of stand-alone erosion and stand-alone corrosion (Equations (1) and (2)). The experimentally determined values of synergy are summarized in Figure 13A, in which the contributions of stand-alone erosion and corrosion are also demonstrated. For all the factors, negative synergy is observed and its value is generally significant for the total wear rate. The highest relative contribution of erosion is observed for the high level of pH (11), whereas the highest relative contribution of corrosion is observed for the low level of V (3 m/s), which is consistent with the literature data of similar systems. The highest and lowest relative contributions of synergy are observed for the low and high levels of pH, respectively. However, the negative value of synergy for all the factors is noteworthy and it would be interesting to study its physical significance.

All the tendencies observed in the experimental data are also present in the E-C NN predictions, which is not always the case for the MFA predictions, as shown in Figure 13B. In particular, the negative sign of synergy is incorrectly predicted for a high level of pH (11) and low level of Cu^{2+} (0 ppm). This difference might not be significant, however, when considering that the value is comparable with experimental error. In this study, only main effects are discussed. In order to explain the origin of the synergy, cross-correlations would have to be examined, but such an analysis is beyond the scope of the present study.

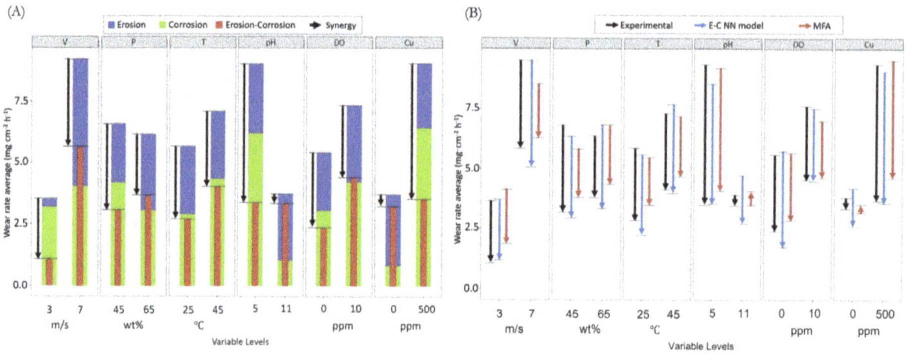

Figure 13. Synergy of wear modes: (**A**) contribution to erosion, corrosion, and synergy in the experimental data. Synergy is represented by an arrow with downward direction corresponding to negative synergy, (**B**) and comparison of synergy contribution to E-C wear rate as measured (black arrows) and as predicted by the E-C NN.

Figure 13 shows that velocity plays a significant role in the wear rate, both in stand-alone erosion and corrosion, as well as in the combined E-C mechanism. The *V* parameter derives from the flow field, which due to it inherent turbulence, affects the kinetic energy conveyed by the particles and also the convection-diffusion mass transfer for corrosion wear. Notably, the impact of particle concentration was not clearly distinguished between the stand-alone erosion and corrosion experiments. This observation can be associated with the fact that particle concentration changes fluid viscosity, consequently influencing the Reynolds number of the system [45], which in turn governs turbulence. Flow-induced corrosion is a critical mechanism in such systems, and this effect was not fully captured by these experiments, leading to a biased interpretation of the particle concentration effect. However, it could explain the negative synergy observed, primarily reducing the corrosion effect. In contrast, the corrosion experiment conducted without particles experienced reduced fluid viscosity, which resulted in augmented flow-induced corrosion and an increase in the corrosion wear rate [46]. In the other experiments, the presence of particles increased fluid viscosity [47,48], leading to a reduction in the Reynolds number (Turbulence level) and, consequently, the corrosion wear rate [46].

The above discussion of synergy shows the importance of correctly modeling the relative contributions of stand-alone effects as they are associated with physical processes that might be studied by analysis of microstructure. However, the high cost of microstructural analysis over the extended design of the experiment can be diminished by the use of E-C NN, as shown in Figure 13B in which superior capacity of the ANN to predict synergy is compared with that of the MFA.

Finally, this research is considered to provide a "toolbox" for effectively predicting E-C wear and understanding its underlying mechanisms. The results emphasize the superiority of ANN over regression models in E-C analysis, offering new alternatives for improving wear prediction accuracy and informing engineering practices aimed at mitigating wear in various systems and applications.

5. Conclusions

In this work, applicability of artificial neural networks (ANN) for predicting E-C wear was explored in comparison to the conventional regression models based on MFA. Previously published experimental data on E-C rate were complemented with new data on stand-alone erosion and stand-alone corrosion. The effect of pre-training was verified with the general conclusion that the pre-trained and fine-tuned ANN outperforms the regression models in accurately predicting the wear rates, indicating that the flexibility of ANN is better suited to represent the complex relationships and patterns that may be omitted by the conventional MFA models. In particular:

- Combination of experimental data with synthetic data generated by the MFA model provides a larger and more diverse dataset, enhancing the training process and improving the generalisation of the applicability and the capabilities of the ANN model in the specific field of E-C studies.
- The ANN shows a superior performance compared to the MFA counterpart as assessed by the mean and median squared errors (MSE). The performance is further improved by including a pre-training step, reducing the mean and median MSE from 3.623 to 2.535, and 0.019 to 0.005, respectively, as compared with 3.694 and 0.345, respectively, of the MFA.
- The ANN trained on E-C data, i.e., the E-C NN, is capable of generalising the importance of experimental parameters without overemphasizing noise in the data, as shown by the absence of global overfitting.
- The errors of the E-C NN predictions are consistently lower than the stand-alone experimental errors, indicating that the model provides highly confident predictions within the factor ranges, outperforming the predictions obtained using MFA.
- Velocity emerges as the predominant factor across all wear mechanisms studied in this work, underscoring the necessity for future focused and deterministic investigations on the specific impact of this parameter in each wear mechanism.
- The synergy effect is highly pronounced compared to the stand-alone wear rate impact. Conducting a detailed study to thoroughly understand and model this particular effect, as well as identifying the key factors that significantly influence it, is of paramount importance.
- Exploring the genuine impact of particle concentration and its influence on fluid viscosity in corrosion wear becomes crucial for a comprehensive understanding of corrosion and E-C wear mechanisms.

In conclusion, the E-C NN proves to be a superior method for predicting wear rates, owing to its comprehensive consideration of all factors and their effects, including direct and synergistic effects. On the other hand, MFA's exclusion of certain parameters to improve linear regression fit limits its ability to fully represent the complexity of erosion-corrosion phenomena. The E-C NN delivers more accurate and reliable predictions, establishing itself as a robust tool for exploring E-C behaviour and advancing tribological research. Moreover, the computational cost of training the ANN model is reasonable, making it viable even with small datasets. This advantage allows for the practical implementation of ANN models in E-C wear analysis, even under data constraints or limited computational resources. As a result, the E-C NN emerges as a promising approach with wide-ranging applications, enabling researchers to gain deeper insights into erosion-corrosion processes and pave the way for future advancements in the field.

Author Contributions: Conceptualisation, M.W. and A.E.-J.; Experiment, J.A.; Software, A.E.-J.; Formal analysis, A.E.-J., I.W. and M.W.; Investigation, A.E.-J. and M.W.; Resources, M.W.; Data curation, A.E.-J.; Writing—original draft preparation, A.E.-J.; writing—review and editing, A.E.-J., M.W. and I.W.; Visualisation, A.E.-J. and M.W.; Supervision, M.W.; Project administration and funding acquisition, M.W. All authors have read and agreed to the published version of the manuscript.

Funding: This research was funded by ANID (Agencia Nacional de Investigación y Desarrollo de Chile) through the grant FONDECYT Regular 1201547 and Pontificia Universidad Católica de Chile through the scholarship Beca VRI.

Institutional Review Board Statement: Not applicable.

Data Availability Statement: The data presented in this study combine results published by Aguirre et al. [35] with new data. The entire set of experimental data is provided as Appendix A.

Acknowledgments: The authors convey special thanks to Alvaro Soto for the inspiration to apply machine learning to the exotic field of E-C and to Alejandro Mac Cawley for his advice in conducting the statistical analysis.

Conflicts of Interest: The authors declare no conflict of interest.

Abbreviations

The following abbreviations are used in this manuscript:

ANN	Artificial neural network
E-C NN	Erosion-corrosion neural network
MFA	Multifactorial analysis
MSE	Meas squared error
RSM	Response surface method

Appendix A

Data of weight loss in $mg{\cdot}cm^{-2}h^{-1}$ measured in all the experimental runs.

Table A1. Erosion data.

Order	V	P	T	pH	DO	Cu^{2+}	Exp. Wear	MFA	E-C NN	F-T E-C NN
1	3	45	25	5	0	0	0.1188	0.1287	0.1139	0.1583
2	7	45	25	5	0	500	1.4600	3.2607	2.4135	2.5356
3	3	65	25	5	0	500	0.2688	0.2327	0.2352	0.3331
4	7	65	25	5	0	0	5.4693	4.9647	3.9248	5.7540
5	3	45	45	5	0	500	0.1584	0.1287	0.1492	0.1764
6	7	45	45	5	0	0	1.5477	3.2607	1.5944	1.5708
7	3	65	45	5	0	0	0.2320	0.2327	0.1816	0.1722
8	7	65	45	5	0	500	4.4563	4.9647	4.3497	4.4579
9	3	45	25	11	0	500	0.0821	0.1287	0.0964	0.1562
10	7	45	25	11	0	0	11.2582	3.2607	11.0948	11.2479
11	3	65	25	11	0	0	0.1075	0.2327	0.1076	0.1748
12	7	65	25	11	0	500	4.3771	4.9647	4.3075	4.3779
13	3	45	45	11	0	0	0.0849	0.1287	0.1021	0.1549
14	7	45	45	11	0	500	3.7751	3.2607	3.7327	3.7596
15	3	65	45	11	0	500	0.2575	0.2327	0.1941	0.1765
16	7	65	45	11	0	0	4.2498	4.9647	6.2028	7.5619
17	3	45	25	5	10	500	2.0836	0.1287	1.9453	2.0045
18	7	45	25	5	10	0	2.7162	3.2607	2.7509	2.7100
19	3	65	25	5	10	0	0.3169	0.2327	0.3019	0.2484
20	7	65	25	5	10	500	5.8144	4.9647	4.7506	5.4991
21	3	45	45	5	10	0	0.0340	0.1287	0.1057	0.1515
22	7	45	45	5	10	500	8.9664	3.2607	8.8790	8.9976
23	3	65	45	5	10	500	0.2462	0.2327	1.7233	1.9999
24	7	65	45	5	10	0	11.0291	4.9647	10.8320	11.0532
25	3	45	25	11	10	0	0.1584	0.1287	0.3864	0.2676
26	7	45	25	11	10	500	3.3161	3.2607	3.2726	3.3056
27	3	65	25	11	10	500	0.1584	0.2327	0.2003	0.1757
28	7	65	25	11	10	0	6.4171	4.9647	6.2448	6.3988
29	3	45	45	11	10	500	0.7102	0.1287	0.6785	0.7017
30	7	45	45	11	10	0	2.1221	3.2607	6.0747	5.0481
31	3	65	45	11	10	0	0.2292	0.2327	0.1953	0.2732
32	7	65	45	11	10	500	5.9106	4.9647	5.7941	5.9130
33	5	55	35	8	5	250	0.8347	1.7535	1.1227	1.0933
34	5	55	35	8	5	250	1.1685	1.7535	1.1227	1.0933
35	5	55	35	8	5	250	1.0808	1.7535	1.1227	1.0933

Table A2. Corrosion data.

Order	*V*	*P*	*T*	*pH*	*DO*	Cu^{2+}	Exp. Wear	MFA	E−C NN	F−T E−C NN
1	3	0	25	5	0	0	0.3537	0.4592	0.3669	0.3526
2	7	0	25	5	0	500	9.4248	8.9749	5.0799	7.6796
3	3	0	25	5	0	500	9.1135	8.9749	5.5686	6.7908
4	7	0	25	5	0	0	0.7753	0.4592	0.7725	0.7774
5	3	0	45	5	0	500	10.1265	11.4943	9.9856	10.1459
6	7	0	45	5	0	0	2.2664	1.0055	2.1457	2.2700
7	3	0	45	5	0	0	0.7074	1.0055	0.6891	0.7111
8	7	0	45	5	0	500	9.2663	11.4943	9.0739	10.9060
9	3	0	25	11	0	500	0.7668	−0.9749	1.5257	0.5824
10	7	0	25	11	0	0	0.3820	0.1906	0.7115	0.6872
11	3	0	25	11	0	0	0.2603	0.1906	0.5175	0.6645
12	7	0	25	11	0	500	0.9846	−0.9749	1.0360	0.9934
13	3	0	45	11	0	0	0.4188	0.7370	0.5629	0.5740
14	7	0	45	11	0	500	2.1079	1.5445	4.5462	3.3760
15	3	0	45	11	0	500	0.9479	1.5445	4.2680	1.6610
16	7	0	45	11	0	0	0.7158	0.7370	1.0281	0.8493
17	3	0	25	5	10	500	9.1503	11.1470	9.0801	9.1590
18	7	0	25	5	10	0	1.0044	0.6722	0.9941	0.9995
19	3	0	25	5	10	0	0.8969	0.6722	1.0509	0.6580
20	7	0	25	5	10	500	9.4819	11.1470	9.4167	9.4943
21	3	0	45	5	10	0	1.0554	1.2186	1.9900	1.2875
22	7	0	45	5	10	500	23.8958	13.6663	11.9901	13.4631
23	3	0	45	5	10	500	10.7659	13.6663	11.4710	12.8372
24	7	0	45	5	10	0	1.1035	1.2186	1.1378	1.1175
25	3	0	25	11	10	0	0.7498	0.4037	0.7171	0.7379
26	7	0	25	11	10	500	0.6564	1.1971	3.9713	2.4994
27	3	0	25	11	10	500	1.7712	1.1971	3.8260	1.5332
28	7	0	25	11	10	0	0.6593	0.4037	0.6668	0.6620
29	3	0	45	11	10	500	3.4972	3.7165	6.3058	3.9462
30	7	0	45	11	10	0	1.0497	0.9500	0.8715	1.3876
31	3	0	45	11	10	0	0.6366	0.9500	0.6113	0.6354
32	7	0	45	11	10	500	0.8941	3.7165	6.9660	4.7817
33	5	0	35	8	5	250	1.2421	3.5252	3.7010	3.7723
34	5	0	35	8	5	250	3.7942	3.5252	3.7010	3.7723
35	5	0	35	8	5	250	2.8719	3.5252	3.7010	3.7723

Table A3. Erosion-Corrosion data.

Order	*V*	*P*	*T*	*pH*	*DO*	Cu^{2+}	Exp. Wear	MFA	E-C NN	F-T E-C NN
1	3	45	25	5	0	0	0.1556	-0.8300	0.3622	0.3340
2	7	45	25	5	0	500	2.1447	4.4140	2.1187	2.1424
3	3	65	25	5	0	500	0.3650	0.6220	0.3823	0.4210
4	7	65	25	5	0	0	4.5582	3.8660	1.6334	2.2664
5	3	45	45	5	0	500	0.3282	2.2500	0.3414	0.3573
6	7	45	45	5	0	0	1.8278	3.4940	2.9187	3.0410
7	3	65	45	5	0	0	0.4074	1.7020	0.2723	0.3409
8	7	65	45	5	0	500	4.3403	4.9460	5.6191	4.7799
9	3	45	25	11	0	500	0.1556	1.1700	0.4032	0.3479
10	7	45	25	11	0	0	1.4996	2.4140	1.4928	1.4864
11	3	65	25	11	0	0	0.2943	0.6220	0.5036	0.5608
12	7	65	25	11	0	500	3.9159	3.8660	2.3881	2.8404
13	3	45	45	11	0	0	0.5008	0.2500	0.4720	0.4893
14	7	45	45	11	0	500	7.9959	5.4940	4.3158	4.5988
15	3	65	45	11	0	500	0.1754	1.7020	0.3715	0.3482
16	7	65	45	11	0	0	8.5590	4.9460	3.0907	2.7559
17	3	45	25	5	10	500	1.3553	2.6000	1.3859	1.3562
18	7	45	25	5	10	0	3.6047	5.3240	3.6684	4.8792

Table A3. *Cont.*

Order	V	P	T	pH	DO	Cu^{2+}	Exp. Wear	MFA	E-C NN	F-T E-C NN
19	3	65	25	5	10	0	2.3371	2.0520	0.9745	0.8509
20	7	65	25	5	10	500	6.6887	6.7760	6.5905	6.7015
21	3	45	45	5	10	0	2.2975	1.6800	2.1945	2.2997
22	7	45	45	5	10	500	10.2623	8.4040	10.2508	10.2911
23	3	65	45	5	10	500	1.5449	3.1320	4.4122	4.3425
24	7	65	45	5	10	0	11.6374	7.8560	11.4157	11.6311
25	3	45	25	11	10	0	1.4430	0.6000	1.3731	1.3913
26	7	45	25	11	10	500	6.7793	7.3240	4.3113	5.5060
27	3	65	25	11	10	500	1.1318	2.0520	1.1991	0.9393
28	7	65	25	11	10	0	6.6520	6.7760	3.2735	3.8736
29	3	45	45	11	10	500	3.8056	3.6800	3.6246	3.8071
30	7	45	45	11	10	0	4.7081	6.4040	4.5801	4.7080
31	3	65	45	11	10	0	0.8149	3.1320	0.8151	0.8023
32	7	65	45	11	10	500	4.9175	7.8560	8.0262	8.7464
33	5	55	35	8	5	250	2.9794	3.6430	2.8227	2.9507
34	5	55	35	8	5	250	2.8351	3.6430	2.8227	2.9507
35	5	55	35	8	5	250	2.9596	3.6430	2.8227	2.9507

References

1. Moore, P. *Ausenco Symposium Outlines Importance of Long Term Mining Slurry Pipeline Management*; Ausenco: South Brisbane, Australia, 2018.
2. *ASTM G119-09*; Standard Guide for Determining Synergism Between Wear and Corrosion. ASTM International: West Conshohocken, PA, USA, 2021. [CrossRef]
3. Javaheri, V.; Porter, D.; Kuokkala, V.T. Slurry erosion of steel—Review of tests, mechanisms and materials. *Wear* **2018**, *408*, 248–273. [CrossRef]
4. Turenne, S.; Fiset, M.; Masounave, J. The effect of sand concentration on the erosion of materials by a slurry jet. *Wear* **1989**, *133*, 95–106. [CrossRef]
5. Neville, A.; Hodgkiess, T.; Xu, H. An electrochemical and microstructural assessment of erosion–corrosion of cast iron. *Wear* **1999**, *233*, 523–534. [CrossRef]
6. Stachowiak, G.W. Particle angularity and its relationship to abrasive and erosive wear. *Wear* **2000**, *241*, 214–219. [CrossRef]
7. Stack, M.; Jana, B. Modelling particulate erosion–corrosion in aqueous slurries: Some views on the construction of erosion–corrosion maps for a range of pure metals. *Wear* **2004**, *256*, 986–1004. [CrossRef]
8. Desale, G.R.; Gandhi, B.K.; Jain, S. Effect of erodent properties on erosion wear of ductile type materials. *Wear* **2006**, *261*, 914–921. [CrossRef]
9. Tian, B.R.; Cheng, Y.F. Electrochemical corrosion behavior of X-65 steel in the simulated oil sand slurry. I: Effects of hydrodynamic condition. *Corros. Sci.* **2008**, *50*, 773–779. [CrossRef]
10. Rauf, A.; Mahdi, E. Studying and comparing the erosion-enhanced pitting corrosion of X52 and X100 steels. *Int. J. Electrochem. Sci.* **2012**, *7*, 5692–5707. [CrossRef]
11. Wood, R.J.K.; Walker, J.C.; Harvey, T.J.; Wang, S.; Rajahram, S.S. Influence of microstructure on the erosion and erosion–corrosion characteristics of 316 stainless steel. *Wear* **2013**, *306*, 254–262. [CrossRef]
12. Islam, A.M.; Farhat, Z.N.; Ahmed, E.M.; Alfantazi, A.M. Erosion enhanced corrosion and corrosion enhanced erosion of API X-70 pipeline steel. *Wear* **2013**, *302*, 1592–1601. [CrossRef]
13. Yu, B.; Li, D.Y.; Grondin, A. Effects of the dissolved oxygen and slurry velocity on erosion-corrosion of carbon steel in aqueous slurries with carbon dioxide and silica sand. *Wear* **2013**, *302*, 1609–1614. [CrossRef]
14. Lindgren, M.; Perolainen, J. Slurry pot investigation of the influence of erodent characteristics on the erosion resistance of austenitic and duplex stainless steel grades. *Wear* **2014**, *319*, 38–48. [CrossRef]
15. Malik, J.; Toor, I.; Ahmed, W.; Gasem, Z.; Habib, M.; Ben-Mansour, R.; Badr, H. Investigations on the Corrosion-Enhanced Erosion Behavior of Carbon Steel AISI 1020. *Int. J. Electrochem. Sci.* **2014**, *9*, 6765–6780. [CrossRef]
16. Islam, M.A.; Alam, T.; Farhat, Z.N.; Mohamed, A.; Alfantazi, A. Effect of microstructure on the erosion behavior of carbon steel. *Wear* **2015**, *332*, 1080–1089. [CrossRef]
17. Zheng, Z.; Zheng, Y. Erosion-enhanced corrosion of stainless steel and carbon steel measured electrochemically under liquid and slurry impingement. *Corros. Sci.* **2016**, *102*, 259–268. [CrossRef]
18. Jiang, J.; Xie, Y.; Islam, A. The Effect of Dissolved Oxygen in Slurry on Erosion–Corrosion of En30B Steel. *J. Bio Tribo-Corros.* **2017**, *3*, 45. [CrossRef]
19. Kuruvila, R.; Kumaran, S.T.; Khan, M.A.; Uthayakumar, M. A brief review on the erosion-corrosion behavior of engineering materials. *Corros. Rev.* **2018**, *36*, 435–447. [CrossRef]

20. Yi, J.Z.; Hu, H.X.; Wang, Z.B.; Zheng, Y.G. On the critical flow velocity for erosion-corrosion in local eroded regions under liquid-solid jet impingement. *Wear* **2019**, *423*, 94–99. [CrossRef]
21. Messa, G.V.; Wang, Y.; Malavasi, S. A discussion of the test procedures of the API 6AV1 standard based on wear prediction simulations. *Wear* **2019**, *426*, 1416–1429. .: 10.1016/j.wear.2019.01.042. [CrossRef]
22. Nicholls, J.R.; Stephenson, D.J. Monte Carlo modelling of erosion processes. *Wear* **1995**, *186*, 64–77. [CrossRef]
23. Haider, G.; Arabnejad, H.; Shirazi, S.A.; Mclaury, B.S. A mechanistic model for stochastic rebound of solid particles with application to erosion predictions. *Wear* **2017**, *376*, 615–624. [CrossRef]
24. Das, S.K. *Application of a Stochastic Modelling Framework to Characterize the Influence of Different Oxide Scales on the Solid Particle Erosion Behaviour of Boiler Grade Steel*; Technical Report; Indian Academy of Sciences: Karnataka, India, 2011.
25. de Moura, B.F.; da Silva, W.B.; de Macêdo, M.C.S.; Martins, M.F. A statistical approach to estimate state variables in flow-accelerated corrosion problems. *Inverse Probl. Sci. Eng.* **2018**, *26*, 966–995. [CrossRef]
26. Vencl, A.; Svoboda, P.; Klančnik, S.; But, A.; Vorkapić, M.; Harničárová, M.; Stojanović, B. Influence of Al_2O_3 Nanoparticles Addition in ZA-27 Alloy-Based Nanocomposites and Soft Computing Prediction. *Lubricants* **2023**, *11*, 24. [CrossRef]
27. Lalwani, V.; Sharma, P.; Pruncu, C.I.; Unune, D.R. Response surface methodology and artificial neural network-based models for predicting performance of wire electrical discharge machining of inconel 718 alloy. *J. Manuf. Mater. Process.* **2020**, *4*, 44. [CrossRef]
28. Ibrahim, S.; Abdul Wahab, N. Improved Artificial Neural Network Training Based on Response Surface Methodology for Membrane Flux Prediction. *Membranes* **2022**, *12*, 726. [CrossRef] [PubMed]
29. Oh, S. Comparison of a response surface method and artificial neural network in predicting the aerodynamic performance of a wind turbine airfoil and its optimization. *Appl. Sci.* **2020**, *10*, 6277. [CrossRef]
30. Patel, K.A.; Brahmbhatt, P.K. A Comparative Study of the RSM and ANN Models for Predicting Surface Roughness in Roller Burnishing. *Procedia Technol.* **2016**, *23*, 391–397. [CrossRef]
31. Abiodun, O.I.; Kiru, M.U.; Jantan, A.; Omolara, A.E.; Dada, K.V.; Umar, A.M.; Linus, O.U.; Arshad, H.; Kazaure, A.A.; Gana, U. Comprehensive Review of Artificial Neural Network Applications to Pattern Recognition. *IEEE Access* **2019**, *7*, 158820–158846. [CrossRef]
32. Hasan, M.S.; Nosonovsky, M. Triboinformatics: Machine learning algorithms and data topology methods for tribology. *Surf. Innov.* **2022**, *10*, 229–242. [CrossRef]
33. Kokol, P.; Kokol, M.; Zagoranski, S. Machine learning on small size samples: A synthetic knowledge synthesis. *Sci. Prog.* **2022**, *105*, 003685042110297. [CrossRef]
34. Barrionuevo, G.O.; Walczak, M.; Ramos-Grez, J.; Sánchez-Sánchez, X. Microhardness and wear resistance in materials manufactured by laser powder bed fusion: Machine learning approach for property prediction. *CIRP J. Manuf. Sci. Technol.* **2023**, *43*, 106–114. [CrossRef]
35. Aguirre, J.; Walczak, M. Multifactorial study of erosion–corrosion wear of a X65 steel by slurry of simulated copper tailing. *Tribol. Int.* **2018**, *126*, 177–185. [CrossRef]
36. *ASTM G1-03 (Reapproved 2017)*; Standard Practice for Preparing, Cleaning, and Evaluating Corrosion Test. ASTM International: West Conshohocken, PA, USA, 2017.
37. R Core Team. *A Language and Environment for Statistical Computing*; R Core Team: Vienna, Austria, 2022.
38. Ulrike Groemping. *FrF2: Fractional Factorial Designs with 2-Level Factors*; Ulrike Groemping: Berlin, Germany, 2023.
39. Lumley, T.; Miller, A. leaps: Regression Subset Selection. 2023. Available online: https://cran.r-project.org/web/packages/leaps/leaps.pdf (accessed on 1 October 2022).
40. Kobayashi, M.; Sakata, S. Mallows' C criterion and unbiasedness of model selection. *J. Econom.* **1990**, *45*, 385–395. [CrossRef]
41. Keyes, T.K.; Levy, M.S. Goodness of Prediction Fit for Multivariate Linear Models. *J. Am. Stat. Assoc.* **1996**, *91*, 191–197. [CrossRef]
42. Pedregosa, F.; Varoquaux, G.; Gramfort, A.; Michel V.; Thirion, B.; Grisel, O.; Blondel, M.; Prettenhofer, P.; Weiss, R.; Dubourg, V.; et al. Scikit-learn: Machine Learning in Python. *J. Mach. Learn. Res.* **2011**, *12*, 2825–2830.
43. Aguirre, J.; Walczak, M.; Rohwerder, M. The mechanism of erosion-corrosion of API X65 steel under turbulent slurry flow: Effect of nominal flow velocity and oxygen content. *Wear* **2019**, *438*, 203053. [CrossRef]
44. Clark, H.M. Particle velocity and size effects in laboratory slurry erosion measurements OR... do you know what your particles are doing? *Tribol. Int.* **2002**, *35*, 617–624. [CrossRef]
45. Grossmann, S.; Lohse, D.; Sun, C. High–Reynolds Number Taylor-Couette Turbulence. *Annu. Rev. Fluid Mech.* **2016**, *48*, 53–80. [CrossRef]
46. Fang, C.S.; Liu, B. Hydrodynamic and temperature effects on the flow-induced local corrosion rate in pipelines. *Chem. Eng. Commun.* **2003**, *190*, 1249–1266. [CrossRef]
47. Stickel, J.J.; Powell, R.L. Fluid mechanics and rheology of dense suspensions. *Annu. Rev. Fluid Mech.* **2005**, *37*, 129–149. [CrossRef]
48. Chen, S.H.; Li, X.F. Effects of particle concentration and physical properties on the apparent viscosity of a suspension of monodisperse concentric core–shell particles. *Eur. J. Mech. B/Fluids* **2020**, *84*, 542–552. [CrossRef]

lubricants

MDPI

Article

Intelligent Tool Wear Monitoring Method Using a Convolutional Neural Network and an Informer

Xingang Xie [1,2], Min Huang [1,2,*], Weiwei Sun [2], Yiming Li [2] and Yue Liu [2]

[1] School of Mechanical Electronic and Information Engineering, China University of Mining and Technology-BEIJING, Beijing 100083, China; bqt2000401003@cumtb.edu.cn
[2] Mechanical Electrical Engineering School, Beijing Information Science and Technology University, Beijing 100192, China; sww@bistu.edu.cn (W.S.); liyimingxf@bistu.edu.cn (Y.L.); yliu@bistu.edu.cn (Y.L.)
* Correspondence: huangmin@bistu.edu.cn

Abstract: Tool wear (TW) is the gradual deterioration and loss of cutting edges due to continuous cutting operations in real production scenarios. This wear can affect the quality of the cut, increase production costs, reduce workpiece accuracy, and lead to sudden tool breakage, affecting productivity and safety. Nevertheless, since conventional tool wear monitoring (TWM) approaches often employ complex physical models and empirical rules, their application to complex and non-linear manufacturing processes is challenging. As a result, this study presents a TWM model using a convolutional neural network (CNN), an Informer encoder, and bidirectional long short-term memory (BiLSTM). First, local feature extraction is performed on the input multi-sensor signals using CNN. Then, the Informer encoder deals with long-term time dependencies and captures global time features. Finally, BiLSTM captures the time dependency in the data and outputs the predicted tool wear state through the fully connected layer. The experimental results show that the proposed TWM model achieves a prediction accuracy of 99%. It is able to meet the TWM accuracy requirements of real production needs. Moreover, this method also has good interpretability, which can help to understand the critical tool wear factors.

Keywords: tool wear; convolutional neural network (CNN); global time feature; informer; BiLSTM

Citation: Xie, X.; Huang, M.; Sun, W.; Li, Y.; Liu, Y. Intelligent Tool Wear Monitoring Method Using a Convolutional Neural Network and an Informer. *Lubricants* **2023**, *11*, 389. https://doi.org/10.3390/lubricants11090389

Received: 27 July 2023
Revised: 6 September 2023
Accepted: 8 September 2023
Published: 11 September 2023

1. Introduction

Tool wear monitoring (TWM) is important to guarantee the manufacturing process's quality and efficiency [1]. The tool wear will affect the product quality, and excessive wear may result in tool damage and the shutdown of the production line, which will cause substantial economic loss. Therefore, developing an effective tool wear condition monitoring method has important practical significance.

Tool wear monitoring approaches mainly involve conventional and deep learning (DL) approaches. Traditional tool wear monitoring methods mainly rely on hand-designed feature extraction algorithms and machine learning models. Standard features include cutting force, a sound signal, and a vibration signal. Then, the relevant features can be extracted from the original signal using a feature extraction algorithm and classified or regressed by a machine learning algorithm. Shi et al. [2] presented a tool wear prediction approach integrating least squares support vector machine (LS-SVM) and principal component analysis (PCA) techniques. Gomes et al. [3] employed the support vector machine (SVM) and vibration and sound signals to monitor tool wear. Chen et al. [4] presented an SVM-based tool wear prediction approach using the Whale Optimization Algorithm (WOA). Gai et al. [5] established a WOA-SVM classification model using fusion features to identify tool wear states. The combination of these optimization algorithms and SVMs described above suffers from related shortcomings. Firstly, the performance of SVMs is highly dependent on the correct choice of parameters. Improper parameter selection can lead to overfitting or underfitting of the model. Secondly, SVM models are not very interpretable, and although SVMs can deal with non-linear problems by using non-linear kernel functions, choosing the right kernel function is not always intuitive. This can make the decision-making process of the model difficult to understand for non-technical people. Finally, although related researchers have used a variety of optimization algorithms to improve the training efficiency of SVMs, these optimization algorithms suffer from the shortcomings of falling into local optimums, sensitivity to initial values, and slow convergence. They cannot deal with complex and diverse tool wear states. Moreover, capturing the tool wear state's dynamic change using a traditional method is challenging due to its limited modeling ability for long-term dependence.

In order to resolve these issues, DL models have attracted extensive attention in TWM. Characterized by powerful nonlinear fitting capabilities and automatic feature learning capabilities, DL models can derive high-level features from raw sensor data and capture complex tool wear state patterns [6]. Tool wear condition monitoring is a critical research area in the manufacturing industry, and many researchers have proposed various methods to solve it [7].

A convolutional neural network (CNN) is a DL model that can extract local features effectively. In TWM, CNN is often utilized to extract the tool wear state's spatial characteristics [8]. CNN can gradually extract the high-level features of the tool wear state through multi-layer convolution and pooling operations. Many studies have successfully applied CNN to classify and forecast tool wear states. For instance, Dai et al. [9] presented a CNN-based TWM approach. Garcia et al. [10] presented a CNN-based in situ TWM approach. Kothuru et al. [11] combined depth visualization and CNN to achieve tool wear state detection. Wu et al. [12] presented an automatic CNN-based tool wear detection approach.

A recurrent neural network (RNN) is a DL model suitable to process sequence data [13]. However, traditional RNNs have deficiencies like gradient disappearance and explosion when dealing with long sequence data. In order to overcome these problems, scholars have proposed improved RNN structures like long short-term memory (LSTM) and gated cycle units (GRU). For example, Xu et al. [14] presented a multi-scale convolutional GRU network to predict tool wear. Liu et al. [15] presented a TWM approach that combines Densetnet and GRU. Chen [16] presented a tool wear prediction approach using parallel CNN and BiLSTM. These improved RNN models can capture the temporal pattern in the tool wear state sequence well and have excellent long-term dependence modeling ability.

Transformer is a self-attention mechanism-based DL model initially utilized for natural language processing tasks. The Transformer encoder models the global context of the input sequence and captures dependencies at different points in the sequence. In recent years, scholars have begun to apply the Transformer to time series data analysis, including tool wear condition monitoring. For example, Liu [17] proposed a new CNN-transformer neural network model for TWM. Liu et al. [18] presented a new transformer-based neural network model for tool wear prediction. The Informer model solves this problem by using a sparse attention mechanism and a hierarchical structure to efficiently deal with long time sequences. The Informer encoder is the core of the Informer model. The main task of the Informer encoder is to capture the patterns and dependencies of the input time series and to encode this information into a fixed-length representation. The Informer encoder introduces the ProbSparse self-attention mechanism, which uses a probabilistic mechanism to capture the patterns and dependencies of the input time series and to encode this information into a fixed-length representation. The main task of the Informer encoder is to capture the patterns and dependencies of the input time series and encode this information into a fixed-length representation. The Informer encoder introduces the ProbSparse self-attention mechanism, which uses a probabilistic mechanism to select the critical time steps, thus reducing the computational complexity. To further reduce the computational burden, the Informer encoder uses a hierarchical structure that divides the time series data into multiple sub-sequences and applies the self-attention mechanism to each sub-sequence independently. Therefore, in this paper, an Informer encoder is chosen to model long-term dependencies and sequentially capture important features in the time series to improve the accuracy and efficiency of tool wear condition monitoring.

In summary, the DL-based tool wear state monitoring method has better feature learning capability and long-term dependence modeling ability than the traditional method. The current work presents a DL network model, CIEBM, which combines a CNN, an Informer encoder, and BiLSTM. The CIEBM model utilizes the advantages of the CNN, Informer encoder, and BiLSTM in feature extraction, long-term dependence modeling, and time series modeling to accurately monitor and predict tool wear state. Compared to traditional methods such as optimization algorithms and SVM, the CIEBM model takes full advantage of different neural networks and is able to automatically learn and extract features from the original data without the need to manually design or select the features. It is also more suitable for tool wear prediction because the CIEBM model is able to capture complex and non-linear relationships in the data due to its multi-layer structure.

The essential novelties are as follows:

(1) This study presents a new TWM approach that combines the advantages of the CNN, the Informer encoder, and BiLSTM. This is the first time these three DL techniques have been combined to monitor tool wear conditions.
(2) This method can extract spatial features from the raw sensor data, capture long-term dependence and time patterns, and learn the feature representation of tool wear state comprehensively to enhance the TWM's precision and reliability.
(3) The presented approach has excellent efficiency and good interpretability, which can help to understand the key factors of tool wear and prepare a valuable reference to prevent and manage tool wear.

The paper is structured as follows: Section 2 focuses on the theory related to the CIEBM model; Section 3 focuses on the structure of the CIEBM model and the parameters related to the network; Section 4 focuses on the experimental procedure and results; and finally, Section 5 presents the conclusions.

2. Methods

2.1. D-CNN

One-dimensional CNN (1D-CNN) is a DL model widely used in time series data analysis and signal processing [19]. Compared with traditional fully connected neural networks, 1D-CNN can efficiently derive local patterns and associated features from time series data using local perception and parameter sharing.

The input data for 1D-CNN is 1D time series data. The input data in the tool wear monitoring can be cutting force, a vibration signal, or a sound signal. Discrete sample points typically represent these time series data, each representing a measured value at a specific point in time. Convolution operators are core components of 1D-CNN and can extract local patterns and associated features from input data [20]. The convolution is performed on the input sequence by a sliding window; the convolution operation between the input data and the convolution kernel in the window is calculated; and the feature mapping is generated. The calculation of the 1D convolution layer can be expressed by Equation (1):

$$y[t] = \sum_{i=0}^{k=1} x[t-i] \cdot w[i] + b \tag{1}$$

where $y[t]$ is the output sequence value at time t, $x[t-i]$ is the output sequence value at time $t-i$, $w[i]$ is the convolution kernel value in position i, and b describes the offset term.

In addition to 1D convolution layers, 1D-CNN usually includes activation functions and pooling layers. Activation functions are utilized to introduce nonlinearities so that the model can fit complex functions. Common activation functions involve ReLU, sigmoid, and tanh functions, as presented in Equations (2)–(4). The pooling layer alleviates the sequence length and improves the model's computational efficiency and robustness. Common pooling operations involve maximum and average pooling, as shown in Equations (5) and (6).

$$f(x) = \max(0, x) \tag{2}$$

$$f(x) = \frac{1}{1+e^{-x}} \tag{3}$$

$$f(x) = \tanh(x) = \frac{e^x - e^{-x}}{e^x + e^{-x}} \tag{4}$$

$$y[t] = \max_{i=t}^{t+p-1} x[i] \tag{5}$$

$$y[t] = \frac{1}{p}\sum_{i=t}^{t+p-1} x[i] \tag{6}$$

where $y[t]$ describes the output sequence value at time t, $x[i]$ describes the input sequence value in position i, and p is the length of the pooling window.

An essential advantage of 1D-CNN when processing sequence data is that local features of the sequence can be automatically extracted without complicated manual feature engineering. Moreover, due to its parameter-sharing characteristics, 1D-CNN can maintain low model complexity and avoid overfitting even when processing long sequence data. However, 1D-CNN has its limitations. Since it mainly focuses on the sequence' s local features, 1D-CNN may ignore the sequence's global features. Furthermore, 1D-CNN cannot handle long-term dependencies in sequences, that is, relationships between elements far apart.

2.2. *Informer Encoder*

Vaswani established the Transformer model in 2017 [21], which has shown remarkable success in natural language processing, image detection, and fault diagnosis. Although Transformer introduces a self-attention mechanism to model long-distance dependencies, its computational complexity increases rapidly for a long input sequence, resulting in a large memory footprint and reduced computational efficiency. In order to solve the mentioned issues, Zhou et al. [22] established the Informer model, as shown in Figure 1. The Informer encoder is the core component of the model and is responsible for feature extraction and representation learning of input sequences. Its core includes ProbSparse Self-Attention and distilling layer operations.

2.2.1. ProbSparse Self-Attention

As shown in Figure 2, ProSparse Self-Attention is one of the key components in the Informer model to sparse self-attention weights, reduce compute and memory overhead, and accommodate the need to handle long sequences. For the input array X, the corresponding Query, Key, and Value vectors can be attained by multiplying various weight matrices, as shown in Equations (7)–(9):

$$\text{Query} = XW_Q \tag{7}$$

$$\text{Key} = XW_K \tag{8}$$

$$\text{Value} = XW_V \tag{9}$$

where X is the input array, and W_Q, W_K, and W_V are weight matrices for linear transformations. First, a dot product is performed on Q and K to obtain an attention score, calculated by Equation (10), which reflects the correlation between each query and key.

$$Score = \frac{QK^T}{\sqrt{d}} \tag{10}$$

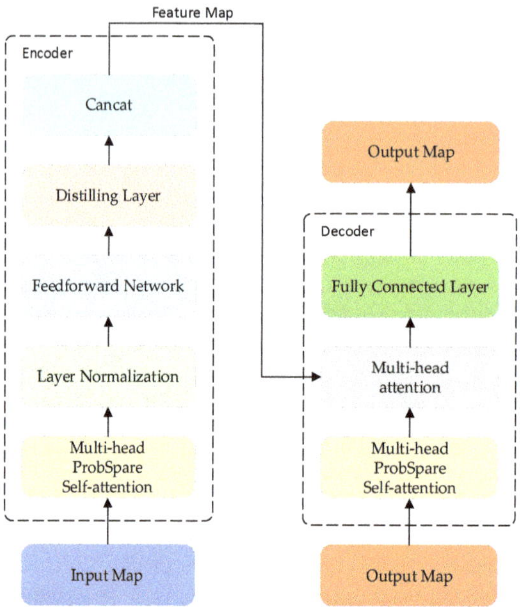

Figure 1. Network structure of the Informer.

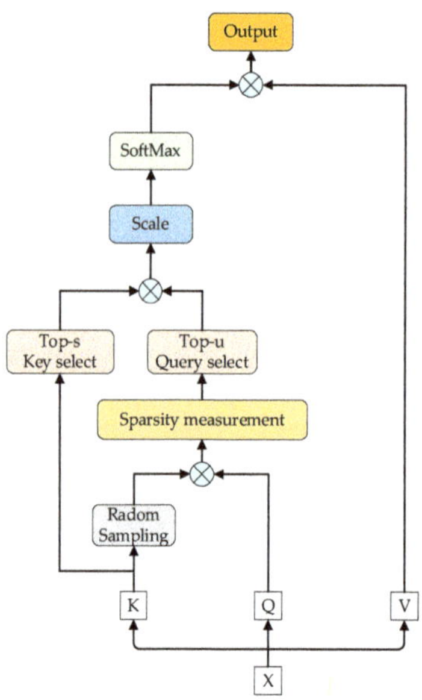

Figure 2. Network structure of ProSparse Self-Attention.

In order to select the important Q, ProSparse Self-Attention calculates the sparsity measurement $M(q_i, K)$ of q_i for the key set K, as shown in Equation (11):

$$M(q_i, K) = \max_j \left\{ \frac{q_i^T k_j}{\sqrt{d}} \right\} - \frac{1}{L_K} \sum_{j=1}^{L_K} \left\{ \frac{q_i^T k_j}{\sqrt{d}} \right\} \quad (11)$$

where k_j is the jth key in K, L_K is the number of keys, and $M(q_i, K)$ is the importance between and key set K, determining the difference between the query and key vectors. According to $M(q_i, K)$, $Top - u$ queries with greater sparsity are selected, where u is the default value representing the number of query vectors to be retained. Accordingly, important query vectors a high correlation to key set K can be screened. For the selected important query vector, the Softmax operation is performed on the dot product score matrix to convert the attention score into a probability distribution, as described by Equation (12).

$$\text{AttentionWeights}(q_i, K) = \text{Softmax}(\text{Score}(q_i, K)) \quad (12)$$

In order to reduce the computation and memory overhead, ProSparse Self-Attention can further sparse the attention weight. For each query vector, only $Top - s$ key vectors with greater attention weights are reserved, where s is the default value representing the number of key vectors to retain. Finally, the sparse attention weight is multiplied by the Value vector, and its summation is employed to obtain the output of ProSparse Self-Attention, as shown in Equation (13).

$$\text{Output}\ (Q, K, V) = \sum_{i=1}^{u} \sum_{j=1}^{s} \text{AttentionWeights}\ (q_i, k_{ij}) v_{ij} \quad (13)$$

where k_{ij} represents the jth reserved key vector of the ith query vector, and v_{ij} represents the value vector corresponding to k_{ij}. Through the above steps, ProSparse Self-Attention realizes the sparseness and selection of attention weights, reduces the calculation and memory overhead, and retains the key information with a high correlation to the query.

2.2.2. Distilling Layer

Figure 3 shows the distilling process. For a too-long input sequence, Probsparse Attention only selects $Top - u$ Query for dot product to form dot product pairs, while the rest of the dot product pairs are set as zero. Therefore, many information items are generated when multiplied by Value. In order to alleviate the information redundancy, a distilling layer is located at the end of the encoder [23], which can highlight the essential features, reduce the long sequences' input complexity, and improve the model's performance [24]. The "distilling" process is advanced from layer j to layer $(j + 1)$ by Equation (14), where $[\cdot]_{AB}$ is the attention block.

$$X_{j+1}^t = \text{MaxPool}(\text{ELU}(\text{Conv1d}[X_j^t]_{AB}))) \quad (14)$$

2.3. *BiLSTM Network*

BiLSTM is a variant of recurrent neural networks extensively utilized in time series data processing [25]. As displayed in Figure 4, compared with traditional one-way LSTM, BiLSTM captures more comprehensive contextual information and timing patterns by running two LSTM layers in both forward and backward orientations on the time series [26].

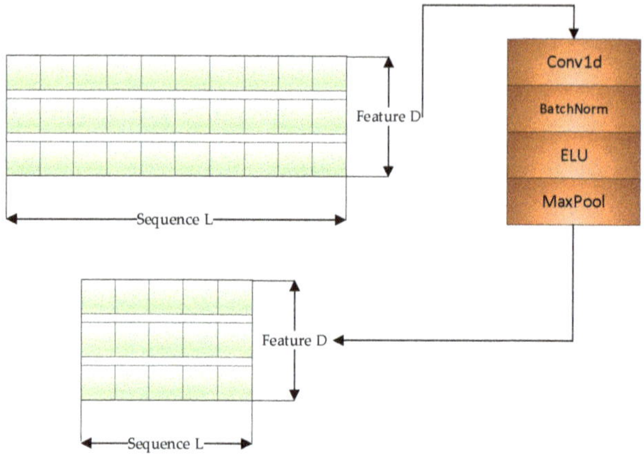

Figure 3. Network structure of the distilling layer.

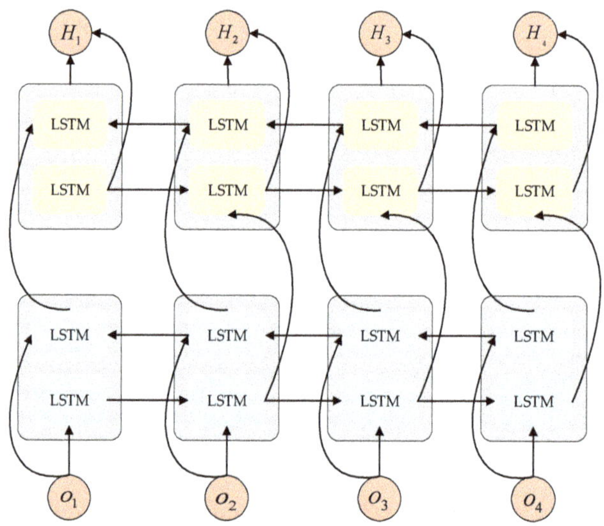

Figure 4. Network structure of BiLSTM.

As illustrated in Figure 5, the LSTM unit is the core component of BiLSTM. The main property of LSTM is to introduce gating mechanisms, including input, forget, and output gates, as well as a cell state, to better control the information flow. The forgetting gate indicates the novel information discarded from the cell state, and the calculation formula is presented in Equation (15). The input gate determines the novel information updated into the cell state, and the calculation formula is presented in Equation (16). The cell states first discard some information through the forgetting gate, and then add new candidate information through the input gate. The computation formulas are described by Equations (17) and (18). The output gate indicates the information the next hidden state should contain, and the calculation formula is shown by Equations (19) and (20).

$$f_t = \sigma\left(W_f \cdot [x_t, h_{t-1}] + b_f\right) \tag{15}$$

$$i_t = \sigma(W_i \cdot [h_{t-1}, x_t] + b_i) \tag{16}$$

$$\widetilde{C}_t = \tanh(W_C \cdot [h_{t-1}, x_t] + b_C) \tag{17}$$

$$C_t = f_t * C_{t-1} + i_t * \widetilde{C}_t \tag{18}$$

$$o_t = \sigma(W_o \cdot [h_{t-1}, x_t] + b_o) \tag{19}$$

$$h_t = o_t * \tanh(C_t) \tag{20}$$

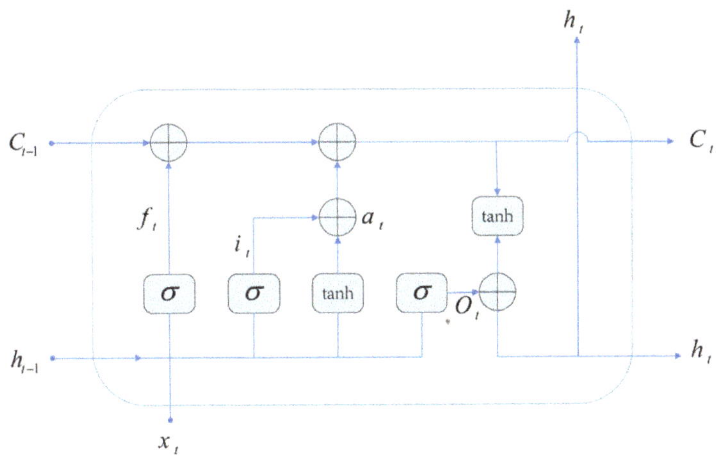

Figure 5. Network structure of LSTM.

3. Proposed Methods

3.1. Frame

This study utilizes the combination of 1D-CNN, Informer encoder, and BiLSTM for TWM. Figure 6 presents the CIEBM tool wear state monitoring algorithm. First, the raw sensor data for tool wear condition monitoring is collected, and the data is pre-processed, including denoising, normalization, and data segmentation. The 1D-CNN neural network is introduced as the basic model for extracting the tool wear state characteristics [27]. CNN can extract local patterns and feature representations related to tool wear from raw sensor data. The Informer encoder is introduced to capture long-term dependencies and global context information in tool wear states. Informer encoders employ self-attention mechanisms to model dependencies between locations in a sequence. Accordingly, features can be correlated from different locations, and important patterns and relationships in the sequence can be captured. The feature sequence extracted by CNN is input into the Informer encoder to obtain a richer feature representation. BiLSTM is introduced to further capture context information in time series data. The Informer encoder's output sequence is taken as the BiLSTM input, and the relationship between sequence timing and different tool wear states is further extracted through the BiLSTM layer.

3.2. Parameter Settings

The model's structural parameters are presented in Table 1.

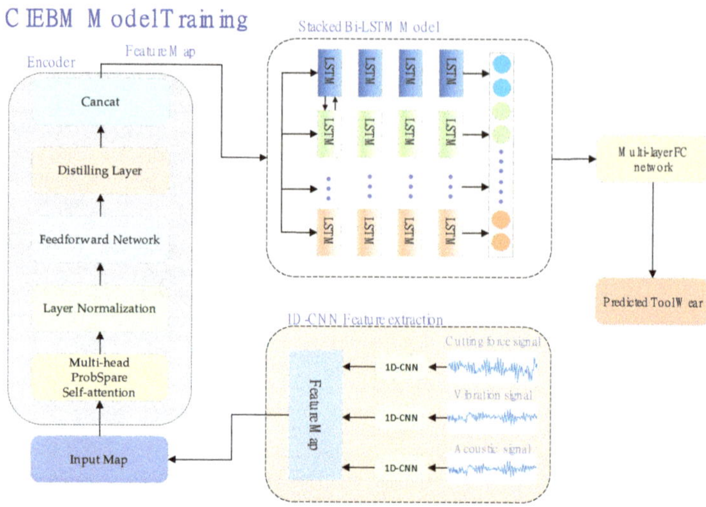

Figure 6. Network structure of CIEBM.

Table 1. The CIEBM structural parameters.

Layer	Output Shape
Conv1D	(20, 16)
MaxPooling	(10, 16)
Informer Encoder	(10, 32)
LayerNormalization	(10, 32)
Attention	(10, 32)
Dropout	(10, 32)
Lstm	(10, 30)
Dropout	(10, 30)
Lstm	(10, 15)
Dropout	(10, 15)
Lstm	(1, 15)
Dropout	(1, 15)
Output	(1, 3)

4. Experiments

4.1. Experimental Sets

The current field of tool wear prediction has largely been experimentally validated using the IEEE PHM2010 Challenge dataset. The current work utilized the IEEE PHM2010 [28] challenge dataset as experimental data to evaluate the precision of the CIEBM model.

The workpiece is first cut out of the raw material, and the surface of the workpiece is treated by face milling to remove the rough surface containing hard particles. A Kistler cutting force sensor, three-way vibration sensor, and acoustic emission sensor are adopted to acquire cutting force, vibration, and noise signals, respectively [29], as shown in Figure 7. The output of these sensors outputs the corresponding voltage signal through the charge amplifier, which is collected by the NI DAQ PCI 1200 board at a frequency of 50KHz. Acquisition of 7 signals (force_x(N), force_y(N), force_z(N), vibration_x(g), vibration_y(g), vibration_z(g), AE_RMS(v)) is carded. Under dry cutting situations, the surface of the stainless-steel workpiece is machined along the *Z*-axis with a 3-slot alloy milling cutter. Table 2 presents the specific processing parameters. Experiments were performed with three tools (C1, C4, and C6), and 315 experiments were accomplished on each tool. After each experiment, a microscope measured the tool wear.

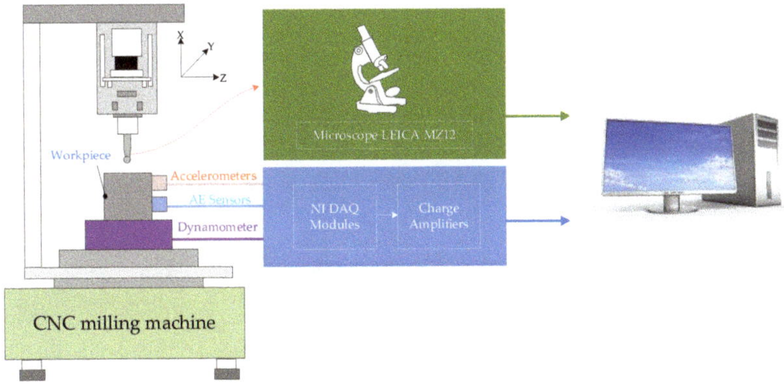

Figure 7. Schematic diagram of the experimental setup.

Table 2. Experimental test parameters.

Parameter	Value
Spindle	10,400 (r/min)
Feed rate	1555 (mm/min)
Depth of cut (y direction, radial)	0.125 (mm)
Depth of cut (z direction, axial)	0.2 (mm)
Sampling rate	50 (kHz)
Workpiece material	Stainless steel (HRC52)

4.2. Data Pre-Processing

Since the experimental process includes feeding and retracting processes, the original signal acquired by the sensor contains some invalid data [30]. Meanwhile, a single time step contains less effective information because the original signal has a high sampling rate. In order to better analyze and monitor the tool wear state, it is necessary to conduct data pre-processing [31]. The data pre-processing comprises the following stages:

Invalid data elimination. Since the cutting process includes feeding and retracting processes, 0.5-s data at the beginning and end should be eliminated to avoid the impact of invalid data on the experiment.

Data segmentation. Since the original data contains less effective information in a single time step, the sensor data of each channel in the original data is divided into five segments on average, and the mean value of each segment is extracted to form a new time series.

Data standardization. Data normalization is performed and transformed through Equation (21) to ensure that the numerical ranges of different features are similar and avoid the excessive influence of specific features on model training.

$$X_{\text{scalled}} = \frac{X-\mu}{\sigma} \tag{21}$$

Dataset division. As presented in Table 3, the tool wear state is categorized into light, moderate, and heavy wear, and one-hot coding is employed to perform the label conversion of the three wear states. The different stages of tool wear are shown in Figure 8. The dataset is divided into training, validation, and test sets. The cross-validation method is adopted for verification. Two datasets are selected from the three for training, and the remaining ones are utilized for verification and evaluation. The ratio of the training dataset to the validation dataset was 9:1. Table 4 records the number of category samples included in c1, c4, and c6.

Table 3. Classification standard for tool wear conditions.

Degree of Wear	Light Wear	Moderate Wear	Heavy Wear
Wear loss (mm)	0–0.12	0.12–0.17	0.17–0.30
One-hot coding	0	1	2

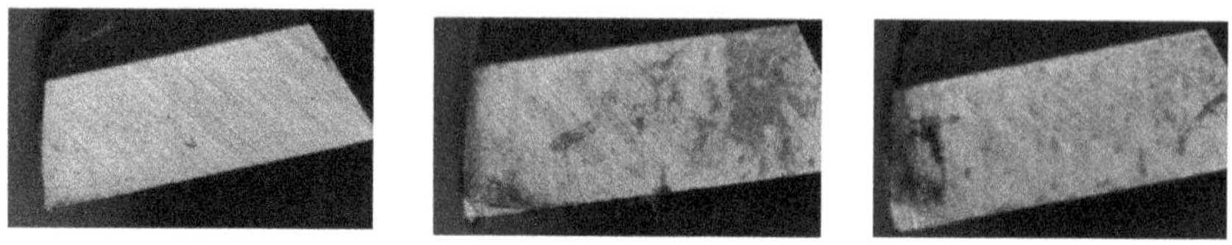

(a) (b) (c)

Figure 8. Diagram of different stages of tool wear: (**a**) light wear; (**b**) moderate wear; (**c**) heavy wear.

Table 4. Number of different types of samples contained in each state.

Tool Number	Category		
	Light Wear	Moderate Wear	Heavy Wear
C1	99	50	146
C4	99	50	146
C6	99	50	146

4.3. Hyperparameter Setting

Hyperparameters immediately influence the model's performance and generalization capability. Different hyperparameter values may result in different complexity and robustness of the model. Hyperparameter selection and optimization is an iterative and experimental process. In this paper, we have tried different combinations of parameters and performed several experiments to finally find the appropriate hyperparameters for the task of tool wear condition monitoring. The best combination of hyperparameters can be found through hyperparameter settings to optimize model performance and improve its generalization ability on new data. Table 5 describes the CIEBM model's hyperparameter settings.

Table 5. Hyperparameter settings of the CIEBM model.

Project	Value
Epoch	150
Batch size	32
Learning rate	0.0001
Dropout	0.2
Objective function	CrossEntropy Loss
Objective function	RMSprop
Activation function	ReLU
Bilstm Stack number	3

4.4. Results

In order to better evaluate the performance of the model, we chose Confusion Matrix, Accuracy, Precision, and Recall to evaluate its performance. The calculations are provided in Equations (22)–(24). As presented in Figure 9, the accuracy of the CIEBM model in dataset identification reaches 99.11% after hyperparameter optimization. The analysis results indicate that the CIEBM model can efficiently detect different wear states of the tool despite the complicated interaction between the tool and the workpiece in the milling process, demonstrating that the CIEBM model has good performance in state recognition.

$$\text{Accuracy} = \frac{TP + TN}{TP + TN + FP + FN} \tag{22}$$

$$\text{Precision} = \frac{TP}{TP + FP} \tag{23}$$

$$\text{Recall} = \frac{TP}{TP + FN} \tag{24}$$

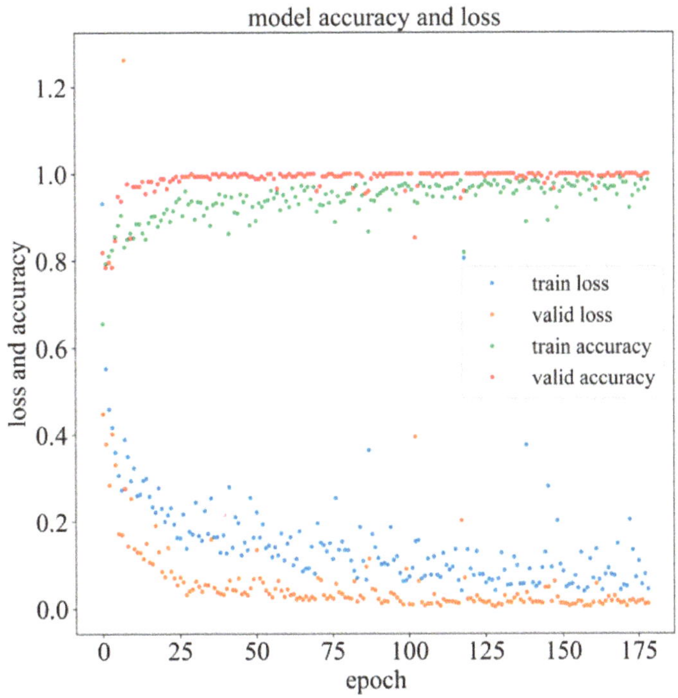

Figure 9. Training loss rate and precision of the CIEBM model.

TP: positive samples are classified as positive samples; *FP*: negative samples are classified as positive samples; *TN*: negative samples are classified as negative samples; *FN*: positive samples are classified as negative samples.

In order to verify the model's precision under different tool datasets, a confusion matrix is utilized to display the classification results. Taking C1 as an example, its confusion matrix is shown in Figure 10. Its horizontal and vertical coordinates are the true and predicted values, respectively. There are four sample classification errors, among which two samples that originally belonged to light wear were wrongly classified as normal wear, one sample that originally belonged to normal wear was wrongly classified as light wear, and one sample that originally belonged to heavy wear was wrongly classified as light wear. As shown in Figures 11 and 12, there are only one and three classification errors in the C4 and C6 datasets, respectively. Light wear has a precision rate of 100% and a recall rate of 97.05%. Moderate wear has a precision rate of 100% and a recall rate of 94%. Heavy wear has a precision rate of 100% and a recall rate of 100%. These metrics show the excellent performance of the CIEBM model. The established model can efficiently extract the features of different tool wear stages and identify and classify the tool wear state.

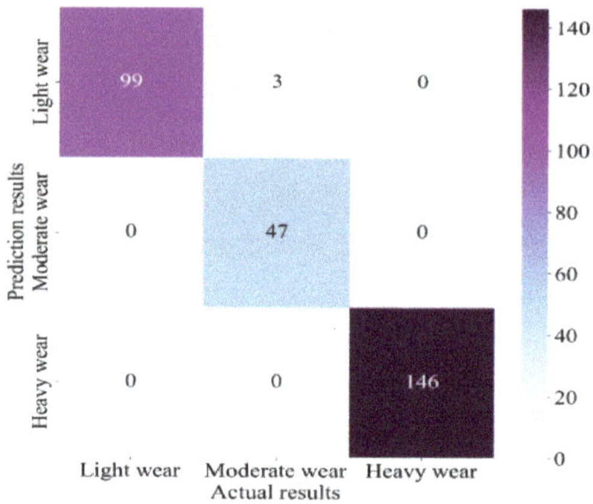

Figure 10. Classification confusion matrix results of the C1 tool wear state.

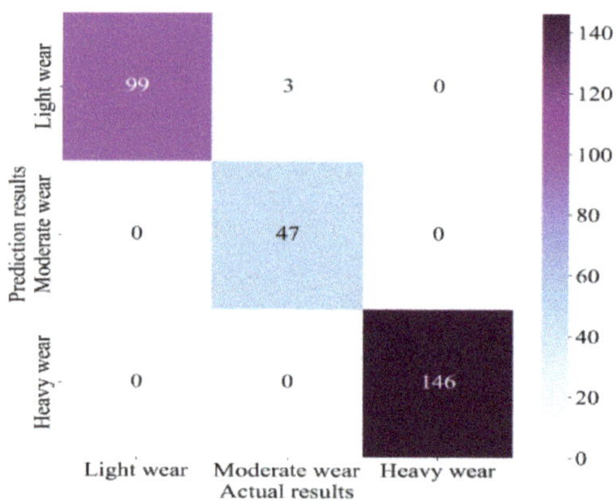

Figure 11. Classification confusion matrix results of the C4 tool wear state.

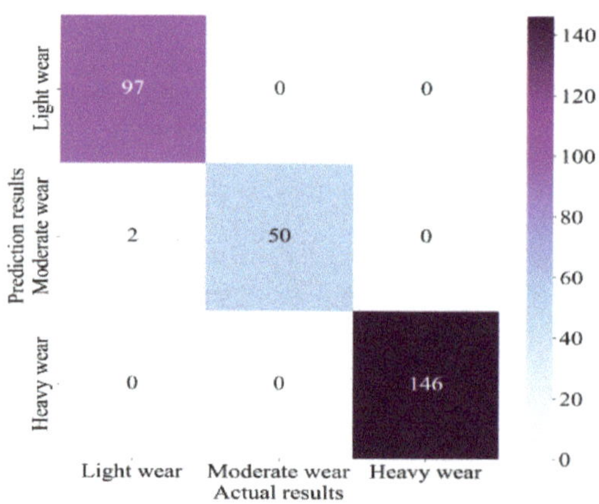

Figure 12. Classification confusion matrix results of the C6 tool wear state.

In order to more intuitively observe each network layer's feature extraction process and demonstrate the ability of the CIEBM model to extract the sensitive features for subsequent state detection effectively, the t-SNE algorithm is utilized for dimension reduction visualization of each network layer, as presented in Figure 13. Figure 13a describes the t-SNE visualization results of the original signal with mixed data and a poor clustering effect. Figure 13b describes the t-SNE visualization results of the CNN layer. The first-type samples have been separated, and there is an aggregation trend among the same type of samples. Figure 13c shows the t-SNE visualization results through the Informer layer. Except for some mixed samples, all samples were classified, and the three types of samples were completely separated. Figure 13d shows the t-SNE visualization results of the Linear layer. All samples are classified, and the clustering effect is obvious. It can be seen that the CIEBM model can effectively identify and classify different tool wear states.

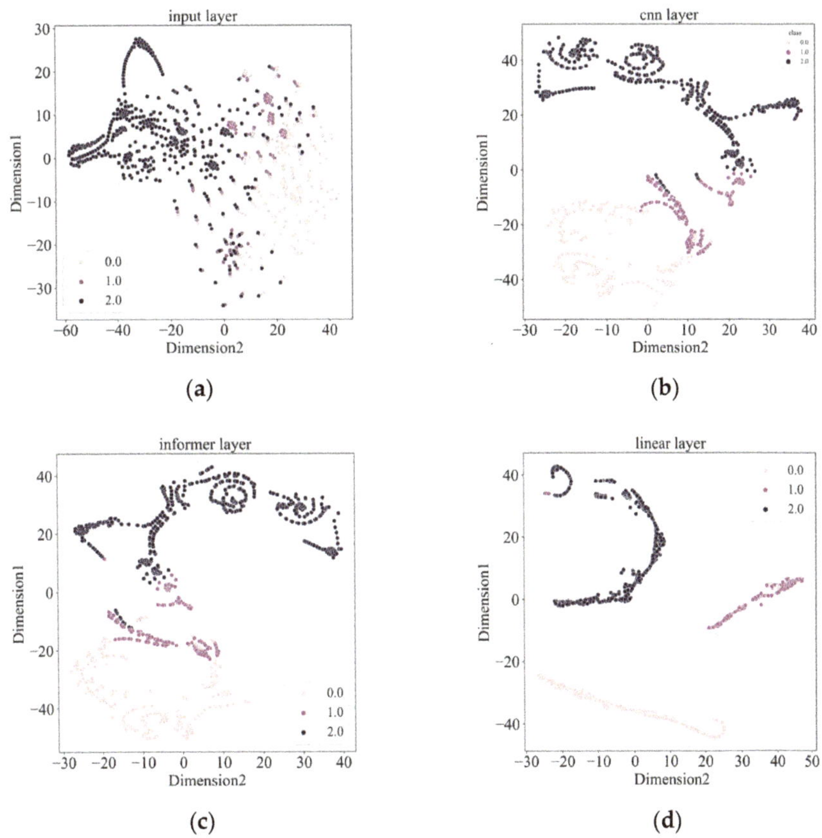

Figure 13. t-SNE visualization results of each network layer of the CIEBM model: (**a**) Input layer visualization results; (**b**) CNN layer visualization results; (**c**) Informer layer visualization results; (**d**) Linear layer visualization results.

4.5. Comparative Analysis

In order to evaluate the advantages of the established CIEBM model, it is compared with the CNN and BiLSTM models in the PHM2010 dataset with the same hyperparameter settings. As shown in Figure 14a,b, the accuracy of the CIEBM model is increased by 17.42% and 2.05% compared with CNN and BiLSTM, respectively. Comparing the convergence rates of different models indicates that the CIEBM model has the maximum convergence rate, indicating that it can extract valuable features from the input data to represent the key information in the data. At the same time, the CIEBM model can learn and adapt the relationship between the features extracted from the training data and the tool wear faster.

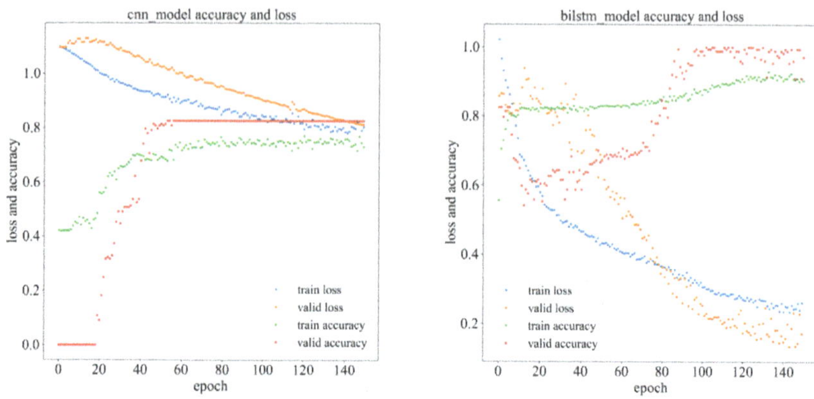

(a) (b)

Figure 14. (**a**) CNN model's loss rate and precision. (**b**) Loss rate and accuracy of the BiLSTM model.

For a more intuitive comparative analysis, Figures 15 and 16 present the confusion matrix of the prediction results of these models. Through the experiment, it can be found that the identification precision of each model in the light and heavy wear is generally higher than that in the moderate wear due to the faster wear change rate of the early wear and heavy wear. As presented in Figure 14, the CNN model is completely wrong in the moderate wear stage with a slower change rate because the CNN is mainly concerned with local patterns and features, and there may be important correlations between each time step in time series data. However, the traditional 1D-CNN mainly focuses on feature extraction from local neighborhoods and cannot utilize global context information. Additionally, underfitting occurs when using only the CNN model due to the lack of feature representation. As presented in Figure 15, the BiLSTM model performs well in time series and can capture long-term dependence. However, it lacks the ability to use global modeling in tool wear time series, considerably increasing the misclassification in the prediction of moderate and heavy wear, especially in the moderate wear stage. This demonstrates the effectiveness of the CIEBM model with the Informer encoder module for global feature modeling.

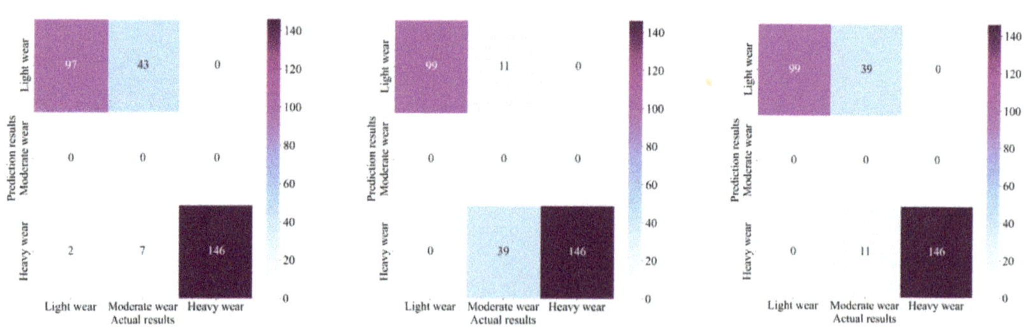

(a) (b) (c)

Figure 15. CNN model confusion matrix results of tool wear status classification: (**a**) C1; (**b**) C4; (**c**) C6.

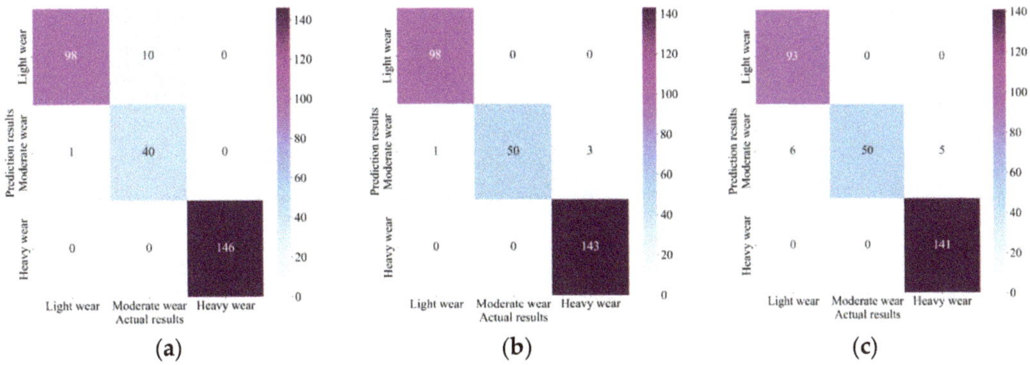

Figure 16. BiLSTM model confusion matrix results of tool wear status classification: (**a**) C1; (**b**) C4; (**c**) C6.

As shown in Figure 17, attention mechanism links are visualized to form attention heat maps to further illustrate the Informer encoder's role, which can understand the key features that the model relies on when making classifications. As revealed from the attention heat map, the Informer encoder can make the CIEBM model better focus on features that are more closely dependent on the tool wear condition monitoring and improve the model's computational performance through the sparse attention mechanism, demonstrating the effectiveness of the introduction of the Informer encoder module.

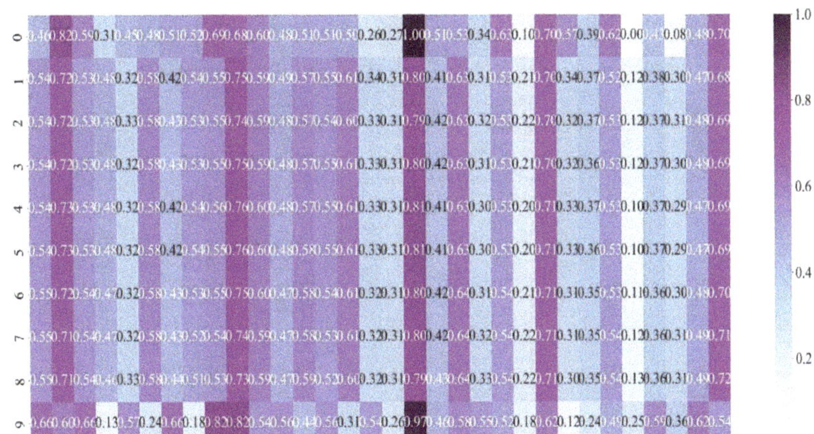

Figure 17. Attention heat map visualization results of the Informer encoder.

5. Conclusions

A tool wear state monitoring approach using CNN, Informer encoder, and BiLSTM was proposed to evaluate its performance on the tool wear state dataset. The experimental results and analysis demonstrate the following results:

(1) Experimental results reveal that the presented TWM approach based on CNN, Informer encoder, and BiLSTM has high accuracy in TWM. All of them reached over 95% in the relevant evaluation indexes, reflecting the excellent performance of the CIEBM model, which can efficiently classify and forecast the tool wear state.

(2) In tool wear monitoring, CNN can extract spatial features from sensor data. Informer encoders can model long-term dependencies and capture global context information with ProbSparse Self-Attention and a feedforward neural network layer.

BiLSTM captures temporal patterns and context information to further improve monitoring accuracy.

(3) Our model is the first to use CNN, an Informer encoder, and BiLSTM together for tool wear condition monitoring, and it is also the first to target global feature modeling based on the non-linearity of the tool wear process to enable the model to better learn the relationship between the features of different wear stages. This is of great importance for further research.

(4) Further analysis shows that our method has an excellent classification impact on normal and different degrees of wear, and the confusion between normal and heavy wear is slight, indicating that the method can effectively distinguish tool states with different degrees of wear.

In summary, the tool wear state monitoring approach using CNN, Informer encoder, and BiLSTM performed well in the experiment. This method has significant application value for TWM in the industrial field. Nevertheless, many details still need to be improved, such as further optimization of the model architecture, hyperparameter adjustment, and dataset size expansion, to enhance the monitoring's precision and robustness.

In future work, the method of combining physical models of tool wear with deep learning will be further investigated. By modeling the physical model, the interpretability of deep learning will be further improved while providing a theoretical basis for optimizing the deep learning network model for the production scenario of tool wear. In the next phase, we will continue to conduct field experiments to study wear under variable working conditions to further improve the generalization ability of the model.

Author Contributions: This article has the help of all the authors. X.X. provided initial ideas and defined research objectives, designed experiments and selected appropriate algorithms, developed codes for the BiLSTM model and performed initial tests, verified experimental results, guaranteed the accuracy of experimental results, and wrote the first draft of the manuscript. W.S. collected, cleaned, and organized the dataset used in the study. Reviewed and edited the manuscript, improving its clarity and coherence. Y.L. (Yingming Li) and Y.L. (Yue Liu) reviewed and edited the manuscript, improving its clarity and coherence. M.H. oversaw the research project, providing guidance and feedback. Secured funding for the research. All authors have read and agreed to the published version of the manuscript.

Funding: The research was funded by the Ministry of Industry and Information Technology's High-end Numerical Control Systems and Servo Motors Project (Grant No. ZTZB-22-009-001).

Data Availability Statement: All data are contained within the article.

Conflicts of Interest: The authors declare no conflict of interest.

Nomenclature

$y[t]$	Calculated output
b	Shift factor
$w[i]$	Weighting coefficient
$x[]$	Sequence input value
$f(x)$	Activation function output
W	Weight matrix
X	Model input
Q/q	Query vector
K/k	Key vector
V/v	Value vector
L	Number of vectors
d	Length of vector
$M()$	Attention score
f_t	Forget gate output
C_{t-1}	Previous cell state

i_t	Input gate output
$\widetilde{C}_t$	Candidate
C_t	New cell state
o_t	Output gate output
h_t	Hidden state
TP	True positive
FN	False negative
FP	False positive
TN	True negative

References

1. Zhang, C.; Wang, W.; Li, H. Tool wear prediction method based on symmetrized dot pattern and multi-covariance Gaussian process regression. *Measurement* **2022**, *189*, 110466. [CrossRef]
2. Shi, D.; Gindy, N.N. Tool wear predictive model based on least squares support vector machines. *Mech. Syst. Signal Process* **2007**, *21*, 1799–1814. [CrossRef]
3. Gomes, M.C.; Brito, L.C.; da Silva, M.B.; Duarte, M.A.V. Tool wear monitoring in micromilling using Support Vector Machine with vibration and sound sensors. *Precis. Eng.* **2021**, *67*, 137–151. [CrossRef]
4. Cheng, Y.; Gai, X.; Jin, Y.; Guan, R.; Lu, M.; Ding, Y. A new method based on a WOA-optimized support vector machine to predict the tool wear. *Int. J. Adv. Manuf. Technol.* **2022**, *121*, 6439–6452. [CrossRef]
5. Gai, X.; Cheng, Y.; Guan, R.; Jin, Y.; Lu, M. Tool wear state recognition based on WOA-SVM with statistical feature fusion of multi-signal singularity. *Int. J. Adv. Manuf. Technol.* **2022**, *123*, 2209–2225. [CrossRef]
6. Xu, D.; Qiu, H.; Gao, L.; Yang, Z.; Wang, D. A novel dual-stream self-attention neural network for remaining useful life estimation of mechanical systems. *Reliab. Eng. Syst. Safe* **2022**, *222*, 108444. [CrossRef]
7. Liu, B.; Li, H.; Ou, J.; Wang, Z.; Sun, W. Intelligent recognition of milling tool wear status based on variational auto-encoder and extreme learning machine. *Int. J. Adv. Manuf. Technol.* **2022**, *119*, 4109–4123. [CrossRef]
8. Liu, Z.; Hao, K.; Geng, X.; Zou, Z.; Shi, Z. Dual-Branched Spatio-Temporal Fusion Network for Multihorizon Tropical Cyclone Track Forecast. *IEEE J. Sel. Top. Appl. Earth Obs. Remote Sens.* **2022**, *15*, 3842–3852. [CrossRef]
9. Dai, W.; Liang, K.; Wang, B. State Monitoring Method for Tool Wear in Aerospace Manufacturing Processes Based on a Convolutional Neural Network (CNN). *Aerospace* **2021**, *8*, 335. [CrossRef]
10. García-Pérez, A.; Ziegenbein, A.; Schmidt, E.; Shamsafar, F.; Fernández-Valdivielso, A.; Llorente-Rodríguez, R.; Weigold, M. CNN-based in situ tool wear detection: A study on model training and data augmentation in turning inserts. *J. Manuf. Syst.* **2023**, *68*, 85–98. [CrossRef]
11. Kothuru, A.; Nooka, S.P.; Liu, R. Application of deep visualization in CNN-based tool condition monitoring for end milling. *Procedia Manuf.* **2019**, *34*, 995–1004. [CrossRef]
12. Wu, X.; Liu, Y.; Zhou, X.; Mou, A. Automatic Identification of Tool Wear Based on Convolutional Neural Network in Face Milling Process. *Sensors* **2019**, *19*, 3817. [CrossRef] [PubMed]
13. Qin, X.; Zhang, W.; Gao, S.; He, X.; Lu, J. Sensor Fault Diagnosis of Autonomous Underwater Vehicle Based on LSTM. In Proceedings of the 2018 37th Chinese Control Conference (CCC), Wuhan, China, 25–27 July 2018; pp. 6067–6072. [CrossRef]
14. Xu, W.; Miao, H.; Zhao, Z.; Liu, J.; Sun, C.; Yan, R. Multi-Scale Convolutional Gated Recurrent Unit Networks for Tool Wear Prediction in Smart Manufacturing. *Chin. J. Mech. Eng.* **2021**, *34*, 53. [CrossRef]
15. Liu, X.; Zhang, B.; Li, X.; Liu, S.; Yue, C.; Liang, S.Y. An approach for tool wear prediction using customized DenseNet and GRU integrated model based on multi-sensor feature fusion. *J. Intell. Manuf.* **2023**, *34*, 885–902. [CrossRef]
16. Cheng, M.; Jiao, L.; Yan, P.; Jiang, H.; Wang, R.; Qiu, T.; Wang, X. Intelligent tool wear monitoring and multi-step prediction based on deep learning model. *J. Manuf. Syst.* **2022**, *62*, 286–300. [CrossRef]
17. Liu, H.; Liu, Z.; Jia, W.; Zhang, D.; Wang, Q.; Tan, J. Tool wear estimation using a CNN-transformer model with semi-supervised learning. *Meas. Sci. Technol.* **2021**, *32*, 125010. [CrossRef]
18. Liu, H.; Liu, Z.; Jia, W.; Lin, X.; Zhang, S. A novel transformer-based neural network model for tool wear estimation. *Meas. Sci. Technol.* **2020**, *31*, 065106. [CrossRef]
19. Lin, H.; Sun, Q. Financial Volatility Forecasting: A Sparse Multi-Head Attention Neural Network. *Information* **2021**, *12*, 419. [CrossRef]
20. Zou, R.; Duan, Y.; Wang, Y.; Pang, J.; Liu, F.; Sheikh, S.R. A novel convolutional informer network for deterministic and probabilistic state-of-charge estimation of lithium-ion batteries. *J. Energy Storage* **2023**, *57*, 106298. [CrossRef]
21. Vaswani, A.; Shazeer, N.; Parmar, N.; Uszkoreit, J.; Jones, L.; Gomez, A.N.; Kaiser, Ł.; Polosukhin, I. Attention is all you need. Advances in Neural Information Processing Systems 30 (NIPS 2017). In Proceedings of the 31st International Conference on Neural Information Processing Systems, Long Beach, CA, USA, 4–9 December 2017; pp. 6000–6010. [CrossRef]
22. Zhou, H.; Zhang, S.; Peng, J.; Zhang, S.; Zhang, W. In Informer: Beyond efficient transformer for long sequence time-series forecasting. In Proceedings of the AAAI Conference on Artificial Intelligence, Virtual, 2–9 February 2021. [CrossRef]
23. Li, R.; Ye, X.; Yang, F.; Du, K.L. ConvLSTM-Att: An Attention-Based Composite Deep Neural Network for Tool Wear Prediction. *Machines* **2023**, *11*, 297. [CrossRef]

24. Xie, X.; Huang, M.; Liu, Y.; An, Q. Intelligent Tool-Wear Prediction Based on Informer Encoder and Bi-Directional Long Short-Term Memory. *Machines* **2023**, *11*, 94. [CrossRef]
25. Li, W.; Liang, Y.; Wang, S. *Data Driven Smart Manufacturing Technologies and Applications*; Springer: Berlin/Heidelberg, Germany, 2021. [CrossRef]
26. Sun, M.; Liu, Z.; Zhang, M.; Liu, Y. Chinese computational linguistics and natural language processing based on naturally annotated big data. In Proceedings of the S14th China National Conference, CCL 2015 and Third International Symposium, NLP-NABD 2015, Guangzhou, China, 13–14 November 2015. [CrossRef]
27. Cheng, Y.; Gai, X.; Guan, R.; Jin, Y.; Lu, M.; Ding, Y. Tool wear intelligent monitoring techniques in cutting: A review. *J. Mech. Sci. Technol.* **2023**, *37*, 289–303. [CrossRef]
28. 2010 phm Society Conference Data Challenge 2010. Available online: https://phmsociety.org/competition/phm/10 (accessed on 6 September 2023).
29. Huang, Q.; Wu, D.; Huang, H.; Zhang, Y.; Han, Y. Tool Wear Prediction Based on a Multi-Scale Convolutional Neural Network with Attention Fusion. *Information* **2022**, *13*, 504. [CrossRef]
30. Du, M.; Wang, P.; Wang, J.; Cheng, Z.; Wang, S. Intelligent Turning Tool Monitoring with Neural Network Adaptive Learning. *Complexity* **2019**, *2019*, 8431784. [CrossRef]
31. Bergs, T.; Holst, C.; Gupta, P.; Augspurger, T. Digital image processing with deep learning for automated cutting tool wear detection. *Procedia Manuf.* **2020**, *48*, 947–958. [CrossRef]

lubricants

Article

Rolling Bearing Fault Diagnosis across Operating Conditions Based on Unsupervised Domain Adaptation

Zhidan Zhong [1,*], Hao Liu [1], Wentao Mao [2], Xinghui Xie [3] and Yunhao Cui [1]

[1] College of Mechanical and Electrical Engineering, Henan University of Science and Technology, Luoyang 471023, China; liuhao3070@163.com (H.L.); 9906412@haust.edu.cn (Y.C.)
[2] School of Computer and Information Engineering, Henan Normal University, Xinxiang 453007, China; maowt@htu.edu.cn
[3] State Key Laboratory of Aerospace Precision Bearings, Luoyang 471000, China
* Correspondence: zzd@haust.edu.cn

Abstract: In practical industrial scenarios, mechanical equipment frequently operates within dynamic working conditions. To address the challenge posed by the incongruent data distribution between source and target domains amidst varying operational contexts, particularly in the absence of labels within the target domain, this study presents a solution involving deep feature construction and an unsupervised domain adaptation strategy for rolling bearing fault diagnosis across varying working conditions. The proposed methodology commences by subjecting the original vibration signal of the bearing to a fast Fourier transform (FFT) to extract spectral information. Subsequently, an innovative amalgamation of a one-dimensional convolutional layer and an auto-encoder were introduced to construct a convolutional auto-encoder (CAE) dedicated to acquiring depth features from the spectrum. In a subsequent step, leveraging the depth features gleaned from the convolutional auto-encoder, a balanced distribution adaptation (BDA) mechanism was introduced to facilitate the domain adaptation of features from both the source and target domains. The culminating stage entails the classification of adapted features using the K-nearest neighbor (KNN) algorithm to attain cross-domain diagnosis. Empirical evaluations are conducted on two extensively used datasets. The findings substantiate that the proposed approach is capable of accomplishing the cross-domain fault diagnosis task even without labeled data within the target domain. Furthermore, the diagnostic accuracy and stability of the proposed method surpass those of various other migration and deep learning approaches.

Keywords: convolutional auto-encoder; balanced distribution adaption; domain adaptation; cross-condition fault diagnosis

Citation: Zhong, Z.; Liu, H.; Mao, W.; Xie, X.; Cui, Y. Rolling Bearing Fault Diagnosis across Operating Conditions Based on Unsupervised Domain Adaptation. *Lubricants* **2023**, *11*, 383. https://doi.org/10.3390/lubricants11090383

Received: 3 August 2023
Revised: 26 August 2023
Accepted: 1 September 2023
Published: 8 September 2023

1. Introduction

Rolling bearings stand as vital constituents within rotating machinery and equipment, with extensive applications in domains such as wind power generation, aerospace, and numerous others [1,2]. Due to their regular exposure to elevated temperatures, heavy loads, dynamic operational conditions, and other demanding and intricate work settings, the reliability of rolling bearings is subject to diminishment. A single instance of failure can engender disruptions in regular industrial production and potentially yield considerable economic repercussions, in some cases, even posing a threat to human life [3]. As indicated in references [4–6], bearing failures contribute to 45–90% of the overall error occurrences in certain machinery and equipment. Thus, undertaking research on intelligent fault diagnosis specifically targeting rolling bearings [7] is of paramount importance.

The vibration signal encapsulates a wealth of information pertaining to the bearing's health condition, forming the foundation for the majority of ongoing research [8]. As stipulated by the literature [9], fault diagnosis techniques can be broadly categorized into two groups: model-based approaches and data-driven methods. Model-based fault diagnosis methodologies often entail the construction of physical models for characterizing bearing defects [10,11]. Nonetheless, this method heavily relies on a priori knowledge of industrial systems and associated components. The actual operation of industrial systems is marked by uncertainties linked to operational conditions, noise, and other factors, and the assumptions underpinning the model fail to accommodate these sources of uncertainty.

Data-driven methodologies commonly leverage techniques encompassing signal processing, statistical feature extraction, and machine learning for fault diagnosis. Singh introduces a fault diagnosis approach grounded in wavelet analysis [12]. Meanwhile, Wang et al. put forth a nonlinear time-frequency flow form learning technique tailored for bearing fault diagnosis [13], its efficacy being substantiated through application to actual bearings. Shazali Osman et al. introduced a novel morphological Hilbert Huang (MH) approach aimed at the early detection of bearing failures [14]. Empirical mode decomposition (EMD) stands as a classical signal processing technique widely applied in the realm of bearing fault diagnosis [15]. Complementing this, variational mode decomposition (VMD) presents a more sophisticated signal-adaptive decomposition methodology [16]. Wang et al. [17] employ VMD in conjunction with cyclic correlation entropy functions for bearing fault diagnosis. Nevertheless, the aforementioned techniques often necessitate the involvement of domain experts during the diagnosis process, a practice prone to intricacies in computation. In related investigations, signal processing serves as the basis for feature extraction, frequently entailing the manual extraction of pertinent attributes encompassing time and frequency domains, as well as signal entropy [18]. Zhang et al. [19] conducted feature extraction from bearing vibration signals, yielding six time and thirteen frequency-domain features tailored for fault diagnosis. Entropy stands as a quantitative tool to delineate the system complexity and has found utility within the domain of vibration signal analysis [20]. Li et al. introduced multiscale dispersion entropy (MDE) for scrutinizing bearing vibration signals and effecting feature extraction [21]. Jiao et al. merged multi-scale sample entropy (MSE) with Energy Moment (EM) for extracting bearing features [22]. Despite attaining a certain degree of accuracy and stability, these methods can only partially accomplish bearing fault diagnosis. Nonetheless, the manual extraction of fault-related features remains an intricate endeavor. This accentuates the pivotal importance of research in automating feature extraction and ensuring the fluid implementation of end-to-end fault diagnosis.

The advancement of artificial intelligence has garnered scholarly attention toward machine learning technology. This genre of technology undertakes data processing to unearth data patterns [23]. Malhi et al. harnessed the principal component analysis (PCA) algorithm to derive bearing signal features, subsequently feeding these features into radial basis function (RBF) networks for fault classification [24]. The support vector machine (SVM) stands as a classical machine learning classifier. Yang et al. utilized intrinsic mode functions (IMF) such as SVM input vectors, with the output indicating bearing failure modes [25]. Pandya et al. accomplished bearing fault diagnosis by relying on acoustic emission signals and a K-nearest neighbor (KNN) classifier. In addition, algorithms such as singular value decomposition (SVD) [26] and the Naive Bayes classifier [27] have also achieved success in bearing fault diagnosis. Notably, while these machine learning-oriented methods operate without human intervention, they are characterized as shallow, possessing limited learning capacities. Moreover, they often struggle to extract high-quality features from intricate, non-linear, and high-dimensional data [22].

Recent years have witnessed the proliferation of deep learning in the realm of fault diagnosis, with deep neural networks displaying formidable prowess in feature extraction [28]. Xia et al. introduced a multi-sensor fusion approach grounded in a convolutional neural network (CNN), proficiently engendering adaptively extracted signal features conducive to end-to-end fault diagnosis [29]. Shao et al. crafted a deep belief network (DBN) to autonomously capture representative features inherent to the original feature set [30]. Sun et al. harnessed a sparse auto-encoder (SAE) to master the features within vibration signals, followed by classification of the extracted features for fault analysis via a deep neural network (DNN) [31]. Kerboua et al. [32] devised a fresh technique for asynchronous motor fault diagnosis, capitalizing on three-dimensional convolutional neural networks. This innovation wields the potential to significantly curtail downtime and optimize production efficiency. Liang et al. [33] formulated a novel approach for diagnosing rolling bearing faults by leveraging the ICEEMDAN Hilbert and ResNet. This method adeptly tackles issues of neutral energy degradation and pivotal feature information loss within deep learning networks.

Deep learning methods possess the capability to autonomously extract fault features, yielding features that effectively capture the essence of the original signal. Nevertheless, the implementation of deep learning methods relies on two key assumptions: (1) the network training data necessitates supervised learning, entailing a substantial volume of labeled data; and (2) both the training and testing data in the source and target domains adhere to an identical distribution. It is crucial to recognize that these two presumptions frequently fall short in real-world scenarios owing to shifts in operational contexts, environmental interference, and assorted external factors. Procuring substantial quantities of labeled data proves highly costly, and data distribution often fluctuates in accordance with working conditions, collectively constraining the advantages of deep learning algorithms. Domain adaptation (DA) emerges as a potent strategy for mitigating the dearth of labeled data and ameliorating imbalanced data distribution, thereby redressing the inherent limitations of deep learning [34,35]. Wei et al. employed transfer component analysis (TCA) to transfer the vibration characteristics of bearings under various operating conditions [36]. Lu et al. utilized maximum mean discrepancy (MMD) for measuring feature distribution deviation across distinct domains, thereby minimizing the gap between the source and target domains [37]. Li et al. amalgamated joint distribution adaptation (JDA) with SVM, employing JDA to assess both the boundary and conditional distributions of data features, while employing SVM as a fault classifier [38]. However, TCA and MMD, for instance, solely focus on the edge distribution of the data. While JDA is capable of adapting both edge and conditional distributions, it neglects the joint contribution of these distributions during adaptation. Balanced distribution adaptation (BDA), proposed by Wang et al., introduces a balancing factor to JDA, enabling dynamic adjustment of the significance of both edge and conditional distributions [39].

In this study, we synergize deep learning and domain adaptation strategies to introduce a novel model for bearing fault diagnosis that relies on both deep learning and unsupervised domain adaptation principles. This model autonomously captures data features, mitigates the challenge of insufficiently labeled data, and aligns data distributions across domains. Initially, an unsupervised convolutional auto-encoder is established for adaptive data feature extraction across both source and target domains. Subsequently, the introduction of balanced distribution adaptation (BDA) serves to minimize the separation between data in the source and target domains. Ultimately, the K-nearest neighbor (KNN) algorithm, trained on the source domain, is employed for the classification of data in the target domain. The primary contributions of this study are summarized below:

(1) We propose a framework that leverages deep learning and unsupervised domain adaptation to address variable working condition problems. This framework autonomously extracts features from raw data, harnesses the advantages of both deep learning and domain adaptation, and facilitates fault diagnosis;

(2) We introduce the BDA algorithm to handle the feature distance metric and migration. This facilitates domain adaptation by balancing the edge distribution and conditional distribution of the data;

(3) We employ TSNE technology to visualize every stage of feature extraction and migration. This approach elucidates the detailed patterns of data transformation and offers interpretability for both the migration and deep learning algorithms.

2. Theoretical Foundation

2.1. Fast Fourier Transform (FFT)

FFT [40] represents an engineered realization of the discrete Fourier transform (DFT), streamlining the computational procedure. Employing the FFT transformation in the time-domain vibration signal of the bearing facilitates the acquisition of frequency domain particulars. Unique fault patterns are encapsulated within distinct frequency bands, rendering the frequency domain signal more adept for fault diagnosis compared to the original vibration signal [41]. Zhang et al. conducted an FFT transformation on vibration signals and derived features from the resultant spectral signals to facilitate bearing fault diagnosis [42]. On the other hand, Mao et al. applied an FFT transformation to the bearing vibration signal, and the ensuing spectral information underwent processing via a generative adversarial network (GAN) [43], which was employed for the early detection of bearing failures.

For a finite length bearing time-domain vibration discrete signal $x(n)$, the spectral function $X(k)$ can be obtained via DFT:

$$X(k) = \sum_{n=0}^{N-1} x(n)\omega_N^{kn}, 0 \leq k \leq N-1 \tag{1}$$

where, N is the sample length, $\omega_N = e^{-j\frac{2\pi}{N}}$.

FFT decomposes the sequence $x(n)$ into two parts, an even sequence $x_2(n)$ and an odd sequence $x_1(n)$, each with the length $\frac{N}{2}$. Thus, we obtain

$$X(k) = \sum_{n=0}^{\frac{N}{2}-1} x_1(n)\omega_N^{2kn} + \sum_{n=0}^{\frac{N}{2}} x_2(n)\omega_N^{(2k+1)n} \tag{2}$$

Extracting the factorization of Equation (2) yields

$$X(k) = \sum_{n=0}^{\frac{N}{2}-1} x_1(n)\omega_N^{2kn} + \omega_N^{kn}\sum_{n=0}^{\frac{N}{2}} x_2(n)\omega_N^{2kn} \tag{3}$$

Since $\omega_N^{2k} = e^{-j\frac{2\pi}{N}2kn} = e^{-j\frac{2\pi}{N/2}kn} = \omega_{N/2}^{kn}$, integrating it into Equation (3) yields

$$\begin{aligned} X(k) &= \sum_{n=0}^{\frac{N}{2}-1} x_1(n)\omega_{N/2}^{kn} + \omega_N^{kn}\sum_{n=0}^{\frac{N}{2}-1} x_2(n)\omega_{N/2}^{kn} \\ &= X_1(k) + \omega_N^{k}X_2(k) \end{aligned} \tag{4}$$

where, $X_1(k)$ is the result of the odd sequence $x_1(n)$ and $X_2(k)$ is the result of the even sequence $x_2(n)$.

2.2. Autoencoder (AE)

AE [44], an established unsupervised learning algorithm, undertakes the task of reconstructing input data into outputs. The utilization of AE and its adaptations has found extensive application within the realm of bearing analysis. Mao et al. employed a stacked denoising auto-encoder to extract shared features among various health states of bearings [45]. Similarly, Jing et al. employed AE to capture spectral features from the signal and coupled it with a Gaussian mixture model for clustering [46]. A representative self-coding structure is illustrated in Figure 1.

AE typically comprises two components: an encoder and a decoder. The encoder is capable of reducing the dimensionality of the input signal and extracting features, while the decoder reconstructs the input signal by employing the extracted features as the input.

If the input to the encoder is assumed, and the resulting feature vector is obtained through encoding, then

$$h = f(\omega X + b) \tag{5}$$

where, f is the activation function, ω is the weight of the network, and b is the network bias.

In the decoding stage, the input of the decoder is h and the output is $\hat{X}$:

$$\hat{X} = f(\omega' h + b') \tag{6}$$

where, f is the activation function, ω is the weight of the network, and b is the network bias.

The auto-encoder network updates the internal parameters by minimizing the reconfiguration error, where l_{AE} is

$$l_{AE} = |X - \hat{X}|^2 \tag{7}$$

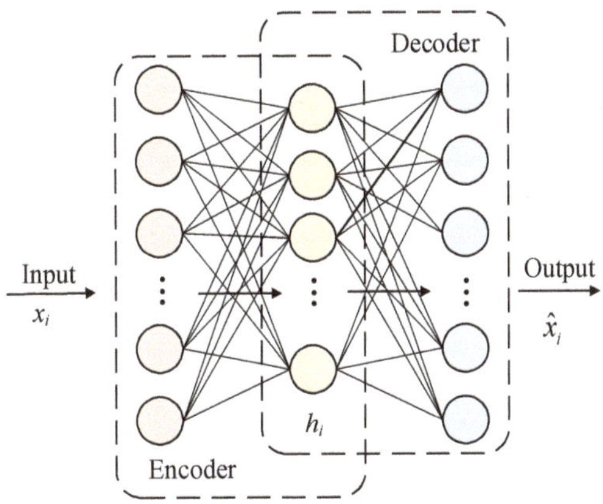

Figure 1. Structure of the AE.

2.3. Convolutional Neural Network (CNN)

CNN [47], a significant component in modern deep learning, is distinguished by its attributes of parameter sharing and translation invariance. These attributes facilitate the extraction of robust features and have contributed to its widespread use in the context of diagnosing faults in rolling bearings [48]. The structure of a one-dimensional convolutional network primarily comprises a one-dimensional convolutional layer, a one-dimensional pooling layer, a fully connected layer, and a classifier, as illustrated in Figure 2.

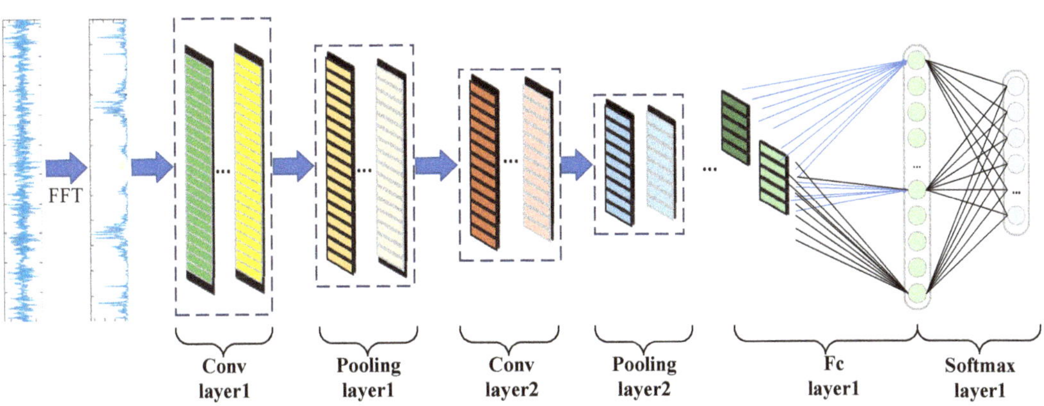

Figure 2. Structure of the CNN.

The convolution layer is composed of a convolutional kernel that executes convolution operations on the input signal and employs a nonlinear activation function to construct features. The resulting output is

$$y_i^l = \sum_{j=1}^{N} x_j^l \otimes k_{ji}^l + b_i^l \tag{8}$$

where, y_i^l represents the jth convolutional computed vector of the lth layer, N is the number of input feature vectors, x_j^l is the jth input feature vector of the lth layer, $\otimes$ represents the convolutional computation, k_{ji}^l is the convolutional kernel of the lth layer with the jth input feature vector, and b_i^l is the ith bias vector of the lth layer.

The output features typically undergo transformation through a nonlinear activation function subsequent to the convolution operation. In this study, the rectified linear unit (ReLU) was employed as the activation function. ReLU is formally defined as

$$a_i^l(j) = f(y_i^l(j)) = \max(0, y_i^l(j)) = \begin{cases} y_i^l(j) & y_i^l(j) \geq 0 \\ 0 & y_i^l(j) < 0 \end{cases} \tag{9}$$

where, $y_i^l(j)$ is the jth output after the ith convolution operation of the lth layer and $a_i^l(j)$ is the activation value of $y_i^l(j)$.

Following ReLU activation, the feature vector is commonly subjected to dimensionality reduction through the utilization of a maximum pooling layer. This operation is computed as

$$p_i^{l+1}(j) = \max_{(j-1)w+1 \leq t \leq jw} \left\{ a_i^l(t) \right\} \tag{10}$$

The fully connected layer can expand the output of the pooling layer to form a one-dimensional feature vector with the activation function ReLU; the softmax classification layer can perform the final multiclassification operation, assuming that the label is $y \in \{1, 2, \ldots, K\}$, and given sample x, its probability of belonging to category k is

$$p(y = k|x; \Theta) = softmax(\theta_k^T x) = \frac{\exp(\theta_k^T x)}{\sum_{i=1}^{K} \exp(\theta_i^T x)} \tag{11}$$

where Θ is all the training parameters in the softmax regression model and $1/\sum_{i=1}^{K} \exp(\theta_i^T x)$ is the normalization function.

2.4. Balanced Distribution Adaptation (BDA)

A labeled source domain space was set as $D_S = \{x_{si}, y_{si}\}_{i=1}^{n}$, and an unlabeled target domain space $D_t = \{x_{tj}\}_{j=1}^{n}$. The same feature space was assumed, but the different edge and conditional distributions were $p(x_s) \neq p(x_t)$ and $p(y_s|x_s) \neq p(y_t|x_t)$.

BDA is a domain adaptive method that adaptively weighs the importance of the edge distribution and conditional distribution between domains by introducing a balancing factor, μ, that minimizes the following distances.

$$D(D_S, D_t) \approx (1 - \mu)(P(x_s), P(x_t)) + \mu(P(y_s|x_s), P(y_t|x_t)) \tag{12}$$

where, $\mu \in [0, 1]$ is a balance factor measuring the importance of the edge distribution and the conditional distribution. When μ is close to 0, the edge distribution is given priority, and when it is close to 1, the conditional distribution is given priority.

It is important to note that since the target domain does not have sample labels, it is not possible to calculate the conditional distribution $p(y_t|x_t)$. Here, the class conditional distribution $p(x_t|y_t)$ is used instead of $p(y_t|x_t)$. The classifier is trained in the target domain and used to predict the pseudo label in the target domain, thus calculating $p(x_t|y_t)$. This process is continuously iterated to improve reliability [46]. The difference between the two distributions was estimated using the maximum mean discrepancy (MMD) Equation (12), which was reduced to

$$\begin{aligned} D(D_S, D_t) \approx (1-\mu)\left\|\frac{1}{n}\sum_{i=1}^{n} x_{si} - \frac{1}{m}\sum_{j=1}^{m} x_{tj}\right\|_{\kappa}^{2} + \\ \mu \sum_{c=1}^{C} \left\|\frac{1}{n_c}\sum_{x_{si}\in D_s^{(c)}} x_{si} - \frac{1}{m_c}\sum_{x_{tj}\in D_t^{(c)}} x_{tj}\right\|_{\kappa}^{2} \end{aligned} \tag{13}$$

where κ is the regenerative kernel Hilbert space, n is the number of samples in the source domain, m is the number of samples in the target domain, n_c is the number of c-class samples in the source domain, and m_c is the number of c-class samples in the target domain.

Continuing to use regularization and matrix transformation techniques, Equation (13) can be formalized as:

$$\begin{aligned} \min \quad & tr\left(A^T X\left((1-\mu)M_0 + \mu\sum_{c=1}^{C} M_c\right)X^T A\right) + \lambda\|A\|_F^2 \\ s.t. \quad & A^T X H X^T A = I, \quad 0 \le \mu \le 1 \end{aligned} \tag{14}$$

where λ is the Frobenius coefficient of the regularization term $\|\cdot\|_F^2$, X is the matrix consisting of x_s and x_t, A denotes the transformation matrix, I is the unit matrix, and H is the central matrix, $H = I - (1/n)$. M_0 and M_c are MMD matrices with the following expressions:

$$(M_0)_{ij} = \begin{cases} \frac{1}{n^2}, & x_i, x_j \in D_s \\ \frac{1}{m^2}, & x_i, x_j \in D_t \\ \frac{1}{-mn}, & otherwise \end{cases} \tag{15}$$

$$(M_c)_{ij} = \begin{cases} \frac{1}{n_c{}^2}, & x_i, x_j \in D_s{}^{(c)} \\ \frac{1}{m_c{}^2}, & x_i, x_j \in D_t{}^{(c)} \\ \frac{1}{-m_c n_c}, & \begin{cases} x_i \in D_s{}^{(c)}, x_j \in D_t{}^{(c)} \\ x_i \in D_t{}^{(c)}, x_j \in D_s{}^{(c)} \end{cases} \\ 0, & otherwise \end{cases} \tag{16}$$

The Lagrangian operator $\phi = (\phi_1, \phi_2, \ldots, \phi_d)$ is introduced and the Lagrangian function of Equations (14) and (15) is obtained by associating the following equation:

$$\begin{aligned} \min \quad & tr\left(A^T X\left((1-\mu)M_0 + \mu\sum_{c=1}^{C} M_c\right)X^T A\right) + \lambda\|A\|_F^2 \\ s.t. \quad & A^T X H X^T A = I, \quad 0 \le \mu \le 1 \end{aligned} \tag{17}$$

Optimization can be viewed as a generalized reformulation problem, such that $\partial L/\partial A = 0$, which yields

$$\left(X\left((1-\mu)M_0 + \mu\sum_{c=1}^{C} M_c\right)X^T + \lambda I\right)A = XHX^T A\phi \tag{18}$$

The optimal mapping transformation matrix A is obtained by solving for it.

3. Proposed Architecture

In order to enhance the precision of bearing fault diagnosis amidst varying operating conditions, this section introduces an unsupervised domain adaptive fault diagnosis framework that relies on autoencoder depth features and balanced distribution adaptation. The architectural layout of the network and core procedural steps are illustrated in

Figures 3 and 4. The detailed parameters governing the convolutional auto-encoder (CAE) network structure are provided in Table 1.

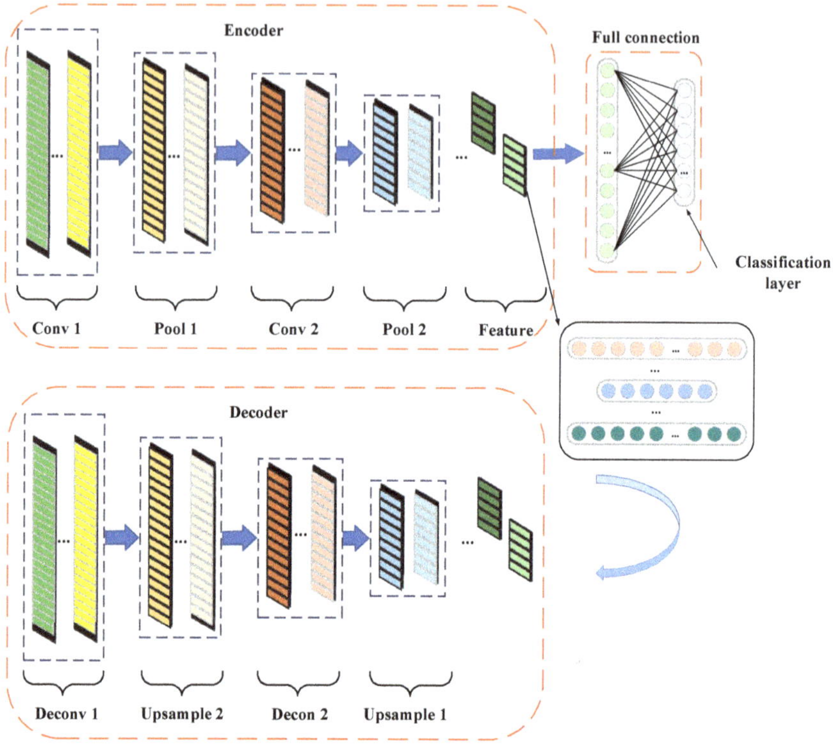

Figure 3. Network structure of the CAE.

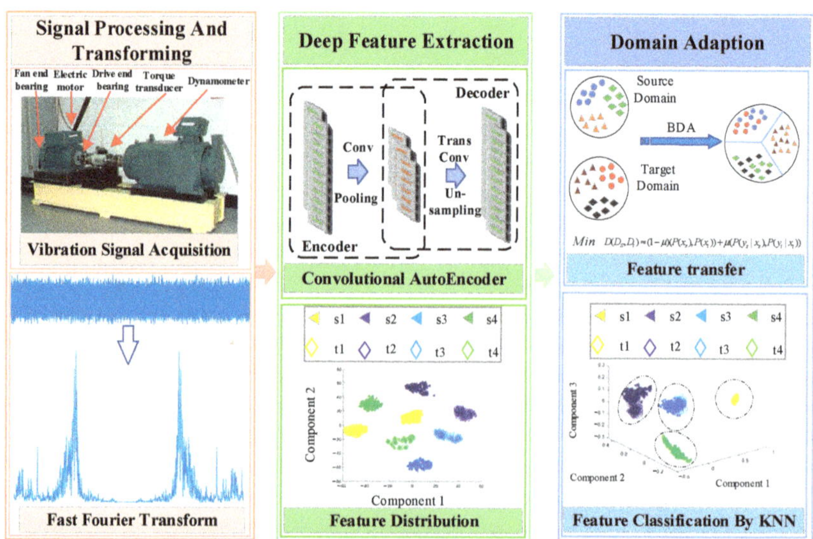

Figure 4. Unsupervised domain adaptation fault diagnosis framework.

Table 1. Detailed configuration of the architecture for the CAE.

Layer	Type	Kernel Size/Stride	Output
Input	Data	/	1 × 1024
Conv1	Convolution1d	32/16	128 × 16
Pool	MaxPool	2/2	64 × 16
Conv2	Convolution1d	3/1	64 × 32
Pool	MaxPool	2/2	32 × 32
Conv3	Convolution1d	3/1	32 × 64
Pool	MaxPool	2/2	16 × 64
Conv4	Convolution1d	3/1	16 × 64
Pool	MaxPool	2/2	8 × 64
Conv5	Convolution1d	3/1	6 × 64
Pool	MaxPool	2/2	3 × 64
Upsample	MaxUnpool	2/2	6 × 64
Deconv1	ConvTranspose1d	3/1	8 × 64
Upsample	MaxUnpool	2/2	16 × 64
Deconv2	ConvTranspose1d	3/1	16 × 64
Upsample	MaxUnpool	2/2	32 × 64
Deconv3	ConvTranspose1d	3/1	32 × 32
Upsample	MaxUnpool	2/2	64 × 32
Deconv4	ConvTranspose1d	3/1	64 × 16
Deconv5	ConvTranspose1d	3/1	1 × 1024

As depicted in Table 1, the initial convolutional layer extracts features directly from the input raw signal without undergoing additional transformations. The principal distinction between the comprehensive architecture of the devised CAE model and that of conventional CNN models resides at the filter level. Specifically, adopting a larger convolutional kernel in the inaugural layer is more adept at attenuating high-frequency noise, while the subsequent convolutional kernels are comparatively diminutive. The incorporation of multiple layers with smaller convolutional kernels enhances the network's depth, consequently facilitating the acquisition of a robust representation of the input signal and expediting the training procedure.

The process is primarily divided into three sequential steps. First, the data undergo a fast Fourier transform to acquire spectral information. Subsequently, convolutional autoencoders are employed to extract depth features from the spectra. Lastly, the BDA algorithm is applied to facilitate domain adaptation. The specific procedures are elucidated as follows:

Step 1: Signal acquisition and preprocessing. Utilize acceleration sensors to capture raw vibration signals from rolling bearings under different operational conditions and fault types. Then, preprocess the signal by selecting 1024 original vibration data points as samples. Apply the fast Fourier transform to obtain the bilateral spectrum, and subsequently, perform z-score standardization on the spectrum.

Step 2: Deep feature extraction and migration. Construct a convolutional autoencoder model and employ FFT spectra as inputs for unsupervised deep feature extraction. Throughout the migration procedure, the encoder architecture of the convolutional autoencoder is capable of generating profound features. The BDA algorithm is then applied to modify the edge and condition distributions of features from the source and target domains, thereby achieving feature migration.

Step 3: Fault classification. The migrated features are shared, and the classification task employs the K-nearest neighbors (KNN) classifier.

4. Experiments and Analysis

In order to showcase the efficacy of the proposed approach, this section opts for the utilization of two extensively employed publicly accessible datasets for conducting cross-service fault diagnosis experiments (CWRU and SEU datasets).

4.1. Variable Load Dataset from the CWRU

4.1.1. Data Description

The CWRU bearing dataset was sourced from the Electrical Engineering Laboratory at the Case Western Reserve University, with the experimental setup as depicted in Figure 5.

Figure 5. CWRU experimental platform.

Single-point damage was inflicted on the bearing arrangement using EDM, resulting in failure diameters of 0.007, 0.014, 0.021, and 0.028 inches, along with distinct failure types: outer ring failure, inner ring failure, and rolling element failure. Vibration data of the bearing were captured utilizing accelerometers across various loads (0 HP, 1 HP, 2 HP, and 3 HP).

This section harnesses the fault data acquired at the drive end (DE), employing a sampling frequency of 12 kHz. The differentiation between source and target domains hinges on three key factors: load, fault diameter, and load. The dataset characterized by a load of 1 HP and a fault diameter of 0.007 inch was designated as the source domain signal, while the dataset marked by a load of 3 HP and a fault diameter of 0.014 inch was earmarked as the target domain signal. The captured data are tabulated in Table 2.

Table 2. Category information about the CWRU dataset.

Class	Fault	Damage Diameter (inch)	Load (HP)
Source1	Normal	0.007	1
Source2	Ball	0.007	1
Source3	Inner	0.007	1
Source4	Outer	0.007	1
Target1	Normal	0.014	3
Target2	Ball	0.014	3
Target3	Inner	0.014	3
Target4	Outer	0.014	3

4.1.2. Signal Pre-Processing

Effective preprocessing of vibration signals enhances the extraction of insightful features via deep networks. Certain researchers have employed the Fast Fourier Transform (FFT) on the raw signals, obtaining their spectra as inputs for deep networks aimed at alleviating the impact of noise [7,49,50]. Figure 6 illustrates time and frequency domain waveforms of bearing vibration data under distinct health states: (1) in the normal state, data predominantly exhibit smooth random waveforms characterized by minor amplitude fluctuations; (2) as indicated by Figure 6, noticeable periodic shocks and irregularities manifest in the data related to inner and outer ring failures, while such attributes are less conspicuous in instances of rolling element failures. Examination of the frequency domain waveforms yields the subsequent findings: (1) normal data demonstrate pronounced energy concentration primarily within the low and mid-frequency bands (below 1 kHz);

(2) pertaining to data from other fault states, a distinct resemblance in the amplitude variation pattern is discernible, characterized by a dominant energy concentration in the high frequency range (2, 4 kHz).

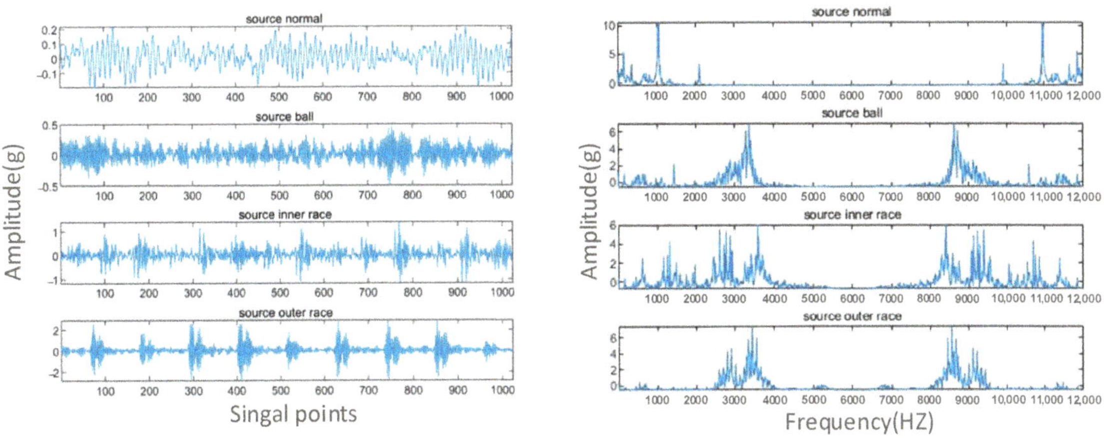

Figure 6. Illustration of vibration signals of the CWRU data set.

4.1.3. Signal Reconstruction and Feature Extraction

The convolutional autoencoder reconstructs input data by acquiring a condensed representation and subsequently generating a reconstruction using this representation. The primary objective of data reconstruction lies in distilling the most valuable information from the input while minimizing noise and redundancy. The outcomes, illustrated in Figure 7, demonstrate that the training of the autoencoder is highly efficacious. The reconstructed signal adeptly emulates the original waveform, showcasing precise reconstruction.

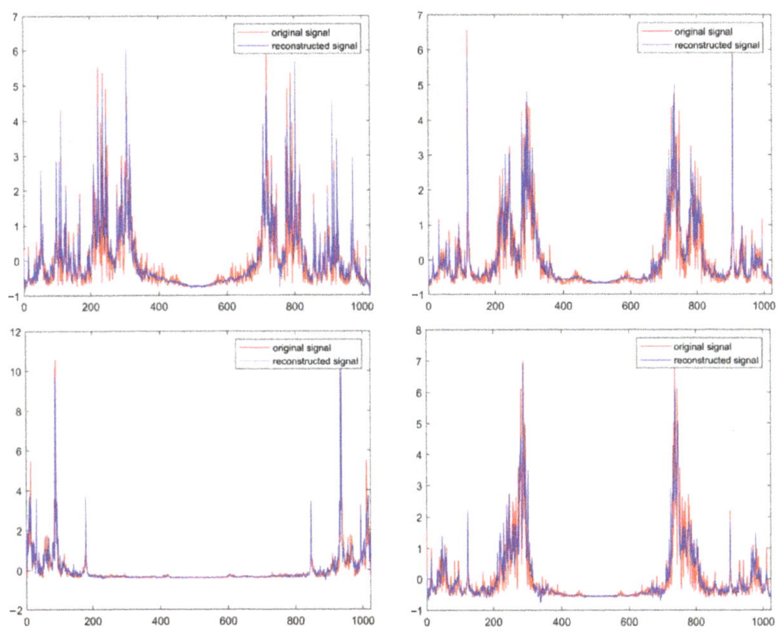

Figure 7. Reconstructed signal and original signal display.

To further elucidate the efficacy of the CAE, we employed the t-sne algorithm to visualize feature distributions. Importantly, in order to underscore the importance of data preprocessing, we fed both the FFT spectrum sample and the original vibration signal sample individually into the CAE. From the encoder, we derived the resultant features and utilized the t-sne technique to diminish their dimensionality and observe their distribution patterns. The original vibration signal is subjected to feature extraction via CAE and subsequently transformed into two- and three-dimensional feature distributions through t-sne, as illustrated in Figure 8.

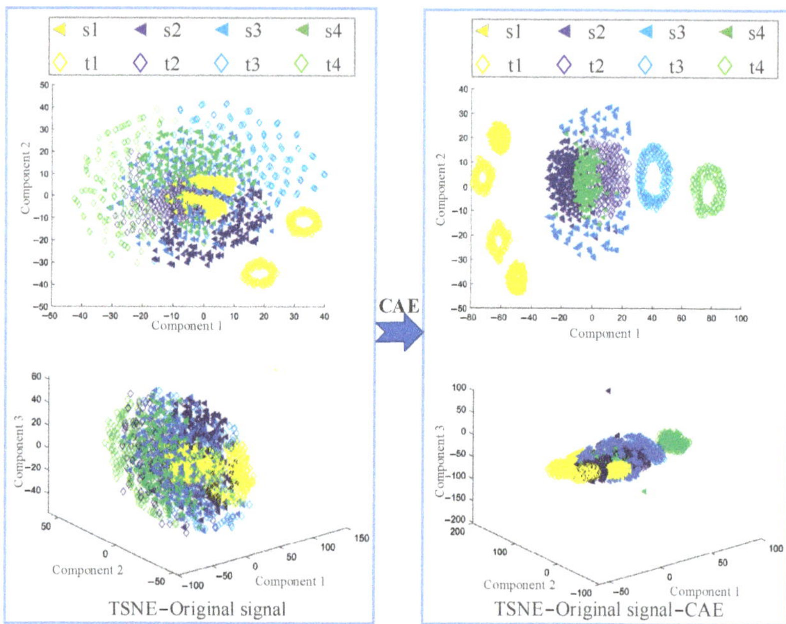

Figure 8. Variation of feature distribution in raw CWRU datasets.

As depicted in the figure, the feature distribution of the original signal (TSNE-Original signal) appears disordered, with various fault types intermingled and lacking distinguishability. This observation implies that the original data fail to elucidate the latent features associated with distinct labels. Following the CAE-based learning of implied features and subsequent dimensionality reduction (TSNE-Original signal-CAE), the features related to different fault types demonstrated an initial level of differentiation. Nonetheless, the distinction between diverse labels remained somewhat indistinct. This suggests that the original vibration signal, subsequent to CAE learning, can manifest the latent features corresponding to different labels to some extent.

The FFT spectrum samples were input into a convolutional auto-encoder (CAE), which employs its encoder to generate latent features. These derived features subsequently undergo t-SNE-based dimensionality reduction, yielding both two- and three-dimensional visualizations, as depicted in Figure 9. Following the encoding step, feature data that share the same labels displayed prominent resemblances, particularly within the two-dimensional space. This outcome yielded a clustering effect of significantly higher quality than the one displayed in Figure 8. By combining insights from Figures 8 and 9, it becomes evident that the CAE possesses a robust capability for profound feature extraction, thereby uncovering the latent characteristics of the signal. Furthermore, the indispensable nature of FFT preprocessing on the original signal is underscored.

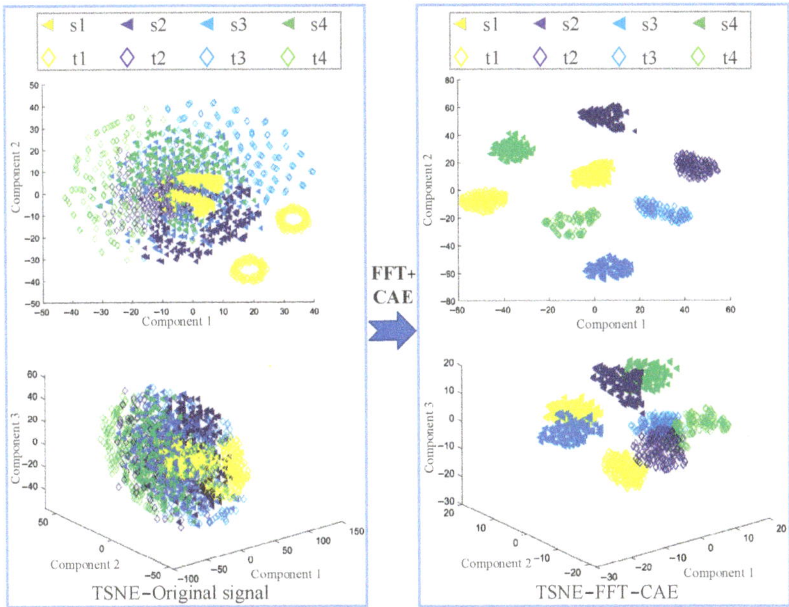

Figure 9. FFT spectral feature distribution variation diagram.

4.1.4. Feature Migration and Analysis

As depicted in Figure 9, diverse fault types within the source domain exhibited a prominent clustering phenomenon, with the two-dimensional visualization particularly accentuating this effect. However, when the same faults manifest across diverse domains, the anticipated clustering patterns did not emerge. This discrepancy poses challenges in effectively diagnosing faults amidst differing operational contexts. Hence, we introduced an unsupervised domain adaptation migration technique grounded in BDA. The objective of this technique is to facilitate the transfer of identical fault characteristics across disparate domains, ultimately augmenting the diagnostic process supported by the classifier.

Source domain: normal and three fault data at 0 HP, fault diameter 0.007 inch $D_S = \{normal, ball, ir, or\}^{0.007}$.

Target domain: normal and three fault data at 3 HP with a fault diameter of 0.014 inches $D_t = \{normal, ball, ir, or\}^{0.014}$.

Task: Migration of untagged data in the target domain using the already tagged source domain data.

The FFT spectrum serves as the input to the CAE, with the encoder generating implied features. Subsequently, the source and target domain features underwent migration through algorithms such as TCA, JDA, CORAL and BDA. These migrated features were then reduced to a three-dimensional space using the t-sne technique, facilitating the observation of clustering effects. The outcomes are presented in Figure 10.

In Figure 10, we present the migration outcomes resulting from the TCA, JDA, CORAL, and BDA migration algorithms, respectively. Contrasting with Figure 8, the TCA algorithm bolstered the clustering efficacy of each label. However, it falls short in effectively bridging the gap between the source and target domains, and its migration outcomes are suboptimal. Comparatively, the JDA algorithm slightly outperformed the TCA algorithm, exhibiting a commendable migration impact on label No. 2. The CORAL algorithm, in contrast, manifested the least successful migration results. The BDA algorithm's migration outcomes, on the other hand, stand out. Under the purview of the BDA algorithm, labels sharing the same category across source and target domains were aptly grouped together, concurrently ensuring distinctiveness among disparate labels. This achievement underscores the method's efficacy in ameliorating data distribution disparities across distinct domains.

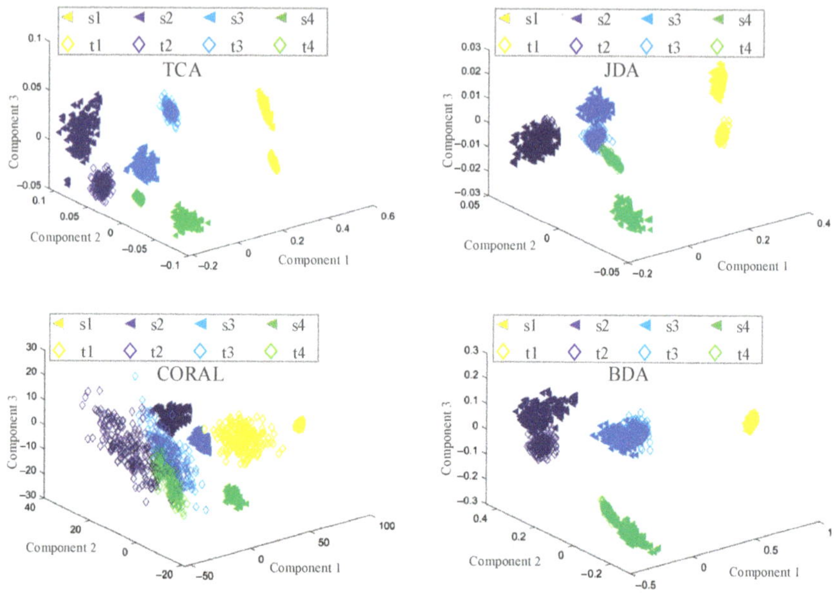

Figure 10. Comparison chart of the effect of different migration algorithms.

4.1.5. Fault Classification and Analysis

In this segment, we enhanced the comparison by integrating various feature types and classification algorithms to underscore the superior attributes of the proposed technique. Initially, FFT spectral features were derived through the CAE. Subsequently, feature migration transpires via TCA, JDA, CORAL, and BDA migration algorithms. Ultimately, the realm of fault classification is navigated through a composite strategy integrating well-established KNN, SVM, and GBDT classifiers. Additionally, we introduced four supplementary approaches: DFCNN, CAE-DTLN [51], 1DRCAE [52] (deep learning algorithms) and DEEP FEATURE-KNN (directly utilizing the KNN algorithm to classify the features extracted by means of the CAE). This amalgamation culminates in a total of eleven distinct methodologies.

(1) TCA-KNN
(2) TCA-GBDT
(3) CORAL-KNN
(4) JDA-KNN
(5) BDA-KNN
(6) BDA-SVM
(7) BDA-GBDT
(8) DFCNN
(9) DEEP FEATURE-KNN
(10) CAE-DTLN
(11) 1DRCAE

The experimental outcomes are showcased in Table 3 and Figure 11, derived from 10 replicate trials. Evaluation metrics encompass the mean accuracy and standard deviation.

Observing Table 3 along with Figures 11 and 12, it becomes evident that the proposed BDA-KNN method attained the highest mean accuracy and nearly the lowest standard deviation. Conversely, CORAL-KNN exhibited the lowest standard deviation but poor accuracy performance. In summary, these findings highlight the superior performance of BDA-KNN. Additionally, it is worth noting that the classification effectiveness substantially surpasses that of the CORAL algorithm when employing BDA, TCA, and JDA migration algorithms. This observation aligns with the feature visualization depicted in Figure 10. A more pronounced clustering effect between source and target domain features correlates with an easier fault diagnosis. Moreover, the majority of migration methods outperformed the deep learning algorithms DFCNN, CAE-DTLN, 1DRCAE, and DEEP FEATURE-KNN, which rely solely on CAE depth features for diagnosis. This underscores the ascendancy of migration learning in cross-service fault diagnosis.

To provide a visually illustrative depiction of the proposed method's performance, Figure 11 presents the confusion matrix based on the CWRU dataset. The figure demonstrates the model's robust diagnostic proficiency across all categories of fault types.

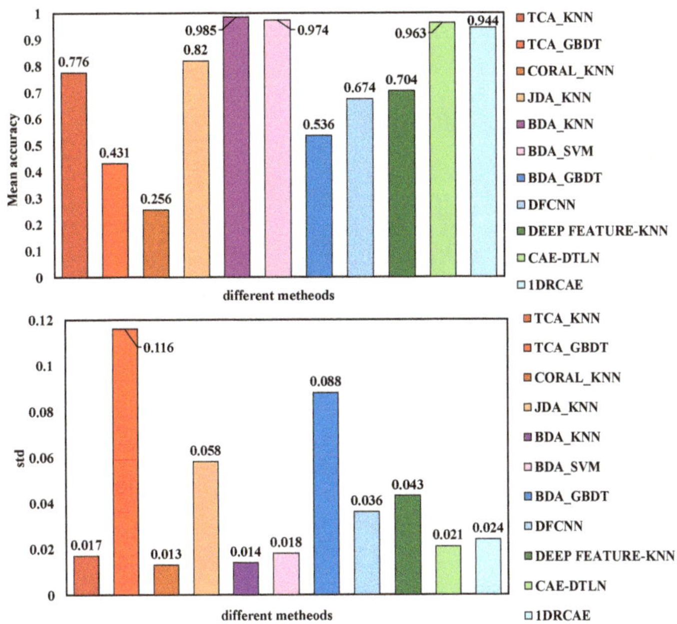

Figure 11. Histograms of the comparative results in Table 3.

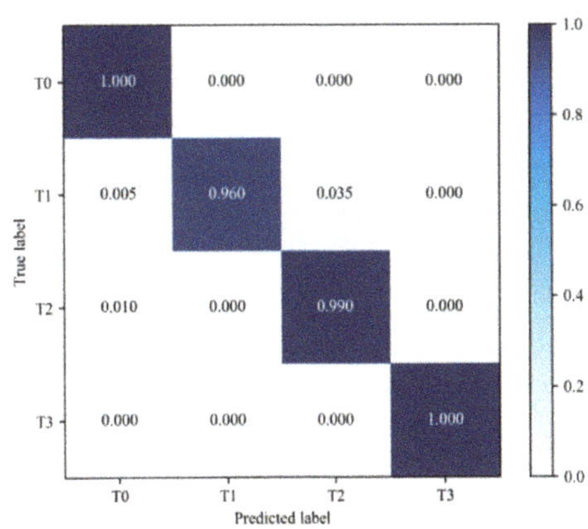

Figure 12. Confusion matrix by the proposed method on the CWRU dataset.

Table 3. Accuracy and standard deviation of different methods on the CWEU dataset.

Evaluating Indicator	TCA_ KNN	TCA_ GBDT	CORAL_ KNN	JDA_ KNN	BDA_ KNN	BDA_ SVM	BDA_ GBDT	DFCNN	DEEP FEATURE-KNN	CAE-DTLN	1DRCAE
Mean accuracy	0.776	0.431	0.256	0.820	0.985	0.974	0.536	0.674	0.704	0.963	0.954
std	0.017	0.116	0.013	0.058	0.014	0.018	0.088	0.036	0.043	0.021	0.024

4.2. Southeastern University Gearbox Dataset

4.2.1. Data Description

Bearing data were acquired from the drivetrain dynamics simulator (DDS), as depicted in Figure 13.

Figure 13. Experimental setup for the gearbox dataset.

As depicted in Table 4, we investigated two distinct operational scenarios, wherein the rotation speed-system load configurations were set to 20 Hz-0 V and 30 Hz-2 V, respectively.

Table 4. Bearing fault types description and data segmentation.

Class	Fault	Condition
Source1	Normal	20 HZ-0 V
Source2	Ball	20 HZ-0 V
Source3	Inner	20 HZ-0 V
Source4	Outer	20 HZ-0 V
Source5	Combination	20 HZ-0 V
Target1	Normal	30 HZ-2 V
Target2	Ball	30 HZ-2 V
Target3	Inner	30 HZ-2 V
Target4	Outer	30 HZ-2 V
Target5	Combination	30 HZ-2 V

4.2.2. Signal Pre-Processing

Figure 14 depicted the time and frequency-domain waveforms of bearing vibration data across distinct health conditions: (1) for normal conditions, the data predominantly exhibit smooth random waveforms characterized by minor amplitude fluctuations; and (2) as illustrated in Figure 14, conspicuous burrs and irregular waveforms manifest in the data linked to inner and outer ring failures, while such features are less noticeable in instances of rolling element failures. The frequency domain waveforms reveal a predominant energy concentration within the high frequency range (4, 6 kHz).

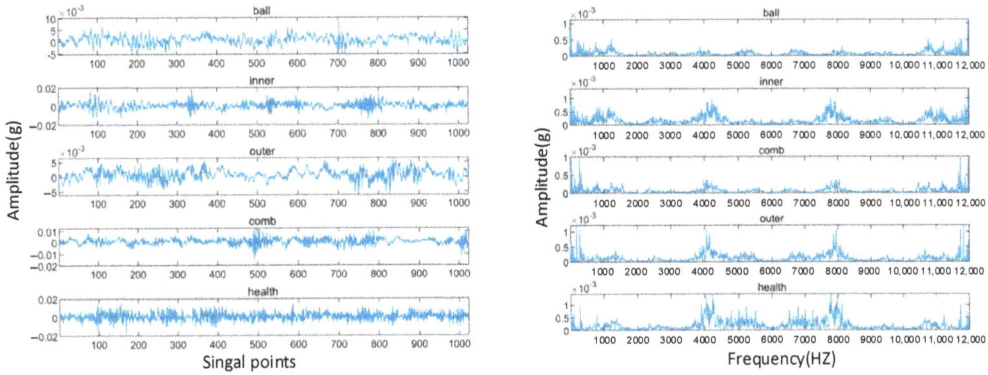

Figure 14. Southeastern University data set visualization.

4.2.3. Signal Reconstruction and Feature Extraction

The convolutional autoencoder reconstructs input data by acquiring a condensed representation and subsequently generating a reconstruction using this representation. The primary objective of data reconstruction lies in distilling the most valuable information from the input while minimizing noise and redundancy. The outcomes, illustrated in Figure 15, demonstrate that the training of the autoencoder is highly efficacious. The reconstructed signal adeptly emulates the original waveform, showcasing precise reconstruction.

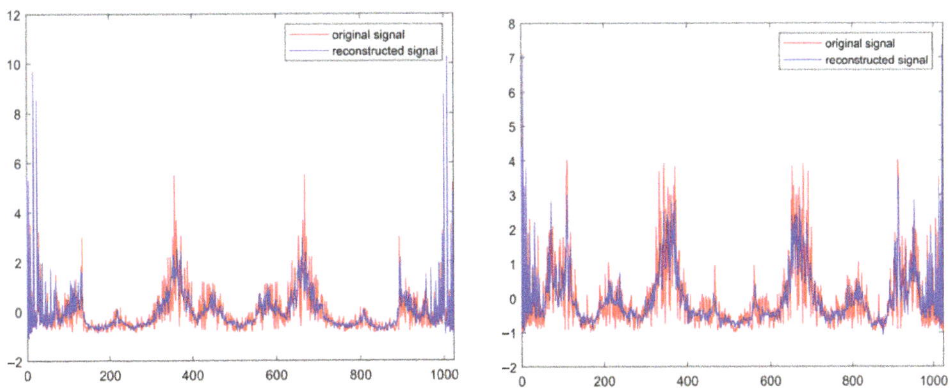

Figure 15. Reconstructed data and original data on the Southeast University data set.

In a bid to bolster the persuasive efficacy, we additionally employ the t-sne algorithm to visualize the implied features extracted by the CAE. Furthermore, to underscore the indispensability of data pre-processing, we juxtaposed the feature distributions of the original data samples with those of the FFT spectrum samples. The ensuing comparison is depicted in Figure 16, wherein the feature distributions, reduced to both two and three dimensions through t-sne, are showcased.

Derived from Figure 16, it becomes apparent that direct dimensionality reduction applied to the original data resulted in a considerable overlap of features across all health states. This outcome underscores the incapacity of the original data to unveil the inherent patterns associated with fault types. On the other hand, feeding the raw data into the CAE for feature extraction, within both two and three-dimensional spaces, reveals nascent distinctions among various fault types. Nevertheless, the lingering challenge of feature overlap remains unaddressed.

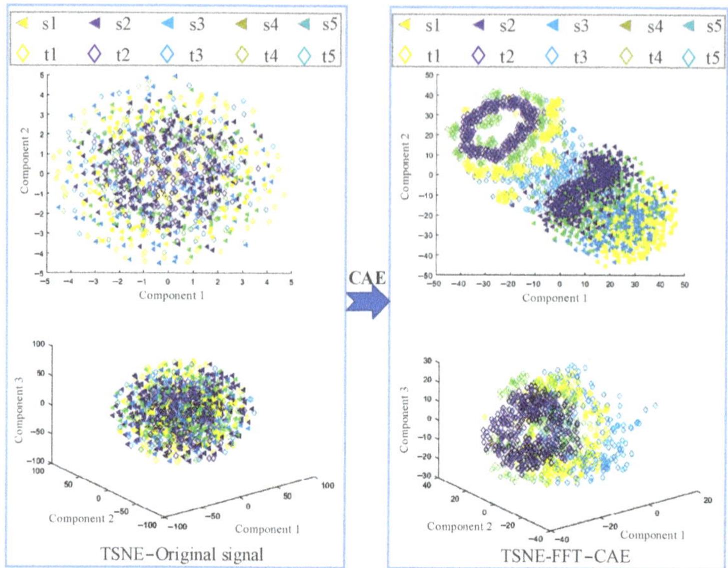

Figure 16. Variation of the feature distribution of the raw data.

As illustrated in Figure 17, the FFT spectrum samples underwent feature extraction within the CAE model, subsequently followed by t-sne visualization. This examination revealed a commendable clustering effect within both two and three-dimensional spaces. The comprehensive insights provided by Figure 15 underscore CAE's adeptness in feature extraction, facilitating the acquisition of the inherent characteristics associated with various fault types. This visualization also underscores the essentiality of FFT pre-processing for the samples. The resultant features, transformed through FFT and subsequently extracted via the CAE, are henceforth denoted as "CAE features".

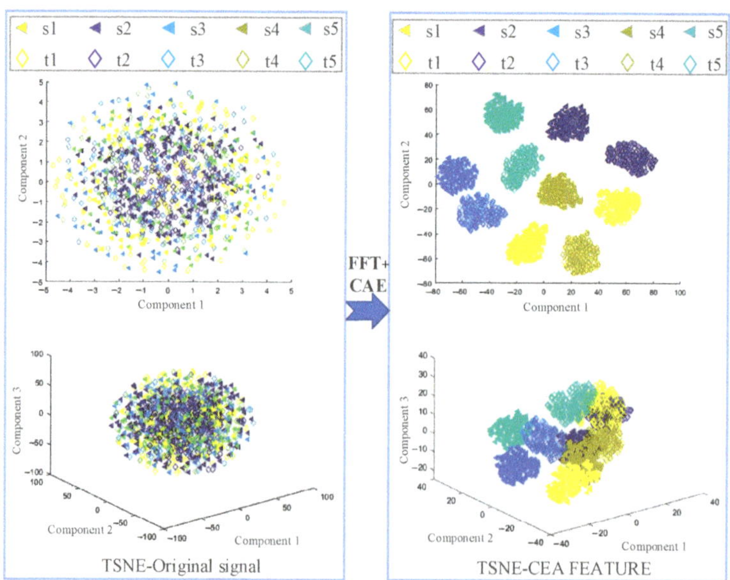

Figure 17. FFT spectral feature distribution variation diagram.

4.2.4. Feature Migration

To substantiate the excellence of the BDA algorithm, four migration learning methodologies TCA, JDA, CORAL, and BDA were employed for migrating CAE features and investigating the ensuing feature clustering phenomenon. This analysis is depicted in Figure 18.

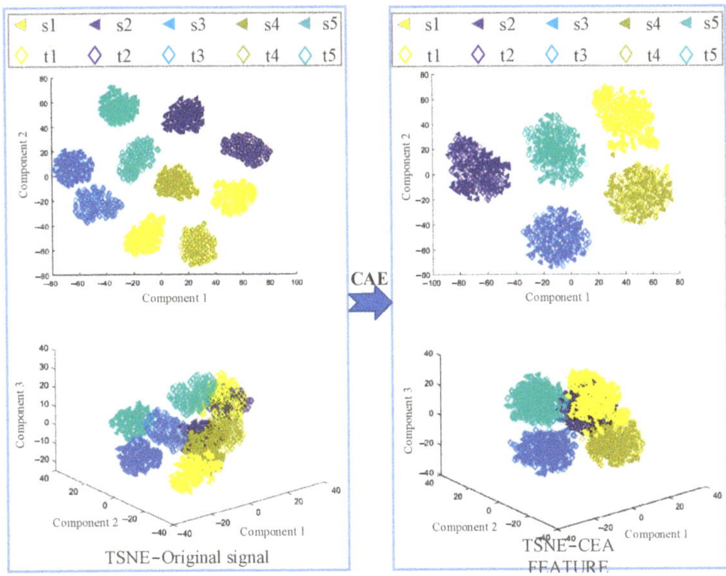

Figure 18. Feature migration results of different migration algorithms.

Observing the figure, it becomes evident that TCA effectively migrates tag types 3 and 4, yet displayed limited efficacy in migrating other tag types. JDA likewise exceled at migrating tag types 3 and 4, while demonstrating subpar performance for the remaining categories. On the other hand, CORAL exhibited the least desirable migration outcomes. BDA, conversely, delivered commendable performance across all tag types, thereby conferring strong discriminability among them. Notably, BDA's proficiency in migrating features between the source and target domains underscores its remarkable domain adaptation capabilities.

4.2.5. Fault Classification and Analysis

In alignment with the CWRU dataset, an identical algorithm was employed for comparison purposes. The experiment was replicated 10 times, with the average diagnostic accuracy and standard deviation (STD) serving as the evaluation metrics. The outcomes are visually depicted in Figure 19 and tabulated in Table 5.

As Table 5 and Figures 19 and 20 demonstrate, it is clear that the proposed BDA-KNN method has achieved the highest mean accuracy and almost the lowest standard deviation. In contrast, CORAL-KNN displayed the lowest standard deviation, but poor accuracy performance. To sum up, these results underscore the superior performance of BDA-KNN. Furthermore, it is noteworthy that the classification effectiveness significantly exceeds that of the CORAL algorithm when employing BDA, TCA, and JDA migration algorithms. This observation is consistent with the feature visualization presented in Figure 18. A stronger clustering effect among source and target domain features corresponds to easier fault diagnosis. Moreover, the majority of migration methods outperformed deep learning algorithms such as DFCNN, CAE-DTLN, 1DRCAE, and DEEP FEATURE-KNN, which rely solely on CAE depth features for diagnosis. This highlights the superiority of migration learning in cross-service fault diagnosis.

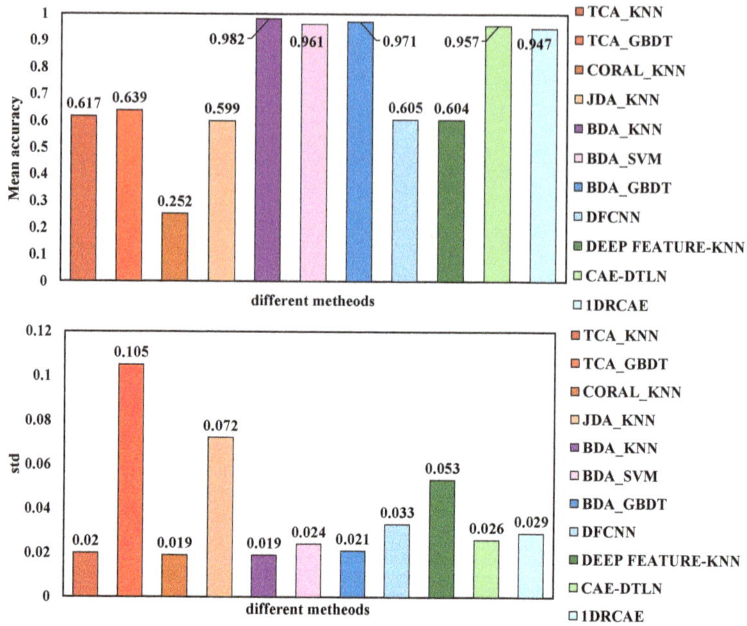

Figure 19. Histograms of the comparative results in Table 5.

Table 5. Accuracy and standard deviation of different methods on the Southeastern University dataset.

Evaluating Indicator	TCA_ KNN	TCA_ GBDT	CORAL_ KNN	JDA_ KNN	BDA_ KNN	BDA_ SVM	BDA_ GBDT	DFCNN	DEEP FEATURE-KNN	CAE-DTLN	1DRCAE
Mean accuracy	0.617	0.639	0.252	0.599	0.982	0.961	0.971	0.605	0.604	0.957	0.947
std	0.020	0.105	0.019	0.072	0.019	0.024	0.021	0.033	0.053	0.026	0.029

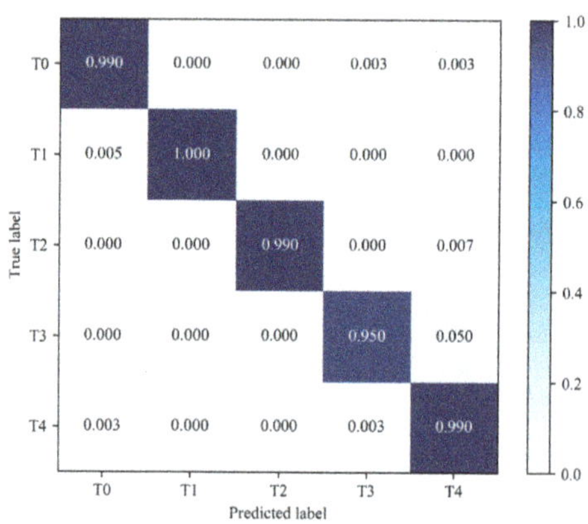

Figure 20. Confusion matrix by the proposed method on the Southeastern University dataset.

For a more visual portrayal of the diagnostic accuracy attributed to the suggested approach, Figure 20 delineates the confusion matrix over the SEU dataset. The visualization demonstrates the model's adeptness in fault diagnosis across all fault types.

In summary, the BDA-KNN method presented in this paper demonstrates a notably high average accuracy and a minimal standard deviation when compared to alternative approaches. Furthermore, this method exhibits remarkable migration results. Utilizing BDA, it effectively combines labels of the same categories from the source and target data sources while preserving the differentiation between distinct labels. This method effectively mitigates data distribution disparities across diverse domains.

5. Conclusions

This article presents an unsupervised domain adaptive approach for the cross-condition diagnosis of bearings. The methodology leverages convolutional auto-encoders (CAEs) to extract intricate features from vibration signals, enabling the subsequent application of the balancing domain adaptation (BDA) algorithm. This algorithm facilitates the unsupervised migration of extracted features between the source and target domains, without necessitating target domain labeling. The BDA algorithm achieves this by intelligently weighing the significance of inter-domain edge distribution and conditional distribution. Ultimately, the diagnosis of rolling bearing faults across varied working conditions is achieved using the K-nearest neighbor (KNN) algorithm. To validate the methodology's efficacy, changes in data distribution during feature extraction and migration were reproduced through the utilization of the t-SNE technique, which further verified the heightened diagnostic prowess of the proposed approach. Furthermore, the performance of the approach was validated across diverse datasets, with the ensuing experimental results concretely confirming the effectiveness and superiority of the method.

Nevertheless, considering that domain adaptation methods are often sensitive to disparities in data distribution, the methods introduced in this paper possessed certain limitations. The domain adaptation approach proposed in this paper was grounded on the assumption of distinct operational circumstances for the same equipment, exhibiting commendable generalization and resilience in the face of varied working conditions for the same equipment. However, there remains a considerable scope for enhancing the model's efficacy in adapting between different devices. Consequently, our future endeavors will be centered on bolstering the generalization capacity and robustness of domain adaptation across diverse devices.

Author Contributions: Methodology, W.M.; Writing—original draft, Z.Z. and H.L.; Writing—review & editing, Z.Z.; Visualization, Y.C.; Project administration, X.X.; Funding acquisition, Z.Z. All authors have read and agreed to the published version of the manuscript.

Funding: This project was supported by the Henan Province Major Science and Technology Special Project (No. 221100220100) and the Henan Provincial Science and Technology Research and Development Joint Fund Project (No. 222103810030).

Data Availability Statement: Experimental data can be downloaded from: https://engineering.case.edu/bearingdatacenter/apparatus-and-procedures and https://github.com/cathysiyu/Mechanical-datasets.

Conflicts of Interest: The authors declare no conflict of interest.

Nomenclature

AE	autoencoder
BDA	balanced distribution adaption
CNN	convolutional neural network
DFT	discrete Fourier transform
FFT	fast Fourier transform
GAN	generative adversarial network
KNN	K-nearest neighbor
MMD	maximum mean discrepancy
T-SNE	T-distributed stochastic neighbor embedding
D_S	source domain space

D_t	target domain space
$X(k)$	spectral function
$\hat{X}$	output of decoder
l_{AE}	reconstruction error
$p(x)$	marginal distribution
$p(y\|x)$	conditional distribution
μ	balance factor
κ	regenerative kernel Hilbert space
H	central matrix
I	unit matrix
A	transformation matrix
ϕ	Lagrangian operator

References

1. Zhang, S.; Wang, B.; Habetler, T.G. Deep Learning Algorithms for Bearing Fault Diagnostics—A Comprehensive Review. *IEEE Access* **2020**, *8*, 29857–29881. [CrossRef]
2. Liu, Z.-H.; Meng, X.-D.; Wei, H.-L.; Chen, L.; Lu, B.-L.; Wang, Z.-H.; Chen, L. A Regularized LSTM Method for Predicting Remaining Useful Life of Rolling Bearings. *Int. J. Autom. Comput.* **2021**, *18*, 581–593. [CrossRef]
3. Neupane, D.; Seok, J. Bearing Fault Detection and Diagnosis Using Case Western Reserve University Dataset with Deep Learning Approaches: A Review. *IEEE Access* **2020**, *8*, 93155–93178. [CrossRef]
4. Rai, A.; Upadhyay, S. A review on signal processing techniques utilized in the fault diagnosis of rolling element bearings. *Tribol. Int.* **2016**, *96*, 289–306. [CrossRef]
5. Benali, J.; Sayadi, M.; Fnaiech, F.; Morello, B.; Zerhouni, N. Importance of the fourth and fifth intrinsic mode functions for bearing fault diagnosis. In Proceedings of the 14th International Conference on Sciences and Techniques of Automatic Control & Computer Engineering—STA'2013, Sousse, Tunisia, 20–22 December 2013; pp. 259–264. [CrossRef]
6. Peng, B.; Bi, Y.; Xue, B.; Zhang, M.; Wan, S. A Survey on Fault Diagnosis of Rolling Bearings. *Algorithms* **2022**, *15*, 347. [CrossRef]
7. Xie, W.; Li, Z.; Xu, Y.; Gardoni, P.; Li, W. Evaluation of Different Bearing Fault Classifiers in Utilizing CNN Feature Extraction Ability. *Sensors* **2022**, *22*, 3314. [CrossRef]
8. Cerrada, M.; Sánchez, R.-V.; Li, C.; Pacheco, F.; Cabrera, D.; de Oliveira, J.V.; Vásquez, R.E. A review on data-driven fault severity assessment in rolling bearings. *Mech. Syst. Signal Process.* **2018**, *99*, 169–196. [CrossRef]
9. Chen, Z.; Xu, J.; Alippi, C.; Ding, S.X.; Shardt, Y.; Peng, T.; Yang, C. Graph neural network-based fault diagnosis: A review. *arXiv* **2021**, arXiv:2111.08185.
10. Liu, Z.-H.; Jiang, L.-B.; Wei, H.-L.; Chen, L.; Li, X.-H. Optimal Transport Based Deep Domain Adaptation Approach for Fault Diagnosis of Rotating Machine. *IEEE Trans. Instrum. Meas.* **2021**, *70*, 1–12. [CrossRef]
11. Li, H.; Lian, X.; Guo, C.; Zhao, P. Investigation on early fault classification for rolling element bearing based on the optimal frequency band determination. *J. Intell. Manuf.* **2013**, *26*, 189–198. [CrossRef]
12. Singh, D.S.; Zhao, Q. Pseudo-fault signal assisted EMD for fault detection and isolation in rotating machines. *Mech. Syst. Signal Process.* **2016**, *81*, 202–218. [CrossRef]
13. Wang, J.; He, Q. Wavelet Packet Envelope Manifold for Fault Diagnosis of Rolling Element Bearings. *IEEE Trans. Instrum. Meas.* **2016**, *65*, 2515–2526. [CrossRef]
14. Osman, S.; Wang, W. A Morphological Hilbert-Huang Transform Technique for Bearing Fault Detection. *IEEE Trans. Instrum. Meas.* **2016**, *65*, 2646–2656. [CrossRef]
15. Lei, Y.; Lin, J.; He, Z.; Zuo, M.J. A review on empirical mode decomposition in fault diagnosis of rotating machinery. *Mech. Syst. Signal Process.* **2013**, *35*, 108–126. [CrossRef]
16. Dragomiretskiy, K.; Zosso, D. Variational Mode Decomposition. *IEEE Trans. Signal Process.* **2014**, *62*, 531–544. [CrossRef]
17. Wang, X.; Sui, G.; Xiang, J.; Wang, G.; Huo, Z.; Huang, Z. Multi-Domain Extreme Learning Machine for Bearing Failure Detection Based on Variational Modal Decomposition and Approximate Cyclic Correntropy. *IEEE Access* **2020**, *8*, 197711–197729. [CrossRef]
18. Wang, F.; Dun, B.; Liu, X.; Xue, Y.; Li, H.; Han, Q. An Enhancement Deep Feature Extraction Method for Bearing Fault Diagnosis Based on Kernel Function and Autoencoder. *Shock Vib.* **2018**, *2018*, 6024874. [CrossRef]
19. Zhang, X.; Wang, B.; Chen, X. Intelligent fault diagnosis of roller bearings with multivariable ensemble-based incremental support vector machine. *Knowl.-Based Syst.* **2015**, *89*, 56–85. [CrossRef]
20. Huo, Z.; Martinez-Garcia, M.; Zhang, Y.; Yan, R.; Shu, L. Entropy Measures in Machine Fault Diagnosis: Insights and Applications. *IEEE Trans. Instrum. Meas.* **2020**, *69*, 2607–2620. [CrossRef]
21. Li, C.; Zheng, J.; Pan, H.; Tong, J.; Zhang, Y. Refined Composite Multivariate Multiscale Dispersion Entropy and Its Application to Fault Diagnosis of Rolling Bearing. *IEEE Access* **2019**, *7*, 47663–47673. [CrossRef]

22. Jiao, W.; Li, G.; Jiang, Y.; Baim, R.; Tang, C.; Yan, T.; Ding, X.; Yan, Y. Multi-Scale Sample Entropy-Based Energy Moment Features Applied to Fault Classification. *IEEE Access* **2021**, *9*, 8444–8454. [CrossRef]
23. Mushtaq, S.; Islam, M.M.M.; Sohaib, M. Deep Learning Aided Data-Driven Fault Diagnosis of Rotatory Machine: A Comprehensive Review. *Energies* **2021**, *14*, 5150. [CrossRef]
24. Malhi, A.; Gao, R. PCA-Based Feature Selection Scheme for Machine Defect Classification. *IEEE Trans. Instrum. Meas.* **2004**, *53*, 1517–1525. [CrossRef]
25. Yang, Y.; Yu, D.; Cheng, J. A fault diagnosis approach for roller bearing based on IMF envelope spectrum and SVM. *Measurement* **2007**, *40*, 943–950. [CrossRef]
26. Shen, F.; Chen, C.; Yan, R.; Gao, R.X. Bearing fault diagnosis based on SVD feature extraction and transfer learning classification. In Proceedings of the 2015 Prognostics and System Health Management Conference (PHM), Beijing, China, 21–23 October 2015.
27. Zhang, N.; Wu, L.; Yang, J.; Guan, Y. Naive Bayes Bearing Fault Diagnosis Based on Enhanced Independence of Data. *Sensors* **2018**, *18*, 463. [CrossRef]
28. Jia, F.; Lei, Y.; Lin, J.; Zhou, X.; Lu, N. Deep neural networks: A promising tool for fault characteristic mining and intelligent diagnosis of rotating machinery with massive data. *Mech. Syst. Signal Process.* **2016**, *72–73*, 303–315. [CrossRef]
29. Xia, M.; Li, T.; Xu, L.; Liu, L.; de Silva, C.W. Fault Diagnosis for Rotating Machinery Using Multiple Sensors and Convolutional Neural Networks. *IEEE/ASME Trans. Mechatron.* **2017**, *23*, 101–110. [CrossRef]
30. Shao, H.; Jiang, H.; Wang, F.; Wang, Y. Rolling bearing fault diagnosis using adaptive deep belief network with dual-tree complex wavelet packet. *ISA Trans.* **2017**, *69*, 187–201. [CrossRef]
31. Sun, W.; Shao, S.; Zhao, R.; Yan, R.; Zhang, X.; Chen, X. A sparse auto-encoder-based deep neural network approach for induction motor faults classification. *Measurement* **2016**, *89*, 171–178. [CrossRef]
32. Kerboua, A.; Kelaiaia, R. Fault Diagnosis in an Asynchronous Motor Using Three-Dimensional Convolutional Neural Network. *Arab. J. Sci. Eng.* **2023**, 1–19. [CrossRef]
33. Liang, B.; Feng, W. Bearing Fault Diagnosis Based on ICEEMDAN Deep Learning Network. *Processes* **2023**, *11*, 2440. [CrossRef]
34. Ma, P.; Zhang, H.; Fan, W.; Wang, C. A diagnosis framework based on domain adaptation for bearing fault diagnosis across diverse domains. *ISA Trans.* **2019**, *99*, 465–478. [CrossRef] [PubMed]
35. Li, X.; Zhang, W.; Ding, Q.; Sun, J.-Q. Multi-Layer domain adaptation method for rolling bearing fault diagnosis. *Signal Process.* **2018**, *157*, 180–197. [CrossRef]
36. Xu, W.; Wan, Y.; Zuo, T.-Y.; Sha, X.-M. Transfer Learning Based Data Feature Transfer for Fault Diagnosis. *IEEE Access* **2020**, *8*, 76120–76129. [CrossRef]
37. Lu, W.; Liang, B.; Cheng, Y.; Meng, D.; Yang, J.; Zhang, T. Deep Model Based Domain Adaptation for Fault Diagnosis. *IEEE Trans. Ind. Electron.* **2016**, *64*, 2296–2305. [CrossRef]
38. Li, M.; Sun, Z.-H.; He, W.; Qiu, S.; Liu, B. Rolling Bearing Fault Diagnosis under Variable Working Conditions Based on Joint Distribution Adaptation and SVM. In Proceedings of the 2020 International Joint Conference on Neural Networks (IJCNN), Glasgow, UK, 19–24 July 2020; pp. 1–8. [CrossRef]
39. Wang, J.; Chen, Y.; Hao, S.; Feng, W.; Shen, Z. Balanced Distribution Adaptation for Transfer Learning. In Proceedings of the 2017 IEEE International Conference on Data Mining (ICDM), New Orleans, LA, USA, 18–21 November 2017.
40. Welaratna, S. *Thirty Years of FFT Analyzers*; Sound and Vibration: San Jose, CA, USA, 1997.
41. Hakim, M.; Omran, A.A.B.; Inayat-Hussain, J.I.; Ahmed, A.N.; Abdellatef, H.; Abdellatif, A.; Gheni, H.M. Bearing Fault Diagnosis Using Lightweight and Robust One-Dimensional Convolution Neural Network in the Frequency Domain. *Sensors* **2022**, *22*, 5793. [CrossRef]
42. Zhang, M.; Yin, J.; Chen, W. Rolling Bearing Fault Diagnosis Based on Time-Frequency Feature Extraction and IBA-SVM. *IEEE Access* **2022**, *10*, 85641–85654. [CrossRef]
43. Mao, W.; Liu, Y.; Ding, L.; Li, Y. Imbalanced Fault Diagnosis of Rolling Bearing Based on Generative Adversarial Network: A Comparative Study. *IEEE Access* **2019**, *7*, 9515–9530. [CrossRef]
44. Hinton, G.E.; Salakhutdinov, R.R. Reducing the Dimensionality of Data with Neural Networks. *Science* **2006**, *313*, 504–507. [CrossRef]
45. Mao, W.; Chen, J.; Liang, X.; Zhang, X. A New Online Detection Approach for Rolling Bearing Incipient Fault via Self-Adaptive Deep Feature Matching. *IEEE Trans. Instrum. Meas.* **2019**, *69*, 443–456. [CrossRef]
46. An, J.; Ai, P.; Liu, C.; Xu, S.; Liu, D. Deep Clustering Bearing Fault Diagnosis Method Based on Local Manifold Learning of an Autoencoded Em-bedding. *IEEE Access* **2021**, *9*, 30154–30168. [CrossRef]
47. Lecun, Y.; Bottou, L.; Bengio, Y.; Haffner, P. Gradient-based learning applied to document recognition. *Proc. IEEE* **1998**, *86*, 2278–2324. [CrossRef]
48. Lei, Y.; Yang, B.; Jiang, X.; Jia, F.; Li, N.; Nandi, A.K. Applications of machine learning to machine fault diagnosis: A review and roadmap. *Mech. Syst. Signal Process.* **2020**, *138*, 106587. [CrossRef]
49. Mao, W.; Liu, Y.; Ding, L.; Safian, A.; Liang, X. A New Structured Domain Adversarial Neural Network for Transfer Fault Diagnosis of Rolling Bearings Under Different Working Conditions. *IEEE Trans. Instrum. Meas.* **2020**, *70*, 1–13. [CrossRef]
50. Liu, Z.-H.; Lu, B.-L.; Wei, H.-L.; Li, X.-H.; Chen, L. Fault Diagnosis for Electromechanical Drivetrains Using a Joint Distribution Optimal Deep Domain Adaptation Approach. *IEEE Sens. J.* **2019**, *19*, 12261–12270. [CrossRef]

51. Qian, Q.; Qin, Y.; Wang, Y.; Liu, F. A new deep transfer learning network based on convolutional auto-encoder for mechanical fault diagnosis. *Measurement* **2021**, *178*, 109352. [CrossRef]
52. Yu, J.; Zhou, X. One-Dimensional Residual Convolutional Autoencoder Based Feature Learning for Gearbox Fault Diagnosis. *IEEE Trans. Ind. Inform.* **2020**, *16*, 6347–6358. [CrossRef]

lubricants

MDPI

Article

Individual Feature Selection of Rolling Bearing Impedance Signals for Early Failure Detection

Florian Michael Becker-Dombrowsky *, Quentin Sean Koplin and Eckhard Kirchner

Department of Mechanical Engineering, Institute for Product Development and Machine Elements, Technical University of Darmstadt, Otto-Berndt-Straße 2, 64287 Darmstadt, Germany; contact@koplin-mail.de (Q.S.K.); kirchner@pmd.tu-darmstadt.de (E.K.)

* Correspondence: florian_michael.becker@tu-darmstadt.de

Abstract: Condition monitoring of technical systems has increasing importance for the reduction of downtimes based on unplanned breakdowns. Rolling bearings are a central component of machines because they often support energy-transmitting elements like shafts and spur gears. Bearing damages lead to a high number of machine breakdowns; thus, observing these has the potential to reduce unplanned downtimes. The observation of bearings is challenging since their behavior in operation cannot be investigated directly. A common solution for this task is the measurement of vibration or component temperature, which is able to show an already occurred bearing damage. Measuring the electrical bearing impedance in situ has the ability to gather information about bearing revolution speed and bearing loads. Additionally, measuring the impedance allows for the detection and localization of damages in the bearing, as early research has shown. In this paper, the impedance signal of five fatigue tests is investigated using individual feature selection. Additionally, the feature behavior is analyzed and explained. It is shown that the three different bearing operational time phases can be distinguished via the analysis of impedance signal features. Furthermore, some of the features show a significant change in behavior prior to the occurrence of initial damages before the vibration signals of the test rig vary from a normal state.

Keywords: condition monitoring; rolling bearing; feature engineering; damage early detection; electrical impedance measurement

Citation: Becker-Dombrowsky, F.M.; Koplin, Q.S.; Kirchner, E. Individual Feature Selection of Rolling Bearing Impedance Signals for Early Failure Detection. *Lubricants* **2023**, *11*, 304. https://doi.org/10.3390/lubricants11070304

Received: 11 May 2023
Revised: 4 July 2023
Accepted: 12 July 2023
Published: 20 July 2023

1. Introduction

Fault-based breakdowns of rotating machinery reduce the reliability, security, and availability of machines [1]. Thus, detecting abnormalities becomes more important to reduce unplanned downtimes. Rolling bearings are one of the most reliable machine elements and are used in a wide range of different rotating machines [2]. They are located in the flux of forces, which means that all changes or harmful abnormalities in rotating machines' behavior interfere with them. Because of that, nearly 20% of all machine failures are based on rolling bearing damages [3]. Monitoring the bearing condition can reduce unplanned downtimes and increase the availability of technical systems. This kind of condition monitoring is the basis for condition-based maintenance or so-called predictive maintenance [2].

The aim of predictive maintenance is to forecast a machine breakdown using condition monitoring and to fulfill necessary maintenance steps at an optimum time slot [2]. A fundamental step for condition monitoring is the detection of failures and the classification of machine element conditions [1,2]. The data acquisition using sensors and sensor systems is essential for different observed parameters to receive information about the monitored system [1].

Early research at the Institute for product development and machine elements shows the applicability of ball bearings as sensors for an in situ load and failure monitoring [4].

This concept uses the electric properties of rolling bearings to calculate the bearing load and gather information about the bearings' condition and operational state [4]. Martin et al. show that the electric impedance signal changes over the lifespan of a ball bearing, and three different phases in the bearing life are distinguished. The occurrence of surface damage is observed in the real and imaginary parts of the impedance signal. Furthermore, it is possible to localize the damage and measure its length by analyzing the impedance data using the characteristic ball-bearing frequencies [5,6]. Maruyama et al. show that measuring the impedance can monitor the lubrication condition [7]. All this displays the opportunities of the electric impedance measurement for ball bearings, which are further investigated in this paper by analyzing the impedance signal itself and features calculated from it to describe the rolling bearing life span.

1.1. Condition Monitoring Using Vibration Data

A common solution for condition monitoring in rolling bearings is measuring the vibration signals resulting from normal and abnormal behavior of the observed components. In the case of pitting, vibrations occur when the damage already harms the surfaces of the contact partners. Overrolling surface damage in the bearing runways or rolling elements leads to a pulse excitation, which is intensified by the elastic material behavior of the components. The resulting vibrations are transferred to the sensor through the structural components, where these are detected and sent to analyzing systems. The signals are prepared for further investigations in the time domain, frequency domain, and time-frequency domain. These data are the basis for receiving information about the system condition and prediction of the remaining operational time [1–3].

In order to predict the remaining life of the bearing, machine learning methods like feature engineering and regression models are used [1,8,9]. The sensors providing the necessary data are not located directly at the monitored component, which is why their signals are a combination of source effects like damages in the bearing runway and transmission path effects influenced by the structural components and their interference [1–3]. This can be a disadvantage because the affectation of the signals can be found in the data used for the machine learning techniques, which leads to uncertainties in the models. Therefore, interfering signals have to be minimized by filtering and other mathematical operation [1,8,9], increasing the complexity of the algorithms. Furthermore, information from the point of interest about the condition of the monitored machine elements is missing.

The impedance is frequency-dependent, which is why it can be investigated in the time and frequency domain [5,6,10]. Since many signal features for vibration data in the time and frequency domain are already commonly used for condition monitoring [1,2,8,9], these existing features will be used for individual feature selection as a feature engineering method in this work.

1.2. Feature Engineering

A feature is a mathematical quantity that describes the attributes and characteristics of a measurement signal. Features are created to decrease the amount of data and to create robust predictors of a specific characteristic of interest [11]. The process of feature engineering is used to create meaningful features with the highest possible quality of information concerning the desired characteristic. Features quantify certain characteristics of a signal. Prominent examples of such features are the mean value and the standard deviation of a set of measurements. Features derived from the time domain describe the temporal behavior of the measurand. Additional features are derived from the frequency domain. It is, therefore, necessary to calculate the frequency spectrum of the impedance signal by applying a discrete Fourier transform. [3]

Feature engineering involves the following steps: First, the measurement signal is preprocessed. Preprocessing the signal enables the reduction of errors such as background noise and errors in the measurement setup [1]. The resulting signal is used to generate features. This can be accomplished by calculating statistical measures like the standard deviation of the signal or by using mathematical methods such as Fourier transform before applying mathematical operations. After generating multiple features, the results are compared to each other to find the features with the most significance regarding the desired information or characteristic. Ranking the individual features according to a specified criterion to select the most valuable ones is called individual feature selection. The criterion quantifies the relevance of each feature [12].

Feature engineering is an important part of further signal-analyzing steps. To fulfill tasks like condition classification and early damage detection, features need to be generated and implemented in machine learning algorithms [1–3].

1.3. Electric Behavior of Rolling Bearings

In an electric circuit, ball bearings show capacitor-like behavior since the electrically conductive components are separated by electrically isolating lubrication films. Depending on the lubrication film thickness, three different electric behaviors are observed and modeled as an equivalent circuit. These behaviors can be modeled in the following way. The Hertzian contact zone is described as a plate capacitor, which is illustrated in Figure 1, so the capacity in the elastohydrodynamic (EHL) contact can be calculated using the capacitor equation [10]:

$$C_{Hz} = \varepsilon_r \varepsilon_0 \frac{A_{Hz}}{h_0}, \quad (1)$$

where A_{Hz} is the Hertzian contact area, h_0 the central lubrication film thickness in the EHL contact, and $\varepsilon_r \varepsilon_0$ the permittivity of the lubricant.

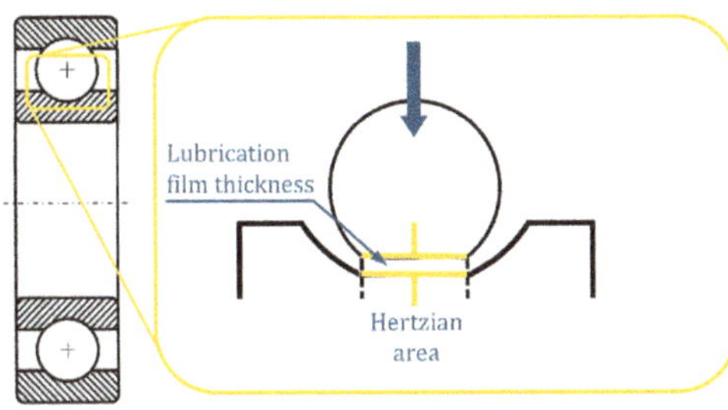

Figure 1. Electric model of the EHL contact in a ball bearing [13].

Gemeinder and Barz enhance the model by initiating a factor k_r considering the influence of the border zone [14,15]:

$$C_{Hz} = k_r \varepsilon_r \varepsilon_0 \frac{A_{Hz}}{h_0}. \quad (2)$$

Schirra shows that the factor k_r is not constant, and Puchtler et al. considered the influence of the unloaded rolling elements in the model [13,16].

The description of the EHL contact as a plate capacitor is only possible when a separating lubrication film exists. In the case of dry friction, direct metallic contact between the rolling elements and the runways leads to a resistive behavior, which means that this condition can be understood as ohmic resistance. An intermediate state can be observed when the lubricant separates the rolling elements and the runway completely so that metallic contacts are avoided. It can be described as an ohmic resistance in parallel connection to a plate capacitor. When the lubrication film is not thick enough, harmful EDM currents occur, damaging the surfaces. This needs to be avoided for sensory usage of the impedance measurement method. Figure 2 gives an overview of the three conditions [10,17–19].

In the case of a sufficiently thick lubrication film, the contact can be modeled as a plate capacitor, whose plate thickness is the lubrication film thickness and whose plate area is the Hertzian area, cf. Figure 1. Film thickness and Hertzian area, and thus also the capacitance, depend on the load. In this study, the complex impedance is measured, which reflects the entire electric behavior of the bearing, including resistive and capacitive terms. In the case of capacitive behavior, the phase angle tends to $-90°$, which can be used as an indicator for lubrication conditions. For phase angles of about $0°$, an ohmic behavior can be observed, and metallic contacts occur [15,18,19].

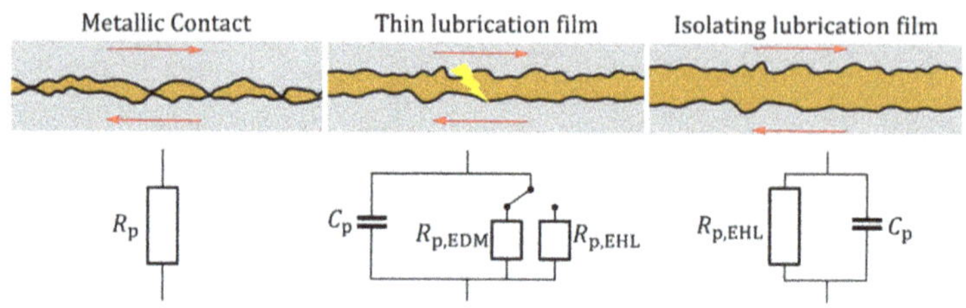

Figure 2. Electric model of the EHL contact as a function of the lubrication film thickness [6].

Because of the usage of the electric properties of rolling element bearings, hybrid bearings, or full ceramic bearings applied in, e.g., electric machinery, cannot be observed using the impedance due to the missing electrical conductivity. For these bearings, classic monitoring approaches have to be used and optimized using feature engineering and other techniques [20,21].

1.4. Research Design

The aim of this work is the further investigation of rolling bearing impedance data from five fatigue tests generated by Martin et al., which already showed the possibility of impedance measurement for rolling bearing observation [5,6]. In their research, only the parameters listed in Table 1 are analyzed, but no additional features were identified or investigated [5,6]. To further identify and analyze additional features is the aim of this contribution. The identified features will build the fundamentals for explainable machine learning algorithms as part of future research. Because impedance measurement for condition monitoring is a new approach in this field, it must be clarified whether the generated signals are appropriate for use in machine learning algorithms like classifiers. Therefore, the focus of this work is to first investigate the opportunities of impedance-based data for condition detection.

Table 1. Signals calculated from the measured complex impedance signal.

Description	Formula	Unit
Real part (effective resistance)	$R = Re(\underline{Z})$	Ω
Imaginary part (reactance)	$X_{LC} = Im(\underline{Z})$	Ω
Absolute value (apparent resistance)	$Z = \sqrt{R^2 + X_{LC}^2}$	Ω
Phase angle	$\varphi = \arctan\left(\frac{X_{LC}}{R}\right)$	rad

To do so, the impedance signals by Martin et al. are filtered and preprocessed to remove outliers. Based on the state of research for condition monitoring using vibration data, time and frequency domain features are calculated from the impedance data and are analyzed. The suitability of these new features will be checked using a normalized label over the operational lifetime of the rolling bearings. A phenomenological explanation of the feature behavior will be provided afterward. The results of the analysis are compared to a different impedance measurement approach with different types of rolling bearings in a validation fatigue test to obtain an indication of a possible generality of the extracted signal information.

2. Materials and Methods

In this section, the used impedance measurement methods and the test rig are presented. After that, the test parameters are introduced.

2.1. Impedance Measurement Methods

Martin et al. used a voltage divider to detect the impedance [5]. The equivalent circuit is shown in Figure 3.

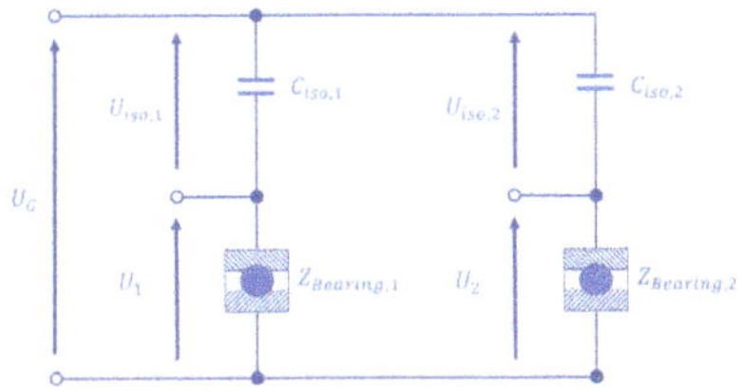

Figure 3. Equivalent circuit of the voltage divider for impedance measurement following [5].

The voltage of the generator and the voltage over the reference impedance are detected. Therefore, the capacity of isolation of the test rig is measured and applied as a reference. In the used configuration, two rolling bearings are observed simultaneously. The impedance is calculated from the known capacity and the measured voltages using the following equations [6]:

$$Z_{Bearing,1} = \left(\frac{U_G}{U_{iso,\,1}} - 1\right) \cdot \frac{1}{j\omega C_{iso,1}}, \tag{3}$$

$$Z_{Bearing,2} = \left(\frac{U_G}{U_{iso,\,2}} - 1\right) \cdot \frac{1}{j\omega C_{iso,2}}, \tag{4}$$

where $Z_{Bearing,i}$ is the complex rolling bearing impedance, U_G is the measured generator voltage, $U_{iso,\,i}$ is the measured voltage over the isolation, and $C_{iso,\,i}$ is the known capacitance of the isolation. The capacitances of the isolations are measured as $C_{iso,1} = 2.2$ nF and $C_{iso,2} = 2.6$ nF. The carrier signal frequency is set to 2.5 MHz, and the sampling rate is 50 MHz. The voltage amplitude is $\hat{U}_G = 2.5$ V [6].

The real part of the measured impedance signals is negative [5,6]. The authors explained this phenomenon as a calculation error because the isolations are assumed ideal. Modeling the isolations as not ideal turns the results into positive real parts, but they were not analyzed further to measure their real behavior [6], which leads to measurement uncertainties.

To avoid these uncertainties, another impedance measurement method has been applied to generate the validation test data. It is based on measurement bridges, using an alternating current as a carrier signal and gauged capacitors for the reference impedance. Figure 4 shows the equivalent circuit.

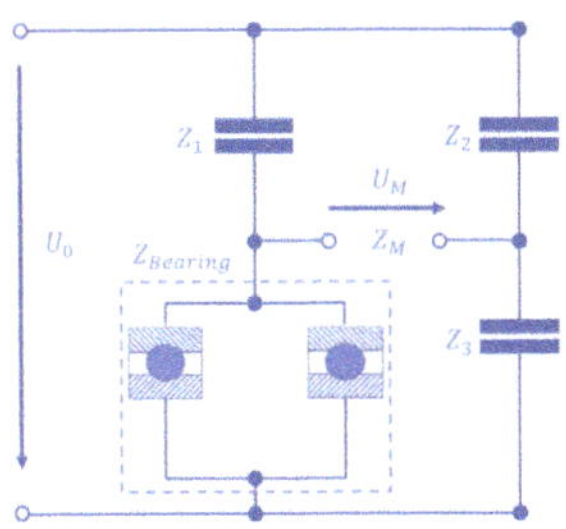

Figure 4. Equivalent circuit of the alternating current measurement bridge for impedance measurement.

The impedance of the bearings in the parallel connection $Z_{Bearing}$ is calculated using the following equation:

$$Z_{Bearing} = Z_1 \frac{Z_3 Z_M + [(Z_2 + Z_3)Z_M + Z_2 Z_3]\frac{U_M}{U_0}}{Z_2 Z_M - [(Z_2 + Z_3)(Z_M + Z_1) + 1]\frac{U_M}{U_0}}. \quad (5)$$

The reference impedance of the capacitors is tagged as Z_1, Z_2, and Z_3. The generator voltage is U_0 and the voltage at the oscilloscope is U_M. The resistance of the oscilloscope is Z_M. An open-short adjustment has been implemented to consider the influence of the measurement lead. To reduce parasitic effects, the carrier signal frequency is set to 25 kHz. The voltage amplitude is identical to the voltage divider. The sampling rate is set to 1 MHz.

This measurement approach has not been used before to detect rolling bearing impedance in fatigue tests. So, an important aspect is if the signals and features of the signals show the same behavior over the bearing operational time. This question will be addressed in this paper.

2.2. Test Rig and Impedance Measurement

All experiments are performed at the rolling bearing test rig of the Institute for product development and machine elements of the Technical University of Darmstadt. It contains four separate test chambers. In each chamber, four rolling bearings are located for observation. The test bench monitors the vibration, the temperature at every bearing, the motor torque, the revolution speed, and the lubricant temperature. Figure 5 shows one of the rig's test chambers. The test bench has an adjustable recirculating oil-lubrication system for each test chamber so that different lubrication conditions can be investigated.

Figure 5. Test chamber of the bearing test rig.

The four bearings in a chamber are placed on the same shaft, which is electrically contacted using a slip ring. The bearing seats consist of two parts, separated by an insulating ceramic layer. A contact pin bypasses the insulation of the electrically observed test bearings Within one of the chambers, two of the four bearings are investigated using impedance measurement methods. The exact configuration for the performed fatigue tests can be read in Martin et al. [5], which is also applied to the new measurement approach tested here for a better comparison. The bearings can be loaded with radial and axial forces by hydraulic actuators.

2.3. Design and Procedure of the Fatigue Tests

The individual feature selection procedure is applied to the data measured in five fatigue tests using the measurement method by Martin. The investigated bearings are angular contact ball bearings of the type 7205B-XL-TVP manufactured by FAG. These tests were executed as part of earlier research at the Institute [6]. For validation purposes, another fatigue test is performed using the alternating current measurement bridge. Therefore, radial ball bearings of the type 6205-C-C3 from the manufacturer SKF are used and stressed under different conditions. A comparison of the bearing loads and test conditions of both measurement setups is displayed in Table 2. As mentioned before, a comparison of the impedance signals of both measurement methods is planned to obtain information about the quality of impedance signals for condition monitoring. All tests run under full lubrication, so the EHL contact and the capacitive electrical behavior in a normal operational stage can be ensured.

Table 2. Test conditions of the two varying measurement setups.

Test Parameter	Investigation Tests [6]	Validation Test
Radial load	3000 N	7884 N
Axial load	28,000 N	3390 N
Dynamic safety	0.95	1.92
Speed	4000 min^{-1}	5000 min^{-1}
Oil temperature	30 °C	60 °C
Time between impedance measurements	1 min	2 min
Length of each impedance measurement	1.34 s	1.5 s
Carrier signal frequency	2.5 MHz	20 kHz
Carrier signal amplitude	5 V peak to peak	5 V peak to peak
Sampling rate	50 MHz	1 MHz

2.4. Preprocessing and Feature Generation

First, the measured impedance data are preprocessed. Four time signals are directly calculated from the measured complex impedance signal, namely the real and imaginary parts and the absolute value and phase angle (see Table 1).

These four signals are further processed. Outliers are removed using a Hampel filter, whose mathematical explanation is described in [22]. A noise filter reduces noise due to the measurement setup and environmental influences. Using wavelets for noise filtering is especially effective when reducing noise while preserving abrupt changes with high-frequency components of the signal [23]. The impedance signal of a damaged rolling bearing is characterized by abruptly occurring peaks in the real and imaginary parts [5]. Therefore, preserving the high-frequency components of the signal is of high importance, which is why a wavelet filter is applied for noise reduction. To prevent misleading signal interpretation due to errors in the measurement setup, the mean value is removed. For this purpose, the mean value of the impedance signal in the run-in stage is subtracted from the signal itself.

In the following step, features are generated. They are derived from the time and frequency domain. For frequency domain features, it is necessary to calculate the frequency spectrum of the impedance signal by applying a discrete Fourier transform. For this purpose, a fast Fourier transform is used [3]. Compared to the time domain signal, the frequency spectrum often contains further information about the signal's properties [24]. In condition monitoring of rolling bearings, the frequency spectrum is particularly important for identifying the location and cause of the initial damage [6,24].

The process of feature generation is established in the field of condition monitoring of ball bearings using the vibration signal. Just like the impedance signal, the vibration signal of a damaged bearing is characterized by periodically occurring peaks during the rollover of the initial damage [1]. Due to this analogy, the state of the art of vibration signal feature engineering is applied to the impedance signal. The generated features listed in Table 3 are taken from studies dealing with vibration signals of rolling bearings [1,25,26]. The features of measurement are calculated for each of the four signals derived from the measured complex impedance signal. In total, this leads to 128 features for each impedance measurement.

Table 3. Features derived from the time and frequency domains.

Number	Formula	Number	Formula
T1	$T_m = \frac{\sum_{i=1}^{N} x(i)}{N}$	F1	$W_1 = W_{mf} = \frac{\sum_{k=1}^{K} s(k)}{K}$
T2	$T_{root} = \left(\frac{\sum_{i=1}^{N} \sqrt{\lvert x(i)\rvert}}{N}\right)^2$	F2	$W_2 = \frac{\sum_{k=1}^{K}(s(k)-W_1)^2}{K-1}$
T3	$T_{rms} = \sqrt{\frac{\sum_{i=1}^{N}(x(i))^2}{N}}$	F3	$W_3 = \frac{\sum_{k=1}^{K}(s(k)-W_1)^3}{K\cdot\left(\sqrt{W_2}\right)^3}$
T4	$T_{max} = max\lvert x(i)\rvert$	F4	$W_4 = \frac{\sum_{k=1}^{K}(s(k)-W_1)^4}{K\cdot W_2^2}$
T5	$T_{sd} = \sqrt{\frac{\sum_{i=1}^{N}(x(i)-T_m)^2}{N-1}}$	F5	$W_5 = W_{fc} = \frac{\sum_{k=1}^{K}(f(k)\cdot s(k))}{\sum_{k=1}^{K} s(k)}$
T6	$T_{skewness} = \frac{\sum_{i=1}^{N}(x(i)-T_m)^3}{(N-1)\cdot T_{sd}^3}$	F6	$W_6 = \sqrt{\frac{\sum_{k=1}^{K}(f(k)-W_5)^2\cdot s(k)}{K}}$
T7	$T_{kurtosis} = \frac{\sum_{i=1}^{N}(x(i)-T_m)^4}{(N-1)\cdot T_{sd}^4}$	F7	$W_7 = W_{rmsf} = \sqrt{\frac{\sum_{k=1}^{K}(f(k)^2\cdot s(k))}{\sum_{k=1}^{K} s(k)}}$
T8	$T_{crest} = \frac{T_{max}}{T_{rms}}$	F8	$W_8 = \sqrt{\frac{\sum_{k=1}^{K}(f(k)^4\cdot s(k))}{\sum_{k=1}^{K}(f(k)^2\cdot s(k))}}$
T9	$T_{clearance} = \frac{T_{max}}{T_{root}}$	F9	$W_9 = \frac{\sum_{k=1}^{K}(f(k)^2\cdot s(k))}{\sqrt{\sum_{k=1}^{K} s(k)\cdot\sum_{k=1}^{K}(f(k)^4\cdot s(k))}}$
T10	$T_{shape} = \frac{T_{rms}}{\frac{1}{N}\cdot\sum_{i=1}^{N}\lvert x(i)\rvert}$	F10	$W_{10} = \frac{W_6}{W_5}$
T11	$T_{impulse} = \frac{T_{max}}{\frac{1}{N}\cdot\sum_{i=1}^{N}\lvert x(i)\rvert}$	F11	$W_{11} = \frac{\sum_{k=1}^{K}((f(k)-W_5)^3\cdot s(k))}{K\cdot W_6^3}$
T12	$T_{pp} = \max(x(i)) - \min(x(i))$	F12	$W_{12} = \frac{\sum_{k=1}^{K}((f(k)-W_5)^4\cdot s(k))}{K\cdot W_6^4}$
T13	$T_{var} = \frac{1}{N}\cdot\sum_{i=1}^{N}(x(i) - T_m)^2$	F13	$W_{13} = \frac{\sum_{k=1}^{K}\left(\lvert f(k)-W_5\rvert^{\frac{1}{2}}\cdot s(k)\right)}{K\cdot\sqrt{W_6}}$
T14	$T_{min} = \min(x(i))$	F14	$W_{14} = \sqrt{\frac{\sum_{k=1}^{K}((f(k)-W_5)^2\cdot s(k))}{\sum_{k=1}^{K} s(k)}}$
T15	$T_{wave} = \frac{\sqrt{\frac{1}{N}\cdot\sum_{i=1}^{N}\lvert x(i)\rvert^2}}{\frac{1}{N}\cdot\sum_{i=1}^{N}\lvert x(i)\rvert}$	F15	$W_{15} = \max\lvert s(k)\rvert$
T16	$T_{peak} = \frac{T_{max}}{\sqrt{\frac{1}{N}\cdot\sum_{i=1}^{N}(x(i))^2}}$		
T17	$T_{LI} = \sum_{i=1}^{N}\sqrt{(x(t_i+\Delta t_s) - x(t_i))^2 + \Delta t_s^2} \approx \sum_{i=1}^{N}\lvert x(t_i+\Delta t_s) - x(t_i)\rvert$ with sampling period Δt_s		

The time series with $i = 1, 2, 3, \ldots, N$ is $x(i)$ while $s(k)$ is a frequency spectrum with $k = 1, 2, 3, \ldots, K$. K is the total number of spectral lines in the spectrum, and $f(k)$ is the frequency value of the k-th spectral line.

Figure 6 summarizes the elaborated steps to derive features from the measured complex impedance signal.

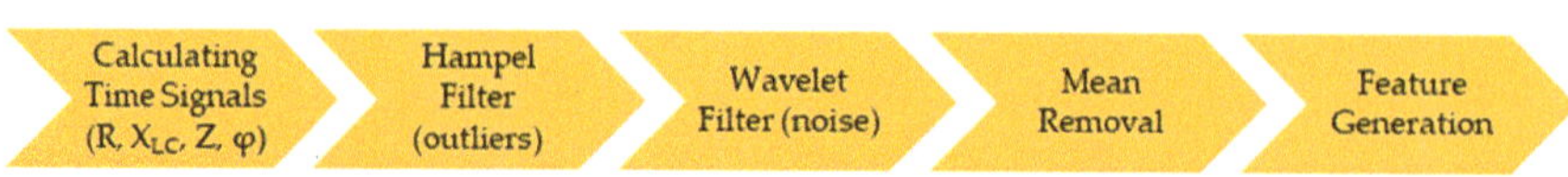

Figure 6. Preprocessing and feature generation process.

2.5. Individual-Feature Selection

The suitability of these generated features for use in condition monitoring is assessed. Therefore, individual-feature selection is applied. This method ranks the features based on a specific criterion [12]. In this case, the criterion is supposed to quantify the ability of a feature to draw conclusions about the condition of the observed rolling bearing.

The conditions of the dynamically stressed bearing are calculated by assuming the hypothesis of linear damage accumulation. Accordingly, the total damage is calculated by summing up the damage portions q_i of each cycle [27]. For the calculation of these damage portions, the duration of the load level t_i and the basic rating life L_{10h} are divided as shown in formula 3.1 [28].

$$q_i = \frac{h_i}{N_{SSZ,i}} = \frac{t_i}{L_{10h} \cdot 60 \frac{\text{min}}{\text{h}} \cdot 60 \frac{\text{s}}{\text{min}}}. \tag{6}$$

The basic rating life is calculated by the speed n_{rpm} of the bearing, its dynamic load capacity C, and life exponent p, as well as the dynamic equivalent load P (see formula 3.2) [29,30]. The dynamic equivalent load depends on the rolling bearing geometry and the radial and axial loads [29]. The test rig records the loads and speed of the bearing during the fatigue tests.

$$L_{10h} = \frac{10^6}{60 \frac{\text{min}}{\text{h}} \cdot n_{rpm}} \cdot \left(\frac{C}{P}\right)^p. \tag{7}$$

The total damage of the bearing can be calculated for the time of each impedance measurement using the recorded operational parameters. The time of initial damage detection of the five fatigue tests scatters a lot. As a result, the total damage of the bearings at the end of the tests differs widely. To obtain a universal measure for the bearing condition, a min-max scaling algorithm normalizes the total accumulated damage (see formula 3.3 [31]). This leads to the normalized accumulated damage, which rises from zero to one during a fatigue test.

$$D^*(m) = \frac{D(m) - D_{min}}{D_{max} - D_{min}}. \tag{8}$$

The criterion for the individual-feature-selection process expresses the relationship between a considered feature and the normalized accumulated damage. This relationship can be quantified by their correlation coefficient [32]. The correlation coefficient, according to Bravais–Pearson, is used to find the strength of the linear relationship between two variables [33]. To consider monotonic, nonlinear relationships, the correlation coefficient, according to Spearman, is used [33]. Features with high correlation coefficients with normalized accumulated damage are considered probable indicators of bearing damage. After calculating the correlation coefficients, each feature is ranked according to its Bravais–Pearson and Spearman correlation coefficient. The final ranking of features is achieved by considering the average rank of a feature resulting from the two mentioned criteria.

The individual feature selection is performed twice at different time intervals. Firstly, all measured data are taken into consideration; thus, the whole lifespan of the tested bearing is observed. Secondly, the last hour before initial damage detection by the test rig is exclusively studied. This enables the observation of the behavior of a feature before initial damage without the influence of the pre-run-in stage. This is of high interest because this stage shows similar effects in the impedance signal to the impedance signal after the initial damage [6].

3. Results

The resulting correlation coefficients of each feature with the label, sorted by their rank in the respective ranking, are displayed in Figure 7. A group of features characterizes the resulting distribution with high correlation coefficients compared to the remaining features.

Since the ten highest-ranked features stand out with a particularly high correlation in both rankings, these features are considered for further investigation.

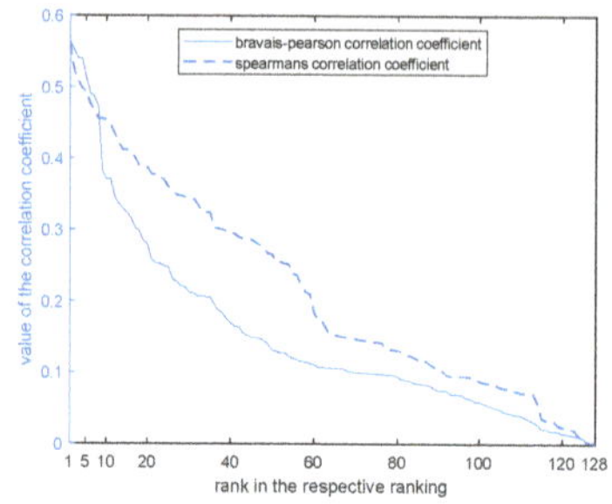

Figure 7. Correlation coefficients of each feature with the label.

The features that appear in both of the chosen subsets are ranked according to their average rank, as described in the previous chapter. The resulting three highest-ranked features are shown in Table 4. Since the procedure has been applied at two different time intervals, two rankings are depicted.

Table 4. Ranking of individual features.

Rank	Whole Lifespan	Last Hour
1.	Feature 88: RMS frequency (F7) of the absolute value	Feature 102: skewness (T6) of the phase angle
2.	Feature 56: RMS frequency (F7) of the imaginary part	Feature 60: skewness of the frequencies (F11) of the imaginary part
3.	Feature 86: central frequency (F5) of the absolute value	Feature 92: skewness of the frequencies (F11) of the absolute value

3.1. Description of Individual Features

In the following chapter, the top-ranked features listed in Table 4 are plotted and described in detail. The observations are further explored in Section 4.

First, the features considering the whole lifespan of the tested bearings are examined. The three highest-ranked features are correlated with each other. Their Bravais–Pearson correlation coefficients are greater than $r = 0.98$, with a deviation of approximately ± 0.009 according to the 95% confidence interval. Thus, only the feature on rank one, namely the root-mean-square (RMS) frequency of the absolute value of the impedance, is representatively examined. The normalized value of this feature over the label is displayed in Figure 8. The feature measurement series of the five fatigue tests using the measurement setup by Martin are depicted in different colors. Three intervals are shown to allow a closer view of the features in the first and last hour of the fatigue test.

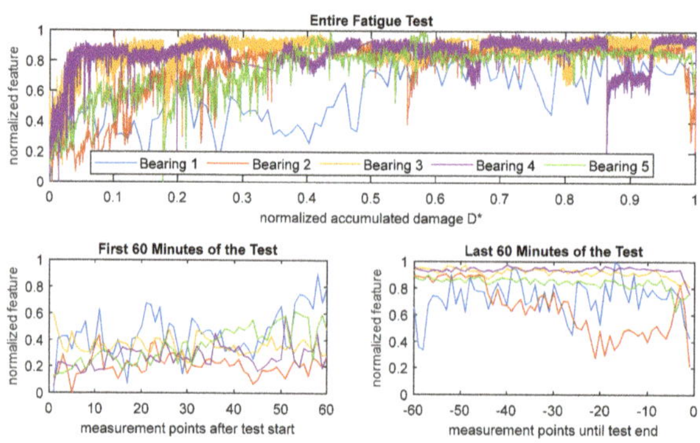

Figure 8. RMS frequency of the absolute value of the impedance.

The behavior of this feature is described in the following paragraph. Starting at a low level in the pro-run-in stage, the feature value increases until shortly before the end of the test. In all fatigue tests, a strong decline of the feature from the last 2 to 45 min before the end of the test is observed. The described three phases across the bearing lifespan are not clearly separated but rather connected to each other by a transition of the feature value. This behavior can be seen most prominently at bearing two.

In Figure 8, some noticeable abnormalities and characteristics are addressed in the following and explained in Section 4. At the beginning of the test, the feature values of the different tests seem to rise at a different pace. Also, there are visible gaps in the graphs, like at bearing 2 in the range of the normalized damage from 0.35 to 0.48 and at bearing 4 from 0.28 to 0.35. Another abnormality can be seen at bearing 3 at approximately $D^* \approx 0.18$ and at bearing 4 at $D^* \approx 0.86$. There, the features suddenly jump to a new level on which they remain for a while.

Now the features, seen as a probable indicator of bearing damage, considering the last hour before the end of the fatigue test, are described. First, the skewness of the phase angle of the impedance is shown in Figure 9.

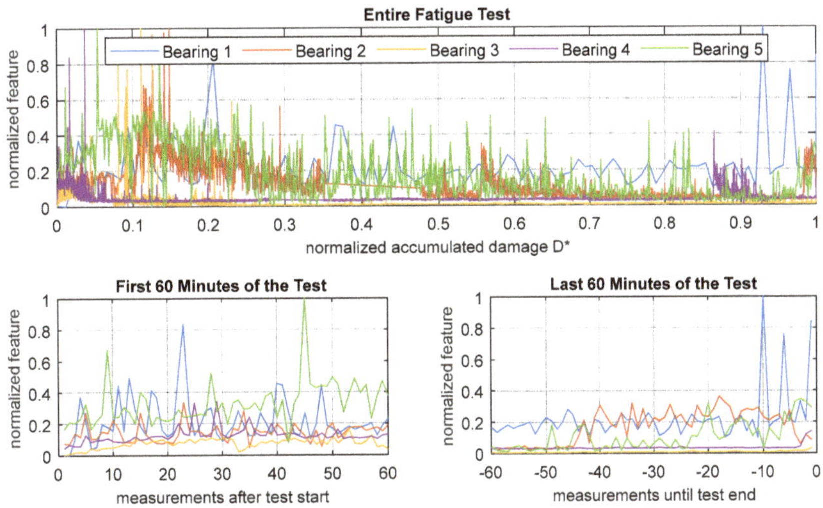

Figure 9. Skewness of the phase angle of the impedance.

At the beginning of the tests, high peaks and noisy behavior are observed. Next, the feature declines to a lower level and significantly less noise. At the fatigue test end, the feature abruptly rises significantly. Again, this feature description indicates three phases of different feature behavior with gradual transitions in between.

The features bearing 2 and 3, considering the last hour of the test, show the same behavior, which is confirmed by the Bravais–Pearson correlation coefficient of $r = 0.99$ with a deviation of approximately ± 0.014 according to the 95% confidence interval. So, only describing the feature on rank two, shown in Figure 10, is sufficient.

The skewness of the frequency values is weighed by the amplitudes of the corresponding frequencies, as shown in Table 3. In the beginning, the fatigue tests show a different behavior. Thus, one can see a significant rise in the feature at the beginning to different levels for all test results. For the rest of the fatigue test, the feature declines steadily. A roughly linear behavior can be observed. The absolute values vary significantly.

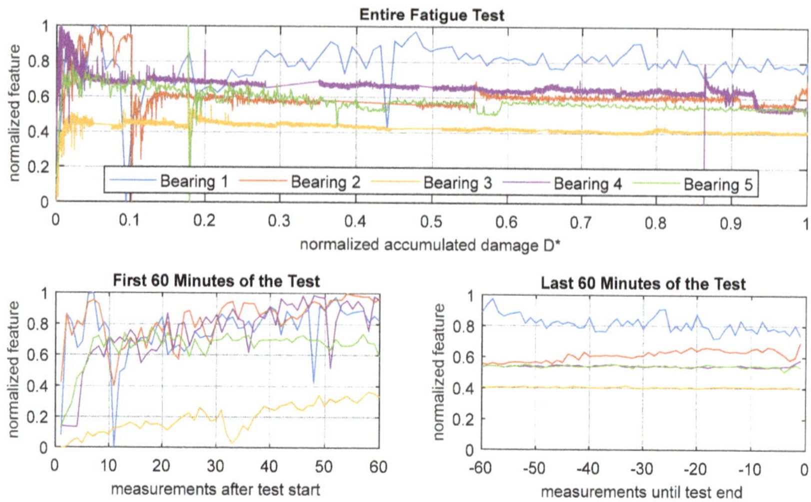

Figure 10. Skewness of the frequency values imaginary part of the impedance.

3.2. *Validation Fatigue Tests*

The results are validated by analyzing the validation fatigue test with the alternating current measurement bridge and different test setup parameters, as described in Section 2.3. Special attention should be drawn to the deviating time between measurements, which is two minutes. At this fatigue test, none of the test bearings but one of the support bearings failed. This support bearing is insulated electrically by ceramic rolling elements, thus not directly influencing the measured impedance. After initial damage detection by the test rig, the test was continued for another 60 min to obtain a higher number of measurements of the damaged bearing. The bearings were not disassembled during the entire test to exclude any fault effect. The features from Table 4 are now shown for the validation test.

In the first considered interval, the whole lifespan, the first two features again have a very high correlation of 0.9995, so, just the highest ranked feature is shown in the following graph in Figure 11.

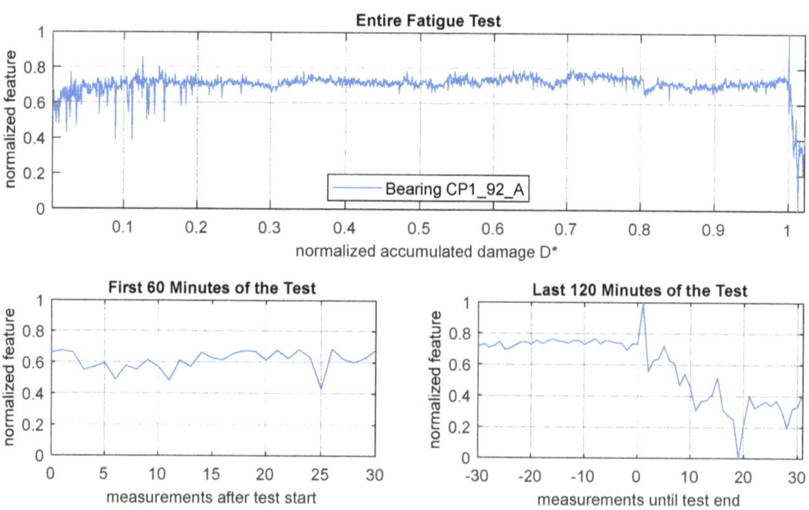

Figure 11. RMS frequency of the absolute value of the impedance (validation test).

The feature behavior corresponds to the behavior previously observed at the fatigue tests by Martin. It starts at a low level, inclines steadily, and suddenly declines at the initial pitting damage. In particular, the decline is clearly visible and very strong. It is particularly interesting that the feature only declines after initial damage detection, whereas in Figure 8, the feature showed abnormalities before the initial damage detection by the test rig. With the continuing load in the validation test, even after initial damage detection, the feature drops way below the level of the pre-run-in stage.

In contrast to the previous features, rank three only possesses a correlation coefficient of $r \approx 0.65$ with ranks one and two, making it necessary to examine this rank additionally. The central frequency of the absolute value of the impedance is depicted in Figure 12. Looking closer at this feature, the difference in the feature's behavior using the measurement setup by Martin becomes obvious: The feature behaves differently in the pre-run-in stage. At the test beginning, the feature is located on the highest level, declines to a lower level in standard operation, and finally drops after initial damage. With this behavior, the feature possibly enables the distinction of run-in and damaged phase using the measurement bridge method. This makes the feature possibly feasible for a distinction between the three bearing life phases that could be observed in the previous chapter.

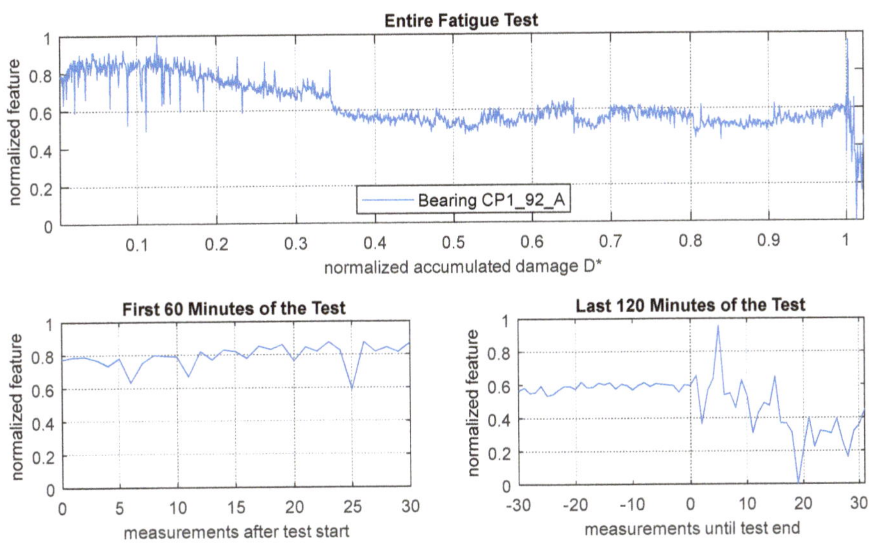

Figure 12. Central frequency of the absolute value of the impedance (validation test).

Looking at the highest-ranked features considering the last hour of the fatigue test, the same phenomena as in the already described measurement series are observed. Nevertheless, the significant indications of bearing damage are again only visible after the initial damage detection and not in advance. The feature on rank one is shown in Figure 13 and confirms the expected feature behavior. Just as in the previous chapter, three phases are clearly visible in the feature behavior.

The features of bearings 2 and 3 again show a very high correlation, and rank two can be seen in Figure 14. The feature again rises at the beginning of the test and declines with further damage progression. However, the decline is much noisier than in Figure 10 and does not show a linear behavior. A significant decrease in the feature can be observed after initial damage detection.

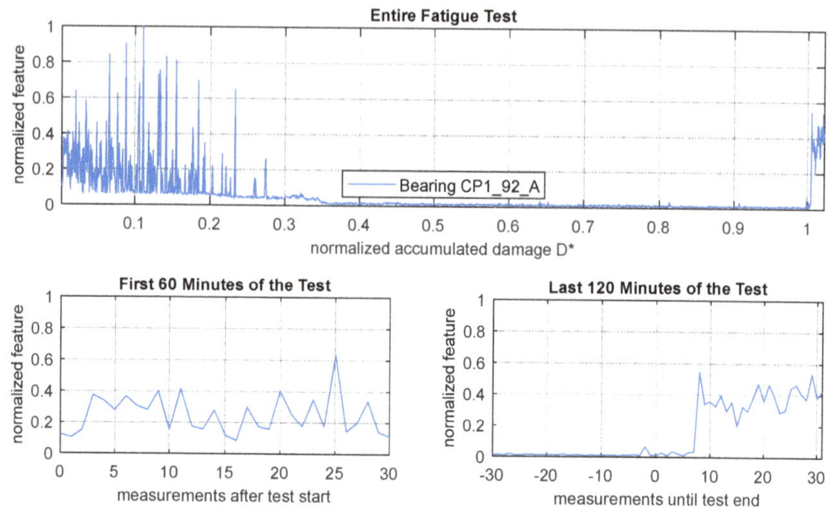

Figure 13. Skewness of the phase angle of the impedance (validation test).

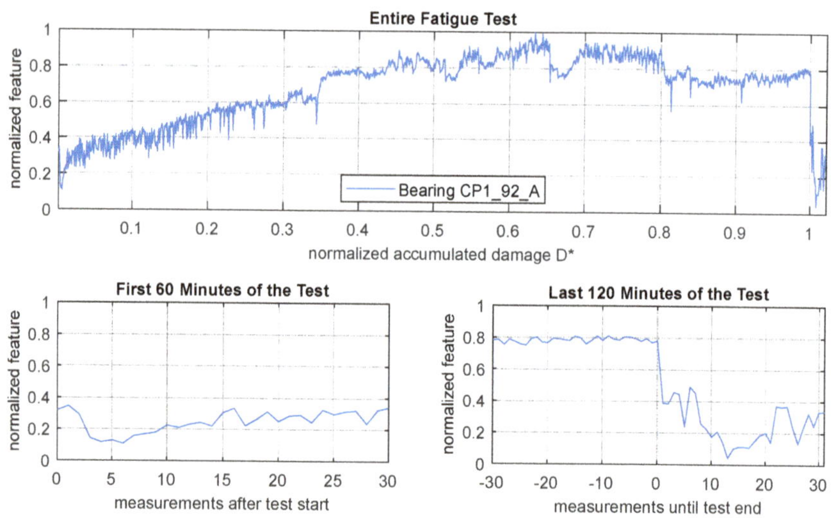

Figure 14. Skewness of the frequency values imaginary part of the impedance (validation test).

3.3. Comparison to Vibration Signals

The vibration signals are automatically recorded by the test rig, and the introduced features of the impedance signals are compared in terms of their suitability as an indicator of rolling bearing damage. The vibration data of bearing 5 show a significant increase in amplitude three minutes prior to the end of the fatigue test. The other tests show a significant increase only 30 s before the test ends (see Figure 15).

In contrast, the most significant features (rank one in each time interval) show a conspicuous behavior way before the vibration signal. As illustrated in Figure 15, initial damage can be recognized by a decrease in the RMS frequency of the absolute value of the impedance. Most bearings show that effect in the last two minutes before the test ends. Bearing 5 shows a significant decrease three minutes before the test ends, and bearing 2, even 40 min before the initial damage detection and test are stopped by the test rig.

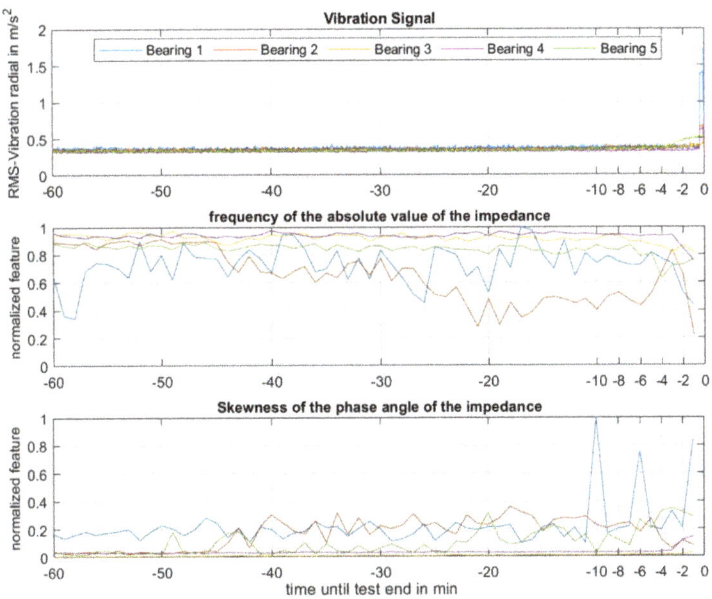

Figure 15. Vibration signals and impedance features in the last 60 min of the fatigue tests.

Likewise, the skewness of the phase angle enables early damage detection. This feature shows an increase in amplitude before the breakdown. Again, the noticeable behavior is seen at bearing 2 40 min before the test ends. Also, bearing 1 shows an increase in feature amplitude nine minutes before the end and bearing 5 more than 45 min before the end of the experiment. The difference in the behavior of the two features regarding damage detection may be affected by different causes, types, or progression of the detected bearing damage.

All in all, the features enable a detection of the bearing damage prior to the vibration signal. Especially the combination of multiple features could be useful to exploit the full potential of the different features for the detection of certain types of bearing damage. Nevertheless, further development of the signal measurement setup and signal processing should be considered to improve the accuracy of damage detection.

4. Discussion

In this section, the results of Section 3 will be interpreted and discussed. The validation is carried out on independent data sets not included in the previously gathered data. Afterward, the results of the validation test will be compared to the results of the investigation tests. In the end, the time gap between the vibration data and impedance features will be explained.

4.1. Phenomenological Explanation

The behavior of the individual features observed in the sections before is interpreted in the following. Along with it, phenomenological explanations of the feature's significance and characteristics are given while taking the available knowledge of the electrical properties of lubricated rolling bearings into account. The explanatory approaches in this chapter obtain a relation between electrical and tribological phenomena to the observed feature behaviors.

Table 4 shows different high-ranked features when looking at the last hour of the fatigue test compared to the whole bearing lifespan. In conclusion, some features are especially significant in the time interval shortly before initial damage, while others possess more information about the damage progression over the whole bearing life. This explanation is plausible since the pre-run-in stage shows similar behavior to the signal before initial damage [6]. Consequently, features that are significant in the last hour of the test may not be significant over the whole lifespan due to the influence of the pre-run-in stage.

From the individual features, the RMS frequency in the spectrum of the absolute value of the measured impedance is interpreted first. According to the chosen criterion, it is the most significant feature regarding overall extracted data during the whole fatigue test. It correlates strongly with the RMS frequency in the imaginary part and the central frequency of the absolute impedance value. Thus, there seem to be phenomena underlying these features that cause their highly similar behavior. This assumption is enforced by the observation of previous investigations that show similar effects in the real and imaginary parts of the bearing impedance when it comes to damage progression [5]. In the following, an approach to explain the characteristic behavior of that feature is introduced.

In the pre-run-in stage, the roughness of the bearings' running surfaces is high; the contact of roughness peaks results in high noise of the impedance signal. The noise often appears with amplitudes of similar magnitude [5]. These seemingly periodically occurring effects lead to low frequencies in the frequency spectrum of the impedance measurement (rough dimension of 1–10 kHz). In the frequency spectrum of impedance measurements in the pre-run-in stage, there are high amplitudes of these frequencies observable (see frequency spectrum at $D^* = 0$ in Figure 16). The same effect is seen in the spectrums of the other tested bearings, too.

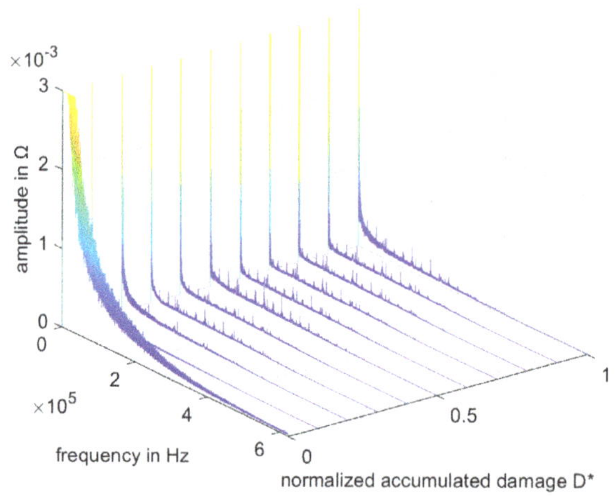

Figure 16. Waterfall diagram of the absolute impedance value of bearing 5.

In the run-in stage, the surface roughness peaks are smoothened, resulting in fewer electrical breakdowns and less noise in the impedance signal. The low frequencies, induced by the high surface roughness, are no longer present, which leads to a higher central and RMS frequency according to the corresponding formula (see Table 3). In Figure 16, this is visible in the decline of amplitudes of the low frequencies with progressing total damage.

The bearing failure due to pitting starts with a crack underneath the contact surface [34]. During the rollover of this beginning bearing damage, impulses in the real and imaginary parts of the impedance signal can occur, similar to those in the pre-run-in stage [5]. These can be caused by the temporary breakdown of the isolating lubricant film during damage rollover. These impulses could again lead to low frequencies in the spectrum. This effect can be observed in Figure 16 in the spectrum right before initial damage detection at $D^* \approx 1$. This again causes a decline in the RMS frequency and the other investigated features, explaining the described behavior before the initial damage detection by the test rig.

Now, after interpreting the rough trend of these features, the described abnormalities in the graphs are addressed as well. The seemingly different paces at which the features rise in the different fatigue tests originate in an approximately constant run-in duration and differing fatigue test times. As a result, the duration of the pre-run-in stage relative to the fatigue test time correlates to the fatigue test time itself, leading to the described effect. Visible gaps in the graphs are caused by a temporary failure of the impedance measurement system. For the duration of the gap, no impedance measurements exist, while the bearing has been damaged continuously, which is captured by the test rig and considered with the calculation of the normalized total damage. Sudden jumps of the feature are caused by a stop of the test rig, including the disassembly of the test rig. These disassembly processes are necessary in order to exchange the failed test bearing and unavoidably lead to inaccuracies in the impedance measurements of the second test bearing. To avoid this effect in future experiments, disassembly during tests should be avoided. This can be achieved by exchanging both test bearings after a detected bearing damage instead of exchanging only the damaged one.

Next, the most significant feature in the last hour before initial damage detection is explained, which is named the skewness of the phase angle of the impedance signal in the time domain. The high skewness at the beginning can be seen as an indicator of the pre-run-in stage since the duration of the high skewness shows the same effects as the duration of the pre-run-in stage described in the previous paragraph.

Before the end of the fatigue test, the skewness again shows distinctly higher values. The high feature values in both the pre-run-in and damaged phase can be phenomenologically explained: A positively skewed distribution describes a graph as shown in Figure 17 [33].

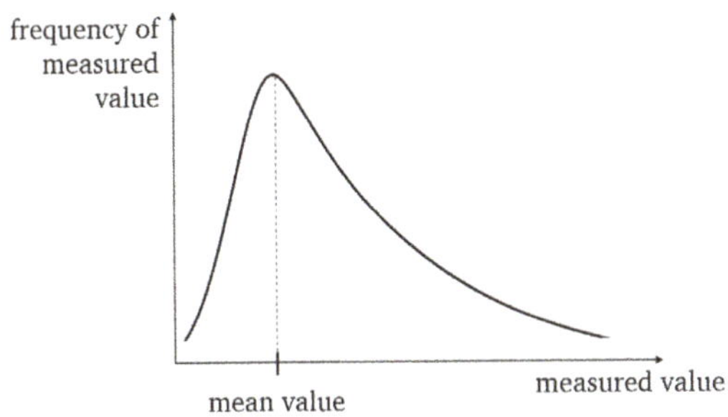

Figure 17. Distribution with positive skewness [35].

In the impedance measurement, the steep rise at the left side of the distribution is caused by the clustering of measurement points with an impedance characteristic for an elastohydrodynamic (EHL) contact. This form of contact is the case in the run-in stage with no electrical breakdowns. Because of the capacitive characteristic of a lubricated bearing with EHL contact [15], the phase angle is approximately $\varphi \approx -90°$, as seen on the left side in Figure 18.

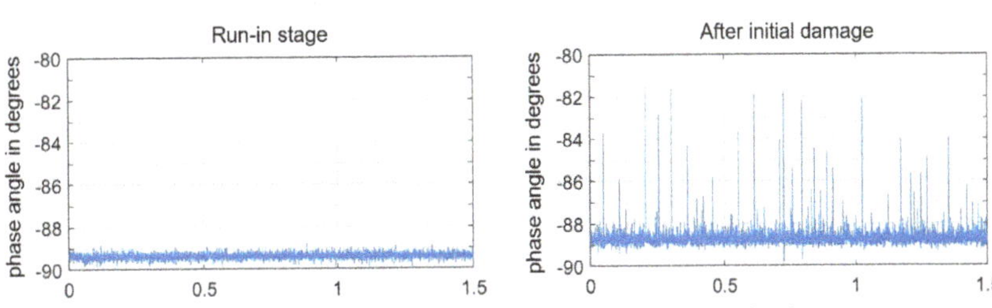

Figure 18. Phase angle of individual impedance measurements of the validation test.

Electrical breakdowns in the pre-run-in and damaged phases lead to signal changes of the impedance with resistive characteristics (see Figure 18 right). Resistive behavior leads to a phase angle closer to zero, which, in this case, results in positive impulses of the phase angle. These positive impulses can be seen in Figure 17 in the form of measurement amplitudes on the right-hand side of the mean value; the more resistive the behavior, the higher the skewness of the phase angle.

The other features in the ranking in Table 4 do not show behavior that clearly shows a distinction between the different test stages. In conclusion, no explanation approach for this behavior is presented in this publication. Nevertheless, the approximately linear behavior of the features in the run-in stage can be useful in damage-progression detection, although the absolute value does not offer as much information about the damage.

In conclusion, the most significant features resulting from the individual feature selection are features in the frequency domain. The features from Table 4 seem to have a relationship with the chosen label. They possess the potential to be useful for early damage detection at rolling bearings and show explainable connections between the bearing impedance and the bearing damage state. It can be said that the information extracted from the impedance signal could be enlarged compared to Martin et al. using this simple feature engineering approach. Therefore, it could be possible to use classification algorithms for more accurate differentiation of different bearing health conditions.

Other impedance-based monitoring approaches focus on the lubrication condition in the EHL contact, e.g., Barz and Maruyama et al. [7,14]. Different methods are used to investigate the lubrication film thickness in rolling element bearings, but they do not include the bearing health condition over its lifespan; this differentiates the approach presented in this paper from the other impedance measurement methods.

4.2. Effects Observed in the Validation Test

In contrast to the findings observed in the other fatigue test, the validation test only showed the expected feature behavior during or after peaks occurring in the vibration signal. Nevertheless, this does not disprove the validity of the impedance features for early rolling bearing detection. This effect is caused by the circumstance of a support bearing failing instead of a test bearing. Thus, none of the bearings, whose impedance has been measured directly, failed, and the impedance consequently remained stable at first.

The support bearing failure may have led to higher stress at the test bearing because of vibrations, impacts, and possible load redistribution. The higher stress of the lubrication film in the test bearings may result in an affinity to metallic contact and, consequently, to resistive behavior. This would explain the phenomena observed in Section 3.2 despite the support-bearing failing. In addition, since the test-bearing impedance is influenced only by a support-bearing failure because of vibrations, the comparatively late impedance feature response in the validation fatigue test can be explained, too.

For both measurement approaches, the same feature was selected. The behavior of the features in both cases was identic with one exception. That means the impedance signal and its features are independent of the measurement approach used to record them. Even if the validation test did not detect damage at a test bearing, failures in other components can be seen in the signal. So, the impedance measurement might be used for condition monitoring not only for rolling bearings but also for other machine elements interacting with the shaft the observed bearings are located at.

Another aspect is that the same features are selected for different rolling bearing types in both tests. The original test was executed using angular groove ball bearings of type 7205. The validation test used deep groove ball bearings of type 6205. Because the feature in both cases showed the same behavior with one exception, it can be said that the impedance is bearing type independent. To explain the signal and feature behavior more precisely, further investigation is needed with a higher variance of bearing types.

4.3. Explanation of Delay between Vibration and Impedance Features

In this section, the possible causes of the delay between the rise of the vibration signal and the observed phenomena in the impedance feature signals shall be examined.

The higher vibrations of a damaged bearing are caused by a crack in the runway surface. During the bearing balls roll over the crack, vibrations are created [1]. However, the bearing impedance might be more sensitive to bearing damage in the early stages. When the crack forms underneath the runway surface, this beginning bearing damage might already have an impact on the electric transition behavior, which would cause changes in the measured impedance signal.

In further research, a comparison of the impedance signal features to the advanced vibration analysis and motor current analysis is necessary. Other papers could show the possibilities of vibration analysis using, e.g., deep feature learning [20]. They are material independent, as mentioned in Section 1.3, which allows a broader application field. In the case of motor current-based condition monitoring, additional sensors are not needed, which is an important cost factor. Impedance-based condition monitoring is applicable for slow-rotating machinery or critical processes and systems [36]. For a higher data quality, disturbance factors have to be identified and analyzed. For system applications, the exact electrical paths through the structure have to be known. First, the results show that disturbance factors have a specific behavior [37] that requires further investigation to be applicable in real applications. A remaining useful life (RUL) prediction for rolling element bearings is not investigated yet. Based on the results discussed before, there is a possibility that the impedance features can be used for RUL prediction. To research this topic, additional fatigue tests are necessary, as well as additional feature engineering methods.

5. Conclusions

The aim of this work was the investigation of impedance signals and their features over the operational time of rolling bearings. The impedance signals have been preprocessed, and individual feature selection was used to extract a higher amount of information from the signals. The features have been analyzed in the time and frequency domain based on the state of research for vibration data. Three phases could be identified in the operative life of a bearing, according to early research. Phenomenological explanations of the feature behavior were derived. In all five fatigue tests, the impedance signal changed before the vibration signals of the test rig sensors showed abnormalities. To clarify this, further research is necessary with a higher amount of fatigue test data. In addition, impedance features in the time-frequency domain have not been investigated yet.

Because uncertainties in the five impedance signals occurred, a more robust measurement approach has been developed and tested in an additional fatigue test. The selected features of both measurement approaches showed the same behavior over the bearing operational life. So, there is the possibility that impedance features map the bearing life independently from the measurement principle. In the validation test, the test bearings did not fail, but the support bearings did. The impedance features changed analog to the vibration signals, which means that the impedance measurement is able to detect damages not only at the observed bearings but also at machine elements located on the same shaft.

In the different test setups, two different bearing types were investigated. Because the impedance shows nearly the same behavior over the bearing's lifetime with one exception, it is possible that the impedance-features behavior is bearing-type independent. Further research is necessary to investigate this phenomenon and explain the feature's behavior.

In summary, it could be shown that impedance measurement can be used for condition monitoring of technical systems. Further research is needed to deepen the understanding of rolling bearing impedance and the features calculated from it. It is also possible to use machine learning algorithms for further investigation. Therefore, more fatigue tests with different operational parameters are necessary. In addition, the changes in bearing type and scale have to be investigated to ensure that the impedance is independent of these factors. In this paper, ball bearings have been used as test bearings. In the future, roller bearings and bearings with line contact, in general, must be examined.

The results presented in this paper show the opportunities for impedance-based condition monitoring. As mentioned in Section 4.3, the technique can be applied for special-use cases where vibration analysis is not sufficient for condition monitoring. The implementation of industrial gearboxes for their observation is already addressed at the Institute. Indicators could be found that the impedance measurement is able to observe not only the rolling element bearings themselves but also the entire gearbox. In this case, further research about the impedance behavior is needed.

Author Contributions: Conceptualization, F.M.B.-D., Q.S.K. and E.K.; methodology, F.M.B.-D. and Q.S.K.; software, F.M.B.-D. and Q.S.K.; validation, F.M.B.-D., Q.S.K. and E.K.; formal analysis, E.K.; investigation, F.M.B.-D. and Q.S.K.; resources, E.K.; data curation, Q.S.K.; writing—original draft preparation, F.M.B.-D. and Q.S.K.; writing—review and editing, E.K.; visualization, Q.S.K.; supervision, E.K.; project administration, F.M.B.-D. and E.K.; funding acquisition, E.K. All authors have read and agreed to the published version of the manuscript.

Funding: The authors thank the Deutsche Forschungsgemeinschaft (DFG, German Research Foundation), which funded the presented research within the project "Early damage detection of rolling bearings by electric impedance measurement". Funded by the Deutsche Forschungsgemeinschaft (DFG, German Research Foundation)—463357020, Gefördert durch die Deutsch Forschungsgemeinschaft (DFG)—463357020.

Data Availability Statement: The authors can be asked for the data used in this paper.

Conflicts of Interest: The authors declare no conflict of interest. The funders had no role in the design of the study; in the collection, analyses, or interpretation of data; in the writing of the manuscript; or in the decision to publish the results.

References

1. Lei, Y. *Intelligent Fault Diagnosis and Remaining Useful Life Prediction of Rotating Machinery*; Elsevier: Oxford, UK, 2016.
2. Randall, R.B. *Vibration-Based Condition Monitoring: Industrial, Aerospace and Automotive Applications*; Wiley: Chichester, UK, 2011.
3. Schaeffler Monitoring Services GmbH. *Condition Monitoring Praxis: Handbuch zur Schwingungs-Zustandsüberwachung von Maschinen und Anlagen*, 1st ed; Vereinigte Fachverlage GmbH: Mainz, Germany, 2019.
4. Schirra, T.; Martin, G.; Vogel, S.; Kirchner, E. Ball Bearings as Sensors for Systematical Combination of Load and Failure Monitoring. In Proceedings of the Design 2018 15th International Design Conference, Dubrovnik, Croatia, 21–24 May 2018; Marjanović, D., Štorga, M., Škec, S., Bojčetić, N., Pavković, N., Eds.; University of Zagreb Faculty of Mechanical Engineering and Naval Architecture: Zagreb, Croatia, 2018; pp. 3011–3022.
5. Martin, G.; Becker, F.M.; Kirchner, E. A novel method for diagnosing rolling bearing surface damage by electric impedance analysis. In *Tribology International 170*; Elsevier: Amsterdam, The Netherlands, 2022. [CrossRef]
6. Martin, G. Die Wälzlagerimpedanz als Werkzeug zur Untersuchung von Oberflächenabweichungen in Wälzlagern. Ph.D. Thesis, Technische Universität Darmstadt, Darmstadt, Germany, 2021.
7. Maruyama, T.; Radzi, F.; Sato, T.; Iwase, S.; Maeda, M.; Nakano, K. Lubrication Condition Monitoring in EHD Line Contacts of Thrust Needle Roller Bearing Using the Electrical Impedance Method. *Lubricants* **2023**, *11*, 223. [CrossRef]
8. Lei, Y.; He, Z.; Zi, Y. A new approach to intelligent fault diagnosis of rotating machinery. *Expert Syst. Appl.* **2008**, *35*, 1593–1600. [CrossRef]
9. Bienefeld, C.; Vogt, A.; Kacmar, M.; Kirchner, E. Feature-Engineering für die Zustandsüberwachung von Wälzlagern mittels maschinellen Lernens. *Tribol. und Schmier.* **2021**, *68*, 5–11. [CrossRef]
10. Prashad, H. *Tribology in Electrical Environments*; Elsevier: Amsterdam, The Netherlands; London, UK, 2006.
11. Alpaydin, E. *Introduction to Machine Learning*, 2nd ed.; The MIT Press: Cambridge, MA, USA, 2010.
12. Soviany, S.; Soviany, C. Feature Engineering. In *Principles of Data Science*; Arabnia, H.R., Daimi, K., Stahlbock, R., Soviany, C., Heilig, L., Brüssau, K., Eds.; Springer: Cham, Switzerland, 2020; pp. 79–103.
13. Schirra, T. Phänomenologische Betrachtung der Sensorisch Nutzbaren Effekte am Wälzlager—Einfluss Unbelasteter Wälzkörper Auf das Elektrische Impedanzmodell. Ph.D. Thesis, Technische Universität Darmstadt, Darmstadt, Germany, 2021.
14. Barz, M. Die Schmierfilmbildung in Fettgeschmierten Schnellaufenden Spindellagern. Ph.D. Thesis, Gottfried Wilhelm Leibniz Universität Hannover, Hannover, Germany, 1996.
15. Gemeinder, Y. Lagerimpedanz und Lagerschädigung bei Stromdurchgang in Umrichtergespeisten Elektrischen Maschinen. Ph.D. Thesis, Technische Universität Darmstadt, Darmstadt, Germany, 2016.
16. Puchtler, S.; Schirra, T.; Kirchner, E.; Späck-Leigsnering, Y.; de Gersem, H. Capacitance calculation of unloaded rolling elements—Comparison of analytical and numerical methods. *Tribol. Int.* **2022**, *176*, 107882. [CrossRef]
17. Harder, A.; Zaiat, A.; Becker-Dombrowsky, F.M.; Puchtler, S.; Kirchner, E. Investigation of the Voltage-Induced Damage Progression on the Raceway Surfaces of Thrust Ball Bearings. *Machines* **2022**, *10*, 832. [CrossRef]

18. Radnai, B.; Gemeinder, Y.; Kiekbusch, T.; Weber, J.; Hering, J.; Arnold, D. Schädlicher Stromdurchgang: Untersuchung des Schädigungsmechanismus und der zulässigen Lagerstrombelastung von Wälzlagern in E-Motoren und Generatoren verursacht durch parasitäre hochfrequente Lagerströme. In *Forschungsvorhaben Nr. 650 I*; Forschungsvereinigungs Antriebstechnik: Frankfurt am Main, Germany, 2015.
19. Muetze, A. *Bearing Currents in Inverter-Fed AC-Motors*; Technische Universität Darmstadt: Darmstadt, Germany, 2003.
20. Saucedo-Dorantes, J.J.; Arellano-Espitia, F.; Delgado-Prieto, M.; Osornio-Rios, R.A. Diagnosis Methodology Based on Deep Feature Learning for Fault Identification in Metallic, Hybrid and Ceramic Bearings. *Sensors* **2021**, *21*, 5832. [CrossRef] [PubMed]
21. Shi, H.; Hou, M.; Wu, Y.; Li, B. Incipient Fault Detection of Full Ceramic Ball Bearing Based on Modified Observer. *International J. Control. Autom. Syst.* **2022**, *20*, 727–740. [CrossRef]
22. Liu, H.; Shah, S.; Jiang, W. On-line outlier detection and data cleaning. In *Computers and Chemical Engineering*; Elsevier: Amsterdam, The Netherlands, 2004; Volume 28, pp. 1635–1647.
23. Pal, S.K.; Mishra, D.; Pal, A.; Dutta, S.; Chakravarty, D.; Pal, S. *Digital Twin—Fundamental Concepts to Applications in Advanced Manufacturing*; Springer: Cham, Switzerland, 2022.
24. Preusche, C. Clusterbasierte Zustandsbewertung von Technischen Systemen zur Unterstützung der Prädiktiven Instandhaltung. Ph.D. Thesis, Technische Universität Darmstadt, Darmstadt, Germany, 2018.
25. Kateris, D.; Moshou, D.; Pantazi, X.-E.; Gravalos, I.; Sawalhi, N.; Loutridis, S. A machine learning approach for the condition monitoring of rotating machinery. *J. Mech. Sci. Technol.* **2014**, *28*, 61–71. [CrossRef]
26. Akpudo, U.E.; Hur, J.-W. Towards bearing failure prognostics: A practical comparison between data-driven methods for industrial applications. *J. Mech. Sci. Technol.* **2020**, *34*, 4161–4172. [CrossRef]
27. Slavič, J.; Mršnik, M.; Česnik, M.; Javh, J.; Boltežar, M. Uniaxial vibration fatigue. In *Vibration Fatigue by Spectral Methods: From Structural Dynamics to Fatigue*; Elsevier: Amsterdam, The Netherlands, 2020; pp. 99–113.
28. Naunheimer, H.; Bertsche, B.; Ryborz, J.; Novak, W.; Fietkau, P. *Fahrzeuggetriebe: Grundlagen, Auswahl, Auslegung und Konstruktion*, 3rd ed.; Springer: Berlin Heidelberg, Germandy, 2019.
29. *ISO-281: Referenznummer ISO 281:2007(E)*; Internationale Organisation für Normung, 2nd ed. International Organization for Standardization: Geneva, Switzerland, 2007.
30. Steinhilper, W.; Sauer, B. *Konstruktionselemente des Maschinenbaus 2*; Springer: Berlin/Heidelberg, Germany, 2012.
31. Akpudo, U.E.; Hur, J.-W. A feature fusion-based prognostics approach for rolling element bearings. *J. Mech. Sci. Technol.* **2020**, *34*, 4025–4035. [CrossRef]
32. Hedderich, J.; Sachs, L. *Angewandte Statistik: Methodensammlung Mit R*, 17th ed.; Springer: Berlin/Heidelberg, Germany, 2020.
33. Fahrmeir, L.; Heumann, C.; Künstler, R.; Pigeot, I.; Tutz, G. *Statistik: Der Weg zur Datenanalyse*, 8th ed.; Springer: Berlin/Heidelberg, Germany, 2016.
34. Dahlke, H. *Handbuch Wälzlager-Technik: Bauarten, Gestaltung, Betrieb*, 1994th ed.; Springer Fachmedien: Wiesbaden, Germany, 1994.
35. Kosfeld, R.; Eckey, H.F.; Türck, M. *Deskriptive Statistik: Grundlagen—Methoden—Beispiele—Aufgaben*, 6th ed.; Springer Gabler: Wiesbaden, Germany, 2016.
36. Becker-Dombrowsky, F.M.; Zaiat, A.; Kirchner, E. Impedanzmessung an Wälzlagern—Servicemodelle zur Überwachung technischer Anlagen. In Proceedings of the VDI-Berichte, 2415, 15. VDI-Fachtagung Gleit- und Wälzlagerungen 2023, Schweinfurt, Germany, 13—14 June 2023; VDI Verlag GmbH: Düsseldorf, Germany. ISBN 978-3-18-092415-1.
37. Becker-Dombrowsky, F.M.; Hausmann, M.; Welzbacher, P.; Harder, A.; Kirchner, E. Systematic identification of disturbance factors on electric characteristics of mechanical gearboxes. In *Forschung im Ingenieurwesen 87*; SpringerNature: Berlin, Germany, 2023; pp. 399–410. [CrossRef]

lubricants

MDPI

Article

Classification of Lubricating Oil Types Using Mid-Infrared Spectroscopy Combined with Linear Discriminant Analysis–Support Vector Machine Algorithm

Jigang Xu [1], Shujun Liu [2,*], Ming Gao [3] and Yonggang Zuo [2]

[1] Unit68709, Qinghai, Haidong 810700, China; xujigang216@163.com
[2] Army Logistic Academy, Chongqing 401331, China
[3] Beijing Aeronautical Technoloogy Research Center, Beijing 100076, China
* Correspondence: jjxyliushujun@163.com

Abstract: To realize the classification of lubricating oil types using mid-infrared (MIR) spectroscopy, linear discriminant analysis (LDA) was used for the dimensionality reduction of spectrum data, and the classification model was established based on the support vector machine (SVM). The spectra of the samples were pre-processed by interval selection, Savitzky–Golay smoothing, multiple scattering correction, and normalization. The Kennard–Stone algorithm (K/S) was used to construct the calibration and validation sets. The percentage of correct classification (%CC) was used to evaluate the model. This study compared the results obtained with several chemometric methods: PLS-DA, LDA, principal component analysis (PCA)-SVM, and LDA-SVM in MIR spectroscopy applications. In both calibration and verification sets, the LDA-SVM model achieved 100% favorable results. The PLS-DA analysis performed poorly. The cyclic resistance ratio (CRR) of the calibration set was classified via the LDA and PCA-SVM analysis as 100%, but the CRR of the verification set was not as good. The LDA-SVM model was superior to the other three models; it exhibited good robustness and strong generalization ability, providing a new method for the classification of lubricating oil types by MIR spectroscopy.

Keywords: mid-infrared spectra; lubricating oil; LDA-SVM; Kennard–Stone algorithm

Citation: Xu, J.; Liu, S.; Gao, M.; Zuo, Y. Classification of Lubricating Oil Types Using Mid-Infrared Spectroscopy Combined with Linear Discriminant Analysis–Support Vector Machine Algorithm. *Lubricants* **2023**, *11*, 268. https://doi.org/10.3390/lubricants11060268

Received: 12 May 2023
Revised: 15 June 2023
Accepted: 17 June 2023
Published: 20 June 2023

1. Introduction

Lubricating oils play a crucial role in industrial practices, serving various functions to ensure the smooth operation of machinery. In the process of mechanical operation, if some parts of the machine do not have the lubrication effect of lubricants, dry friction will occur, causing machine damages. According to experimental data, considerable heat generated by dry friction in a short period of time can melt the metal and even damage the machine. The major working principle is as follows: Lubricating oil which exists between working parts of a machine produces the membrane that can reduce the resistance of the parts in actual work by wrapping an oil film on their surface. Oil films are produced by lubricating oil. Toughness and strength are important indicators for lubricants to play a role. The main aims of gear lubrication are to diminish friction, increase efficiency, reduce wear and contact fatigue of the interacting tooth surfaces, and improve durability [1]. According to the literature [2,3], the gear transmission systems with and without lubrication are very different. A major reduction in energy waste and emissions of mechanical systems can be seen with the optimized performance of lubricating oil [4–6].

Lubricating oil mainly comprises basic oil, which governs its basic properties, and additives that enhance the performance of basic oil, providing certain new functions [7]. As seen from data shown in [8–12], lubricants with different types of additives are supposed to lead to different effects.

It is challenging to distinguish the types of lubricating oil solely from their appearance because of the similarities of their constituents: basic oil and small additives. In the process of using lubricating oil, once the label is defaced or lost, it will lead to misuse, which will lead to engine failure, equipment failure, performance gradation, and even accidents. The lubricating oils and the unknown additive types and contents are qualitatively classified and analyzed using physical and chemical methods. Traditional methods, such as Raman spectroscopy [13], physical and chemical characterization, and gas chromatography, are time-consuming and expensive. The composition of lubricating oil is complex, with various types of additives and wide-ranging mid-infrared (MIR) spectroscopy features. Different additives have their own characteristic peaks in the MIR spectra, but because the characteristic peaks seriously overlap, it is challenging to distinguish different lubrication oils directly using MIR, and chemometric methods are required. In recent years, MIR spectroscopy has been widely used in the determination of oil concentration in water [14], molecular structure analysis of new and in-use engine oils [15], analysis of oil sludge [16], determination of soot content in engine oil [17], qualitative and quantitative analysis of sulfur content in crude oil [18], and the detection of oil pollution [19].

Recent research studies on both crude oil and lubricating oil through the method of infrared spectroscopy combined with chemometrics, such as the chemometric strategy based on pattern recognition which has been developed for clustering and the classification of crude oils of Iran, can be seen in the literature [20]. GC-FID and FT-IR fingerprints were considered for fingerprint analysis, and the potential of PCA/HCA for clustering and PLS-DA/CP-ANN for classification were studied. A hybrid optimization method for feature band selection of the middle infrared spectrum based on binary particle swarm optimization (BPSO) and the genetic algorithm (GA) has been developed by Xia Yanqiu et al. [21]. Firstly, the basic classification model of oil additive species recognition by the K nearest neighbor algorithm (KNN) and random forest algorithm (RF) is established. Then, the GA-BPSO hybrid optimization algorithm is used to screen the characteristic band region in the whole band range of the spectrum. O. Galtier et al. [22] compared the results which were obtained by several chemometric methods, SIMCA, PLS2-DA, PLS2-DA with SIMCA, and PLS1-DA, in two infrared spectroscopic applications, which were optimized by selecting spectral ranges containing discriminant information. In the first application, mid-infrared spectra of crude petroleum oils were classified according to their geographical origins. In the second application, near-infrared spectra of French virgin olive oils were classified in five registered designations of origins (RDOs). In both cases, the PLS1-DA classification indicated a 100% good result. An extreme learning machine was used to train and test the model constructed by the infrared spectral data of the mixed additives, and the greedy algorithm and genetic algorithm were used to optimize the input band, while the optimization results were compared. The test results showed both effective identification of the type and prediction of the content of lubricant additives [23]. Owing to the characteristic that the MIR spectroscopy of lubricating oils provides both linear and nonlinear information, the linear discriminant analysis–support vector mechanism (LDA-SVM) model is proposed, which uses LDA for supervised dimensionality reduction, SVM for classification, and provides a theoretical basis for the rapid classification of lubricating oils.

2. Materials and Methods

2.1. Materials

2.1.1. Samples

A total of 120 Lubricating oil samples (Figure 1) from different lubricating oil manufacturers were analyzed using MIR spectroscopy to identify their types: gear oil, $n = 13$; diesel oil, $n = 41$; gasoline engine oil, $n = 12$; general engine oil, $n = 33$; hydraulic oil, $n = 21$.

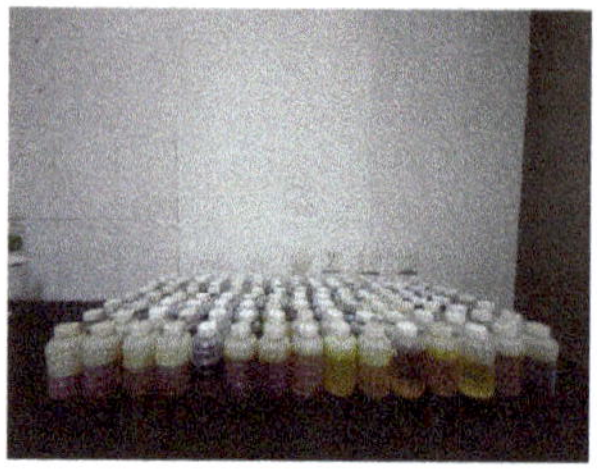

Figure 1. 120 Lubricating oil samples.

2.1.2. Experimental Instruments and Parameters

Instrument: Tensor27 Fourier transform infrared spectrometer produced by BRUKER (Mannheim, Germany), in Figure 2.

Figure 2. BRUKER Tensor27.

Measurement method: transmission method;
Optical path: 0.1 mm;
Measurement parameters: resolution 4 cm^{-1};
Beam range: 600–4000 cm^{-1};
Spectral averaging times: 16 times;
Windows and beam splitters: ZnSe.
Original MIR spectral data of samples are shown in Figure 3.

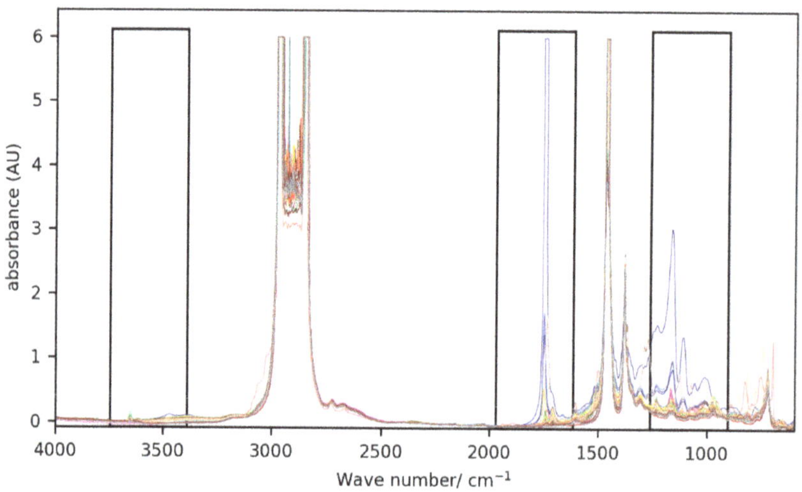

Figure 3. Original MIR spectral data of samples and the selected ranges for modeling.

2.2. Methods

2.2.1. Spectral Data Pre-Processing

Spectral data pre-processing was mainly performed to select the spectral data range and eliminate electrical noise, sample background light, and stray light from the spectral data. The pre-processing method of spectral data greatly influences the stability and generalization ability of the model. In this study, the spectral data pre-processing method was as follows:

(1) Wave number range. Different types of lubricating oils have characteristic peaks in the photon region and fingerprint region of the MIR spectrum, according to the characteristics of the lubrication oil spectrum. The spectral data used in this study consisted of three ranges: 3743.7–3386.9, 1969.3–1612.4, and 1259.5–902.7 cm^{-1} [7]. Figure 3 shows the MIR spectrum of the original data of the experimental samples. The spectral data in the three black boxes were selected for modeling.

(2) Smooth processing. The Savitzky–Golay convolution smoothing method was used to remove random noise in the spectrum and improve the signal-to-noise ratio.

(3) Multiplicative scatter correction (MSC). MSC was used to eliminate the spectral differences caused by different scattering levels, thereby enhancing the correlation between the spectra and data. Assuming the spectrum $x(1 \times m)$, the MSC algorithm was as follows: ① the average spectrum $\overline{x}$ of the samples was calculated; ② linear regression was performed on x and $\overline{x}$, $x = b_0 + \overline{x}b$, and the least squares method was used to determine b_0 and b; ③ $(x - b_0)/b_0$.

(4) Normalization. Also known as vector normalization, for a spectrum, first its average absorbance value was calculated, the average value from the spectrum was subtracted, and then the sum of the squares of the spectrum was divided. Normalization can eliminate spectral variations caused by small optical path differences. The normalization calculation formula was as follows:

$$x'_k = \frac{x_{ik} - \overline{x}}{\sqrt{\sum_{i=1}^{n} x_{ik}^2}} \tag{1}$$

$\overline{x}$ is mean of the vector, x_{ik} is a value of normalization, x'_k is the result of normalization.

Figure 4 shows a flow chart of spectral data pre-processing. The spectral ranges were optimized and selected first and subsequently smoothed; then, MSC and finally normalization were performed.

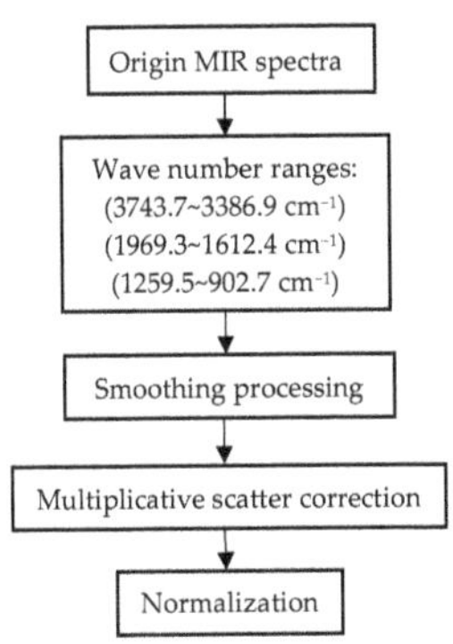

Figure 4. Spectral data pre-processing flow.

2.2.2. Dimensionality Reduction Using LDA Algorithm

LDA, proposed by Fisher in 1936, is a supervised dimensionality reduction technology and is widely used in feature extraction. The LDA algorithm predominantly involves projecting the sample data with large dimensions to the best classification vector area to identify the data and narrow the feature range, and after the projection, it ensures that the data have a large inter-class distance and small intra-class distance; that is, the samples can be well separated within this range. Each sample of its dataset has a class output. This is different from principal component analysis (PCA). LDA uses the Fisher discriminant criterion, so it is also known as Fisher's linear discriminant. The LDA algorithm is widely used in the field of pattern recognition [24–28].

(1) Principle of LDA. Assuming d-dimensional (d features) spectral samples $X = [X_1, \ldots, X_n] \in R^{n \times N}$, $X_i (i = 1, \ldots, N) \in R^n$ represents the i-th sample, and N represents the total number of samples. $X_{ij} \in R^n (i = 1, \ldots, c; j = 1, \ldots, N_I)$ represents the j-th sample in class i, N_i represents the number of samples of the i-th class, and c represents the number of sample classes. The mean of all samples is:

$$\overline{x} = \frac{1}{N} \sum_{i=1}^{N} x_i \tag{2}$$

Let the sample mean of the i-th class be $\overline{x_i} (i = 1, 2, \ldots, c)$, then we have

$$\overline{x} = \sum_{i=1}^{c} \frac{N_i}{N} \overline{x}_i \tag{3}$$

Dimensionality reduction using LDA is used to reduce high-dimensional spatial feature information to a low-dimensional feature space according to the existing category information. The LDA results show that samples of the same type are clustered together, and samples of different types are separated as much as possible. The inter-class and intra-class distances are expressed in the form of discrete matrices, and the change matrix W_{opt} was solved using Fisher's criterion. Fisher's criterion is expressed as follows:

$$J(W) = \mathrm{arg}max \frac{|W^T S_b W|}{|W^T S_w W|} \tag{4}$$

As in (4), S_b is an inter-class discrete matrix, and its specific expression is:

$$S_b = \sum_{i=1}^{c} \frac{N_i}{N} (\overline{x}_i - \overline{x})(\overline{x}_i - \overline{x})^T \tag{5}$$

As in (4), S_w is an intra-class discrete matrix, and its expression is:

$$S_w = \sum_{i=1}^{c} \sum_{j=1}^{N_i} \frac{1}{N} \frac{N_i}{N} (x_{ij} - \overline{x})(x_{ij} - \overline{x})^T \tag{6}$$

Equation (4) is the generalized Rayleigh entropy of matrix S_b relative to matrix S_w. Using the properties of the generalized Rayleigh entropy, the optimal solution for calculating $J(W)$ is $W_{opt} = (w_1, w_2, w_3 \ldots, w_d)$, where $w_1, w_2, w_3 \ldots, w_d$ are the eigenvectors corresponding to the first d non-zero eigenvalues of $S_w^{-1} S_b$.

(2) The steps of LDA are as follows:

① Intra-class divergence matrix S_w was calculated;

② Inter-class divergence matrix S_b was calculated;

③ Matrix $S_w^{-1} S_b$ was calculated;

④ The largest d eigenvalues of $S_w^{-1} S_b$ and the corresponding eigenvectors $(w_1, w_2, \ldots, w_d)$ were calculated to obtain the optimal solution W_{opt};

⑤ $z_i = W_{opt}^T x_i$ was calculated for each sample x_i in the sample set;

⑥ The output sample set $D = \{(z_1, y_1), (z_2, y_2), \ldots, (z_m, y_m)\}$ was obtained.

2.2.3. SVM Algorithm

SVM is a classification technology proposed in 1963 by the AT&T Bell laboratory research group led by Vapnik. SVM is a pattern recognition method based on statistical learning theory, which is mainly used in the field of pattern recognition [29,30]. It provides numerous unique advantages for solving small sample, nonlinear, and high-dimensional pattern recognition, and it can be extended to other machine learning problems such as function fitting. The SVM mechanism involves finding an optimal classification hyperplane that meets the classification requirements so that the hyperplane can maximize the blank areas on both sides of the hyperplane while ensuring classification accuracy. SVM can achieve the optimal classification of linearly separable data.

Taking two types of data classification as examples, given a sample set (x_i, y_i), $i = 1,2,\ldots,l$, $x \in R^n$, $y \in \{\pm 1\}$, with the hyperplane denoted as $(w \cdot x) + b = 0$, to correctly classify all samples and have a classification interval, the following constraints are required:

$$\min_{w,b} \frac{1}{2}\|w\|^2 \tag{7}$$

$$y_i[(w \cdot x_i) + b] \geq 1;\ i = 1,2,3\ldots l \tag{8}$$

This is a convex quadratic programming problem that was solved using the Lagrange function:

$$\mathrm{L}(w,b,a) = \frac{1}{2}\|w\| - a(y((w \cdot x) + b) - 1) \tag{9}$$

The optimal solution was determined by finding the maximum value:

$$\boldsymbol{a}^* = (a_1^*, a_2^*, a_3^*, \ldots, a_l^*)^{\mathrm{T}} \tag{10}$$

The optimal weight vector w^* and the optimal bias b^* were calculated as follows:

$$w^* = \sum_{j=1}^{l} a_j^* y_j x_j \tag{11}$$

$$b^* = y_i - \sum_{j=1}^{l} y_j a_j^* (x_j \cdot x_i) \tag{12}$$

For the linear inseparable case, the kernel method was used. The main idea was to project the input vector to a high-dimensional feature vector space and construct the optimal classification surface in the feature space. The linear discriminant function was constructed in the high-dimensional space, and the commonly used kernel functions were as follows:

① Linear kernel function: $K(x, x_i) = \langle x, x_i \rangle$;

② Polynomial kernel function: $K(x, x_i) = [\gamma(x \cdot x_i) + coef]^d$, where d is the order of the polynomial, and coef is the bias coefficient;

③ RBF kernel function: $K(x, x_i) = \exp\left(-\gamma\|x - x_i\|^2\right)$, where γ is the width of the kernel function;

④ Sigmoid kernel function: $K(x, x_i) = \tanh(\gamma(x \cdot x_i) + coef)$, where γ is the width of the kernel function and coef is the bias coefficient.

2.3. *Construction of Calibration Set and Validation Set*

2.3.1. K/S Algorithm

The K/S algorithm [31] can provide the best expression of the difference between samples and select more representative samples. The K/S algorithm was used to select the sample set, and the steps were as follows: (1) The Euclidean distance between the two samples was calculated, and the two samples with the largest distance were selected for the calibration set. (2) The distance between each remaining sample and the selected calibration set was calculated, and the two farthest and nearest samples were determined and selected for the calibration set. (3) Step (2) was repeated until the number of the selected calibration samples was equal to the predetermined number. (4) The remaining samples were the samples of the validation set.

2.3.2. Specific Construction of Calibration Set and Validation Set

The calibration set and verification set were constructed by the K/S algorithm with a ratio of 6:4 for the spectral data of gear oil, diesel oil, gasoline oil, general oil, and hydraulic oil samples. The specific sample distribution is listed in Table 1, and the statistical distribution of MIR spectral data of samples in the calibration set and prediction set is listed in Table 2.

Table 1. Composition of calibration and validation set.

Sample Types	Calibration Set	Validation Set	Sum of Sample
Gear oil	8	5	13
Diesel engine oil	25	16	41
Gasoline engine oil	8	5	13
All-purpose engine oil	20	13	33
Hydraulic oil	13	9	22
Total number of samples	74	46	120

Table 2. Statistical distribution of MIR spectral data of samples in calibration set and prediction set.

Sample (Unit)	Data Sets	Number of Samples	Maximum	Minimum	Mean	Standard Deviation
Lubricating oils	Calibration set	74	6.0	−0.065	0.070	0.163
	Validation set	46	1.732	−0.063	0.064	0.117

2.4. LDA-SVM Algorithm Steps

Step 1: Data pre-processing. The spectral range was optimized, the signal-to-noise ratio was improved, and the influence of stray light was eliminated;

Step 2: The K/S algorithm was used to divide the sample data to ensure the representativeness of the calibration set and validation set;

Step 3: Supervised dimensionality reduction was performed on the calibration set using LDA, and the optimal vector W_{opt} was calculated;

Step 4: The dimensionality reduction result was provided as the input of SVM, and the grid search method was used to automatically search and calculate the optimal parameters of SVM, when the kernel functions were linear, poly, RBF, and sigmoid;

Step 5: The dimensionality reduction result of the validation set was calculated through the optimal vector W_{opt} obtained in step 3;

Step 6: The optimal parameters were used to predict the validation set through SVM.

2.5. Experimental Design

As shown in Figure 5, the original infrared spectrum data of the lubricating oils were pre-processed, the data were divided into calibration and validation sets by the K/S algorithm, and the calibration set was input into four models: 1. The PLS-DA model was used to calculate the percentage of correct classification (%CC) of the calibration set under different latent variable numbers, and the principal component number with the highest correct rate was selected. 2. The LDA model, when the matrix was decomposed with singular value decomposition (SVD), least square (lsqr), eigenvalue decomposition (eigen), and the %CC of the calibration set and validation set were calculated, and the optimal results were selected. 3. When the principal component number of PCA was 2–40, the results of dimensionality reduction were used as the input of SVM. The grid search method was used to search the hyperparameters automatically to obtain the optimal solutions of the kernel functions when they were linear, poly, RBF, and sigmoid. 4. The principal component number of LDA was 2, 3, or 4, the dimension reduction results were taken as the input of SVM, and the grid search method was used to search the hyperparameters automatically to obtain the optimal solutions of kernel functions when they were linear, poly, RBF, and sigmoid.

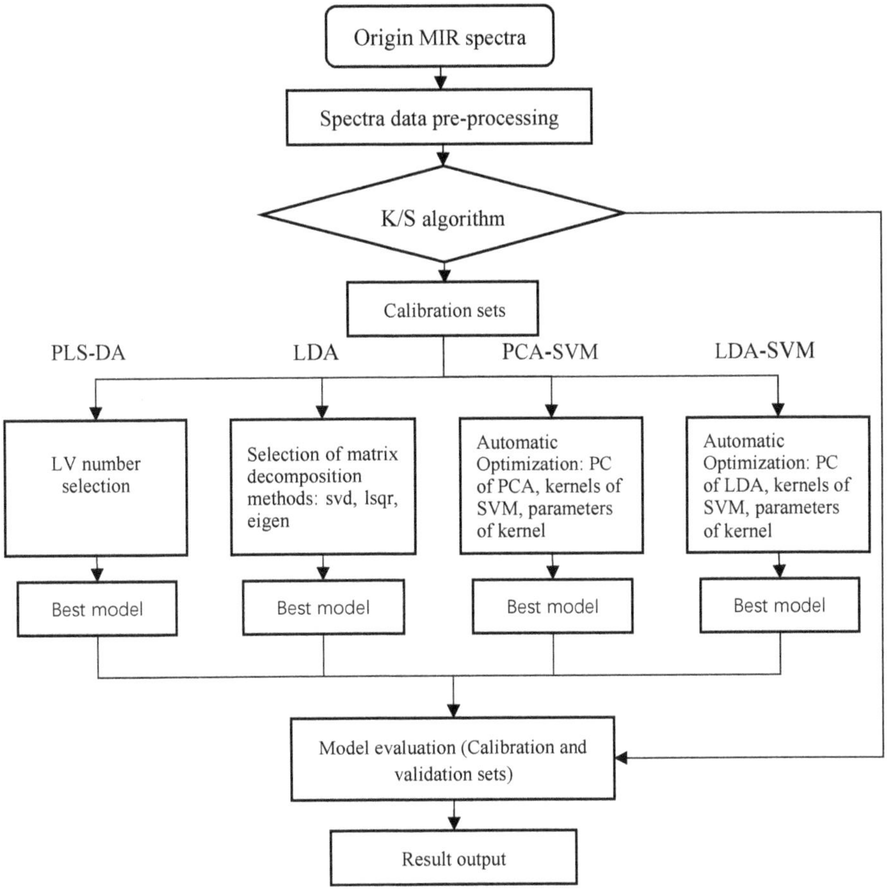

Figure 5. Experimental design.

The %CC was the criterion used to compare classification results.

$$\%CC = N_c/(N_c + N_{ic}) \tag{13}$$

where N_c and N_{ic} represent the numbers of incorrect and correct identifiers, respectively.

PLS-DA, LDA, PCA-SVM, and LDA-SVM models were built using the Keras and Scikit-learn machine learning library. They were developed based on Python 3.7.0, and the data mining and data analysis tools adopted Scikit-learn 0.23.2. The programming platform is based on Jupiter Notebook 4.4.0 and runs on the Windows 10 operating system.

3. Results and Discussion

3.1. PLS-DA Model

The number of latent variables is an important parameter in the PLS-DA model; when the number of latent variables is small, it leads to insufficient feature extraction, and when the number of latent variables is large, it leads to noise information. The %CC of the calibration and validation sets is shown in Figure 6. The number of latent variables ranges from 2 to 74, and the %CC of the calibration set increases with the number of latent variables; when the number of latent variables is >36, the cyclic resistance ratio (CRR) remains unchanged at 100%. The %CC of the validation set fluctuated greatly with the number of latent variables, and when the number of latent variables was 22, the %CC reached its maximum, 78%. The PLS-DA model was over-fitted by comparing the results of calibration and validation sets. When the number of latent variables was 22, the sum of the %CC of the calibration and validation sets reached the maximum value.

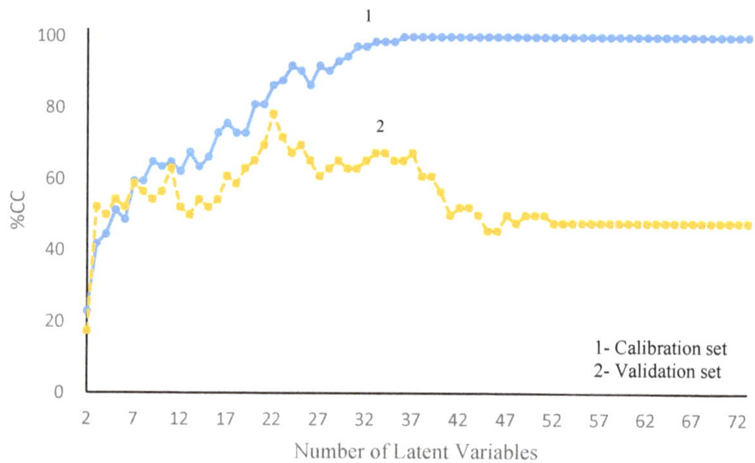

Figure 6. %CC for calibration and validation sets under different numbers of latent variables with PLS-DA model.

3.2. LDA Model

The %CC of the calibration and validation sets is listed in Table 3; different matrix decomposition algorithms have a certain influence on the results of the LDA model. When the matrix decomposition algorithms were used by SVD, the %CC of the calibration and validation sets was 100% and 95%, respectively. When the matrix decomposition algorithms were used by lsqr and eigen, the %CC of the calibration and validation sets was 95% and 98%, respectively. By comparing the three decomposition algorithms, we observe that SVD decomposition algorithms are favorable, where the sum of the %CC of the calibration and validation sets reaches the highest value.

Table 3. %CC for calibration and validation sets under different decomposition methods with LDA model.

Decomposition Method	Calibration Sets (%CC)	Validation Sets (%CC)
SVD	100	95
sqlr	95	97
eigen	95	97

3.3. PCA-SVM Model Recognition Results

PCA is an unsupervised dimensionality reduction technique. The main factors affecting the PCA-SVM model are as follows: principal component number, kernel function, and kernel function parameters. The kernel functions of SVM are linear, poly, RBF, and sigmoid when the principal component number ranges from 2 to 42, and grid search is used for automatic hyperparameter search. As shown in Figure 7a,c, the principal component numbers negligibly influence the linear and RBF kernel functions. When the kernel functions are linear, the %CC of the calibration and validation sets are 89% and 85%, respectively, When the kernel functions are RBF, the %CC of the calibration and validation sets is 100% and 93%, respectively. As shown in Figure 7b, when the kernel function is poly, the %CC of the calibration and validation sets increases first and then decreases. When the principal component number is 16, the %CC of the calibration and validation sets is 100% and 89%, respectively. As shown in Figure 7d, the %CC of the calibration and validation sets increases with an increase in the principal component number and finally stabilizes. When the principal component number is 30, the %CC of calibration and validation sets is 91% and 89%, respectively. Comparing the results of the different kernel functions, the best prediction result of the PCA-SVM model is achieved using the RBF kernel function, and the %CC of the calibration and validation sets is 100% and 93%, respectively.

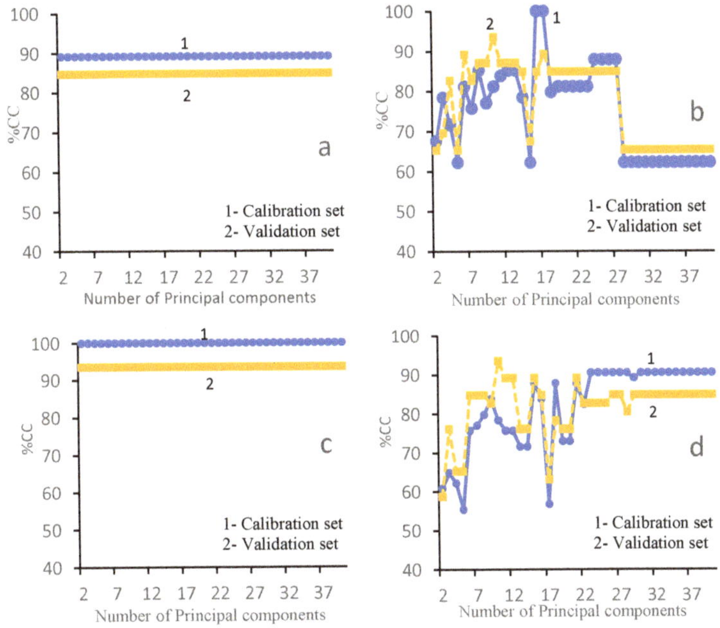

Figure 7. %CC for calibration and validation sets of PCA-SVM model for the following: (**a**) number of principal components used in linear kernel function; (**b**) number of principal components used in poly kernel function; (**c**) number of principal components used in RBF kernel function; (**d**) number of principal components used in sigmoid kernel function.

3.4. LDA-SVM Model

LDA is a dimensionality reduction technique. The main factors that affect the classification results of the LDA-SVM model are as follows: the principal component number, kernel function, and kernel function parameters. When the principal component number is 2, 3, or 4, and the kernel functions of SVM are linear, poly, RBF, and sigmoid, respectively, grid search is used for automatic hyperparameter search to obtain the optimal solutions. The %CC of the calibration and validation sets is listed in Figure 8. The %CC of the calibration and validation sets increases with an increase in the principal component number of LDA, and when the principal component number is 4, the %CC of the calibration and validation sets becomes maximized. Comparing the results of different kernel functions, the best prediction result of the PCA-SVM model is exhibited by the poly kernel function, and the %CC of the calibration and validation sets is 100%.

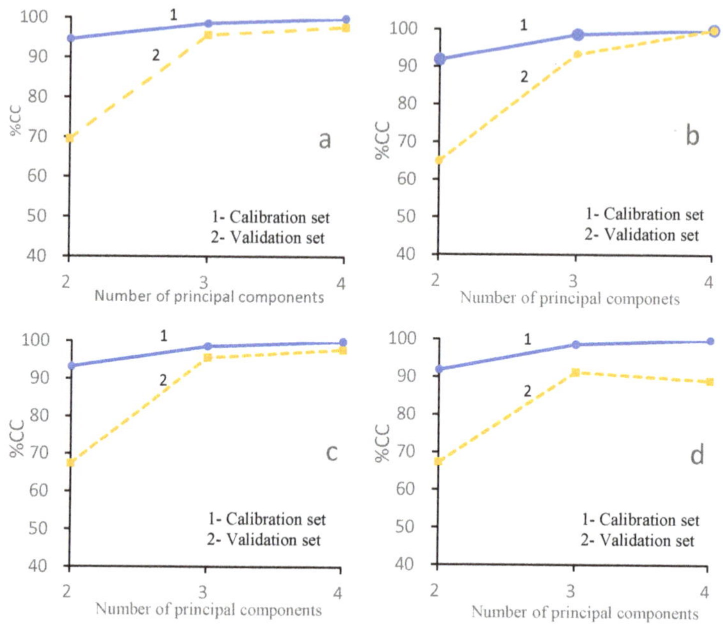

Figure 8. %CC for calibration and validation sets as LDA-SVM model of the following: (**a**) number of principal components used in linear kernel function; (**b**) number of principal components used in poly kernel function; (**c**) number of principal components used in RBF kernel function; (**d**) number of principal components used in sigmoid kernel function.

3.5. Comparison of Model Classification Results

The classification results of PLS-DA, LDA, PCA-SVM, and LDA-SVM are listed in Table 4. The PLS-DA model exhibits the worst recognition ability, the over-fitting phenomenon is serious, and the CRR of calibration and validation sets is poor. When classified using the LDA and PCA-SVM model, the CRR of the calibration set achieved 100%, but the CRR of the validation set is unfavorable; the LDA-SVM has the best recognition, and the CRR of the calibration and validation sets is 100%.

Table 4. Correct classification of calibration and validation sets of different models.

Model	Parameter	Calibration Sets (%CC)	Validation Sets (%CC)
PLS-DA	LV = 22	86%	78%
LDA	Decomposition method = SVD	100%	95%
PCA-SVM	PC = 2, kernel = RBF	100%	94%
LDA-SVM	PC = 4, kernel = poly	100%	100%

4. Conclusions and Future Scope

A classification model based on LDA-SVM was proposed. In this model, LDA was used for the dimensionality reduction of the MIR spectrum of lubricating oils, the samples of the same class were clustered together, and the samples of different classes were separated as far as possible. The results of dimensionality reduction were input to SVM. The results demonstrated that LDA-SVM exhibited higher recognition accuracy and robustness than PLS-DA, LDA, and PCA-SVM models. LDA-SVM is a suitable tool to identify lubricating oil types via MIR spectra.

In the next work, a semi-supervised learning method and an interval selection algorithm will be combined to study the improved LDA-SVM algorithm for oil classification.

Author Contributions: Methodology, J.X.; software, J.X.; formal analysis, J.X.; data curation, M.G.; writing—original draft preparation, S.L.; writing—review and editing, Y.Z. All authors have read and agreed to the published version of the manuscript.

Funding: This research received no external funding.

Data Availability Statement: Not applicable.

Acknowledgments: We thank zhiyun polish (https://www.zhiyunwenxian.cn, accessed on 3 June 2023) for its linguistic assistance during the preparation of this manuscript.

Conflicts of Interest: The authors declare no conflict of interest.

References

1. Liu, H.; Liu, H.; Zhu, C.; Parker, R.G. Effects of lubrication on gear performance: A review. *Mech. Mach. Theory* **2020**, *145*, 103707. [CrossRef]
2. Feng, K.; Pietro, B.; Wade, A.S.; Robert, B.R.; Zhan, Y.C.; Ren, J.; Peng, Z. Vibration-based updating of wear prediction for spur gears. *Wear* **2019**, *426–427*, 1410–1415. [CrossRef]
3. Feng, K.; Ji, J.C.; Ni, Q. A novel gear fatigue monitoring indicator and its application to remaining useful life prediction for spur gear in intelligent manufacturing systems. *Int. J. Fatigue* **2023**, *168*, 107459. [CrossRef]
4. Talbot, D.; Kahraman, A.; Li, S.; Singh, A.; Xu, H. Development and validation of an automotive axle power loss model. *Tribol. Trans.* **2016**, *59*, 707–719. [CrossRef]
5. Fernandes, C.M.; Battez, A.H.; González, R.; Monge, R.; Viesca, J.; García, A.; Martins, R.C.; Seabra, J.H. Torque loss and wear of fzg gears lubricated withwind turbine gear oils using an ionic liquid as additive. *Tribol. Int.* **2015**, *90*, 306–314. [CrossRef]
6. Krantz, T.; Tufts, B. Pitting and bending fatigue evaluations of a new case-carburized gear steel. In Proceedings of the ASME 2007 International Design Engineering Technical Conferences and Computers and Information in Engineering Conference, Las Vegas, NV, USA, 4–7 September 2007; p. 215009.
7. Tian, G.Y.; Chu, X.L.; Yi, R.J. *Lubrication Oil infrared Spectral Analysis Technology*; Chemical Industry Press: Beijing, China, 2014; Volume 20, pp. 119–121.
8. Martins, R.; Seabra, J.; Magalhães, L. Micropitting of austempered ductile iron gears: Biodegradable ester vs. mineral oil. *Rev. Assoc. Port. Anál. Exp. Tensões.* **2006**, *13*, 55–65.
9. Adebogun, A.; Hudson, R.; Breakspear, A.; Warrens, C.; Gholinia, A.; Matthews, A.; Withers, P. Industrial gear oils: Tribological performance and subsurface changes. *Tribol. Lett.* **2018**, *66*, 65. [CrossRef]
10. Brandão, J.; Meheux, M.; Ville, F.; Seabra, J.; Castro, J. Comparative overview of five gear oils in mixed and boundary film lubrication. *Tribol. Int.* **2012**, *47*, 50–61. [CrossRef]
11. Bhaumik, S.; Prabhu, S.; Singh, K.J. Analysis of tribological behavior of carbon nanotube based industrial mineral gear oil 250 cSt viscosity. *Adv. Tribol.* **2014**, *2014*, 341365. [CrossRef]
12. Song, W.; Yan, J.; Ji, H. Fabrication of GNS/MoS_2 composite with different morphology and its tribological performance as a lubricant additive. *Appl. Surf. Sci.* **2019**, *469*, 226–235. [CrossRef]
13. Zhao, Z.Y. *The Lubricant Quality Near-Infrared and Raman Spectroscopy Testing Methods*; East China Jiaotong University: Shanghai, China, 2013.
14. Gao, Z.F.; Zeng, L.B.; Shi, L.; Li, K.; Yang, Y.Z.; Wu, Q.S. Development of a portable Mid-Infrared Rapid Analyzer for oil concentration in water Based on MEMS Linear sensor Array. *Spectrosc. Spectr. Anal.* **2014**, *34*, 1711–1715.
15. Yu, H.W.; Wang, X.X.; Li, J.X.; Zhang, Y.X. Study on the structures of New engine oil and used Engine oil by multi-demensional infrared Spectroscopy. *LuBricating Oil* **2021**, *36*, 37–41.
16. Li, H.; Li, J.F.; Xie, L.Q.; Ren, Z.P.; Ding, Y.Q. Analysis of organic component distribution and inorganic mineral composition of tank bottom sludge in change qing oil field. *Anal. Instrum.* **2021**, *52*, 52–59.
17. Zhang, F.Y.; Lang, X.J.; Zhang, D.H.; Liu, J.; Zhang, Y. Determination of soot content in-service Engine oil by Fourier Transform infrared Specrometry. *LuBricating Oil* **2021**, *35*, 45–48+59.
18. Mohammadia, M.; Khorrami, K.M.; Hamid, V.; Karimi, A.; Sadrara, M. Classification and determination of sulfur content in crude oil samples by infrared spectrometry. *Infrared Phys. Technol.* **2022**, *127*, 104382. [CrossRef]
19. Douglas, R.K.; Nawar, S.; Alamar, M.C.; Coulon, F.; Mouazen, A.M. The application of a handheld mid-infrared spectrometry for rapid measurement of oil contamination in agricultural sites. *Sci. Total Environ.* **2019**, *665*, 253–261. [CrossRef]
20. Fatemeh, S.H.N.; Hadi, P. Pattern recognition analysis of gas chromatographic and infrared spectroscopic fingerprints of crude oil for source identification. *Microchem. J.* **2020**, *153*, 104326. [CrossRef]
21. Xia, Y.Q.; Wang, C.; Feng, X. GA-BPSO Hybrid Optimization of Middle Infrared Spectrum Feature Band Selection of Lubricating Oil Additive Type Identification Technology. *Tribology* **2022**, *42*, 42–152.
22. Galtier, O.; Abbas, O.; Le, D.Y. Comparison of PLS1-DA, PLS2-DA and SIMCA for classification by origin of crude petroleum oils by MIR and virgin olive oils by NIR for different spectral regions. *Vib. Spectrosc.* **2011**, *55*, 132–140. [CrossRef]
23. Xia, Y.Q.; Xu, D.W.; Feng, X.; Cai, M.R. Identification and Content Prediction of Lubricating Oil Additives Based on Extreme Learning Machine. *Tribology* **2020**, *40*, 97–106.

24. He, Z.X.; Wu, M.T.; Zhao, X.Y.; Zhang, S.Y.; Tan, J.R. Representative null space LDA for discriminative dimensionality reduction. *Pattern Recognit.* **2021**, *111*, 107664. [CrossRef]
25. Şahin, D.Ö.; Kural, E.O.; Akleylek, S.; Kılıç, E. Permission-based Android malware analysis by using dimension reduction with PCA and LDA. *J. Inf. Secur. Appl.* **2021**, *63*, 102995. [CrossRef]
26. Amiri, V.; Nakagawa, K. Using a linear discriminant analysis (LDA) based nomenclature system and self-organizing maps (SOM) for spatiotemporal assessment of groundwater quality in a coastal aquifer. *J. Hydrol.* **2021**, *603*, 127082. [CrossRef]
27. Xiong, Y.; Cheng, C.H.; Wu, J.H. Research on Premise Selection Technology Based on Machine Learning Classification Algorithm. *Netinfo Secur.* **2021**, *21*, 9–16.
28. Lu, W.P.; Yan, X.F. Balanced multiple weighted linear discriminant analysis and its application to visual process monitoring. *Chin. J. Chem. Eng.* **2021**, *36*, 128–137. [CrossRef]
29. Han, S.; Li, N.; Xue, L.; Hasi, W.L.J. Study on Classification and Identification of Arsenic Mineral Drugs by Raman Spectroscopy Combined with PCA-SVM. *J. Anal. Sci.* **2022**, *38*, 224–228.
30. Chen, C.H.; Zhong, Y.S.; Wang, X.Y.; Zhao, Y.K.; Dai, F. Feature Selection Algorithm for Identification of Male and Female Cocoons Based on SVM Bootstrapping Re Weighted Sampling. *Spectrosc. Spectr. Anal.* **2022**, *72*, 1173–1178.
31. Li, H.; Wang, J.X.; Xing, Z.N.; Shen, G. Influence of Improved Kennard/Stone Algorithm on the Calibration Transfer in Near-Infrared Spectroscopy. *Spectrosc. Spectr. Anal.* **2011**, *31*, 362–365.

lubricants

MDPI

Article

Machine Learning Composite-Nanoparticle-Enriched Lubricant Oil Development for Improved Frictional Performance—An Experiment

Ali Usman [1,2,*], Saad Arif [3], Ahmed Hassan Raja [4], Reijo Kouhia [1], Andreas Almqvist [5] and Marcus Liwicki [2]

1 Structural Engineering, Faculty of Built Environment, Tampere University, 33720 Tampere, Finland
2 Department of Computer Science, Electrical and Space Engineering, EISLab, Division of Machine Learning, Luleå University of Technology, 97187 Luleå, Sweden
3 Department of Mechanical Engineering, HITEC University, Taxila 47080, Pakistan
4 Department of Mechanical Engineering, COMSATS University Islamabad, Wah Cantt. 47040, Pakistan
5 Department of Engineering Sciences and Mathematics, Division of Machine Elements, Luleå University of Technology, 97187 Luleå, Sweden
* Correspondence: ali.usman@ltu.se

Abstract: Improving the frictional response of a functional surface interface has been a significant research concern. During the last couple of decades, lubricant oils have been enriched with several additives to obtain formulations that can meet the requirements of different lubricating regimes from boundary to full-film hydrodynamic lubrication. The possibility to improve the tribological performance of lubricating oils using various types of nanoparticles has been investigated. In this study, we proposed a data-driven approach that utilizes machine learning (ML) techniques to optimize the composition of a hybrid oil by adding ceramic and carbon-based nanoparticles in varying concentrations to the base oil. Supervised-learning-based regression methods including support vector machines, random forest trees, and artificial neural network (ANN) models are developed to capture the inherent non-linear behavior of the nano lubricants. The ANN hyperparameters were fine-tuned with Bayesian optimization. The regression performance is evaluated with multiple assessment metrics such as the root mean square error (RMSE), mean squared error (MSE), mean absolute error (MAE), and coefficient of determination (R^2). The ANN showed the best prediction performance among all ML models, with 2.22×10^{-3} RMSE, 4.92×10^{-6} MSE, 2.1×10^{-3} MAE, and 0.99 R^2. The computational models' performance curves for the different nanoparticles and how the composition affects the interface were investigated. The results show that the composition of the optimized hybrid oil was highly dependent on the lubrication regime and that the coefficient of friction was significantly reduced when optimal concentrations of ceramic and carbon-based nanoparticles are added to the base oil. The proposed research work has potential applications in designing hybrid nano lubricants to achieve optimized tribological performance in changing lubrication regimes.

Keywords: machine learning; friction; lubrication; nanoparticles; tribology; artificial neural network; Bayesian optimization

Citation: Usman, A.; Arif, S.; Raja, A.H.; Kouhia, R.; Almqvist, A.; Liwicki, M. Machine Learning Composite-Nanoparticle-Enriched Lubricant Oil Development for Improved Frictional Performance—An Experiment. *Lubricants* **2023**, *11*, 254. https://doi.org/10.3390/lubricants11060254

Received: 27 April 2023
Revised: 4 June 2023
Accepted: 6 June 2023
Published: 9 June 2023

1. Introduction

Metal-on-metal interfaces are found abundantly in engineering applications. Some examples are mechanical seals, bearings, pistons/plungers, and gears. These interfaces are prone to wear for various loading conditions. For instance, an intuitive mapping of the wear mechanism of metallic and non-metallic materials with lubricating conditions was graphically presented by Lim et al. [1]. A lubricant may be utilized to establish a thin lubricating film to separate the interfacial surfaces and reduce friction and wear. However, the lubricating film developed by traditional and non-traditional lubricants may not be sustained during operation due to the high loads and relative speed of the mating surfaces [2,3]. Recent advances in nanoparticle (NP)-based lubricant additives have shown promising results in reducing the coefficient of friction (CoF) and wear of highly loaded interfaces operating in the boundary lubrication regime and increasing the load-carrying capacity of full-film hydrodynamic lubricated interfaces. Single-NP-based lubricant oil blends have been evaluated extensively for the last two decades. However, optimizing oil blends for more than one additive particle is needed to address the varying demands of the tribo-pairs for varying lubricating domains.

Several studies have investigated the use of NP as lubricant additives to improve the antiwear and frictional performance of lubricating oils. For example, CuO/Al_2O_3 [4] and Bi_2O_3 [5] NPs were found to reduce friction and wear scar diameter (WSD), whereas CeO_2 was shown to facilitate the frictional performance of polyamide-imide/polytetrafluoroethylene lubricating coatings [6] and engine oils [7]. When used in combination with ZDDP, CeO_2 NP was found to improve antiwear performance even further [8]. Cu [9] and CuO [10], and TiO_2 [11] NPs were also found to improve the thermal conductivity and rheological properties of lubricating oils, respectively. The addition of SiO_2 NP was reported to increase the load-carrying capacity of soya bean and sunflower oil [12], whereas the addition of CuO NP to coconut oil resulted in the lowest friction and a polishing effect on worn surfaces [13].

Mirzaamiri et al. [14] introduced nanodiamonds to water, resulting in a 70% reduction in friction and an 88% reduction in wear that was attributed to the ball-bearing effect of the nanodiamond. Wu et al. [15] added sulfonated graphene to water, increasing viscosity by 25.8% and reducing the WSD and CoF by 74% and 15.7%, respectively. Xu et al. [16] studied the effect of graphene nanosheet (GNS) and Ag hybrids on phenolic composites, reporting that a 9 wt% GNS/Ag hybrid reduced the friction coefficient and wear rate by 40% and 72%, respectively, due to strong molecular interactions. Wang et al. [17] found that thicker copper coated with molybdenum disulfide had a lower friction coefficient but exhibited more severe wear. Yu et al. [18] reported that hydrated silica tribofilm reduced the CoF of MoAlB ceramic to 0.12. Pham et al. [19] showed that SiO_2 enhances the anti-oxidation of lubricants. Simonovic et al. [20] found that the wear of WSC-coated ceramic is reduced under low loads and more WS_2 monolayers are present; however, abrasive wear occurs at loads above 8 N. Xu et al. [21] investigated materials containing 1% kyanite with the best braking performance. Chen et al. [22] compared Si_3N_4-based and carbon-rich MLG-based MLG/Si_3N_4 ceramics and found that the combination of MLG and Si_3N_4 improved wear resistance and reduced the CoF. Fahad et al. [23] studied base oil containing modified TiO_2/CuO NPs, which improved the viscosity index and load-carrying capacity. Sharma et al. [24] found that mixing alumina/graphene (GnP) hybrid NPs reduced cutting tool wear and nodal temperature. Huang et al. [25] found that GO–Al_2O_3 hybrid NPs provided better friction and wear performance than pure GO or Al_2O_3 due to the GO layer preventing surface asperities from direct contact and the Al_2O_3 tribo-layer acting as a load bearer to polish asperities.

Besides ceramic and carbon-based NPs, various studies [26–28] have also investigated the tribological performance of ferrous-NP-based lubricants. Oliveira et al. [26] additized PAO 8 oil with Fe_2O_3 NP to evaluate the lubricant performance for reduced friction and wear. The boundary lubrication resulted in increased scuffing resistance and reduced wear rates by up to 27% for high loads due to the intrinsic properties of metallic oxides. Another study [27] investigated the effect of coated magnetic NPs dispersed in trimethylolpropane trioleate base oil. The Nd and Fe_3O_4 NPs in 0.015 wt% concentration significantly reduced the CoF and WSD by 29% and 67%, respectively, in comparison with the base oil. Alvi et al. [28] enhanced the tribological performance of drilling fluids with iron oxide-based NP. Fe_2O_3 NP in a 0.019 wt% concentration reduced the CoF by 47% and 45% with dispersion in bentonite and KCl-based base fluids, respectively. This indicates that hybrid lubricant blends can outperform previously formulated lubricants; however, application- and operating-condition-dependent optimization is needed. It is a delicate task involving many independent parameters and requiring a highly robust optimization scheme.

Machine learning (ML)-based methodology has shown the capability to handle many multi-featured input parameters and target the desired outcome with high accuracy and precision. Bhaumik et al. [29] presented a method for designing multiple NP-based biolubricants using a multi-layered artificial neural network (ANN). The ANN-based model was optimized with a genetic algorithm and the additized biolubricant showed a decrease in the CoF of 45–50% compared with mineral oils. Humelnicu et al. [30] used a feed-forward ANN to obtain the minimum CoF for blended diesel fuel by optimizing the concentrations of two vegetable oils. A CoF of 0.00156 was achieved using 4% sunflower oil, based on the results from the ANN computations. Haldar et al. [31] designed an ANN-based regression estimator to predict the dynamic viscosity of multi-walled carbon nanotubes (MWCNT) and SiO_2-based nano lubricant in a 20:80 ratio. The perfect estimation was found within a 2.62% maximum deviation by comparing experimental data with the model predictions. Recently, Esfe et al. [32] used a quasi-Newton algorithm based on a multi-layered ANN to predict the viscosity of a hybrid nano lubricant with high precision. The trained Levenberg–Marquardt (LM)-based regression learner achieved a mean squared error (MSE) of 6.15×10^{-4} while estimating the observed behavior of a hybrid lubricant blend of SAE40 oil additized with MWCNT and Al_2O_3 at a 10:90 concentration ratio. Table 1 summarizes studies that effectively employed ML-based data-driven approaches to model the inherent non-linearities of nano lubricants.

Table 1. List of similar studies on ML-based approaches for tribological performance prediction.

Ref.	Methodology	Input/Output Parameter	Base Oil/Additive	Performance
[29]	ANN, GA	Load, speed, concentration/ CoF	NCO, CMO/ GRT, MWCNT, GRPHN, ZnO	CoF ↓ by 45–50% WSD ↓ by 87.5%
[30]	FF-ANN	Concentration/ CoF	Regular diesel fuel/ Sunflower oil, Rapeseed oil	CoF: 1.56×10^{-3} with 4% sunflower oil
[31]	ANN	Temperature, volume fraction, shear rate/Viscosity prediction	SAE68 hydraulic oil/ MWCNT, SiO_2	R^2: 0.998 RMSE: 2.135415
[32]	LM-based MLP	Temperature, volume fraction, shear rate/Viscosity prediction	SAE40/ MWCNT, Al_2O_3	R: 0.9999 MSE: 6.15×10^{-4} $-2\% < \text{MOD} < 2\%$
[33]	DT, RF, GLM, ANN	Temperature, volume fraction/ Kinematic viscosity prediction	SAE30, Hydrex100, EP90/ Al_2O_3, CeO_2	R^2: 0.861 (SAE30) R^2: 0.971 (Hydrex100) R^2: 0.973 (EP90)
[34]	LM-ANN	Temperature, volume fraction, shear rate/Viscosity prediction	SAE50/ MWCNT, Al_2O_3	MSE: 3.58 R: 0.999

GA: genetic algorithm; NCO: neat castor oil; CMO: commercial mineral oil; GRT: graphite; GRPHN: graphene; FF: feed-forward; MLP: multi-layer perceptron; MOD: margin of deviation; DT: decision trees; RF: random forest; GLM: generalized linear model; ↓ shows a decrease.

The studies reviewed above show that NP enrichment can be used to control the dynamic and static properties of the lubricant. However, the mechanisms that govern the changes in lubrication performance are complex and not yet fully understood. Moreover, there exist many parameters that affect the outcome of NP addition to the base lubricant. Due to the complexity of the mechanism and the numerous design parameters of hybrid lubricant blends containing multiple NPs, there is a gap in the literature presenting studies on the possibilities of obtaining performance improvements. In this work, random forest trees (RFT) and support vector machines (SVM)-based regression models are initially developed to capture the NP-based lubricant behavior. In the final approach of designing computational methods, multi-layered ANNs are developed and trained to predict the performance of multiple-NP-based lubricants and their hybrids to minimize the CoF. The details of the experiments, training dataset, modeling, and results are discussed and the capabilities of the ML-based techniques are compared. The CoF is analyzed for varying operating conditions and the evolution of lubricating regimes is analyzed for individual and hybrid oils. Optimization of the NP concentrations for varying lubricating regimes is also evaluated.

2. Design of Experiment

The experimental dataset was created using a pin-on-disc tribometer for experiments with different NP-based lubricants as shown in Figure 1. A commercially available oil, 5W30 by Shell plc, was used in this study to create NP-based blends, and a comparison is drawn with the same oil without NP. The experiments were carried out with varying values of the parameters involved. The parameters under consideration were the NP concentrations in weight percentages (wt%) for silicon dioxide (SiO_2) and nano graphite (NG) with varying load (Newton, N) and speed (revolutions per minute, RPM). The single output CoF was recorded for each experiment. The experiments were conducted at two load levels and five speed levels for all the lubricants comprising the plain oil (PO) without NP and PO with both NPs individually. For each NP, two levels of concentration, along with the above load and speed levels, are used because the load and speed both influence the lubrication regime. Similarly, the NP concentration affects the oil viscosity, which in turn is an important parameter controlling the lubricating regime experienced by the tribo-pair. Therefore, five factors and the corresponding three levels have been adopted to explore the pure hydrodynamic, mixed, and boundary lubrication of the tribo-interface for varying design parameters and to explore the effect of the combination of NPs on the said lubricating regimes.

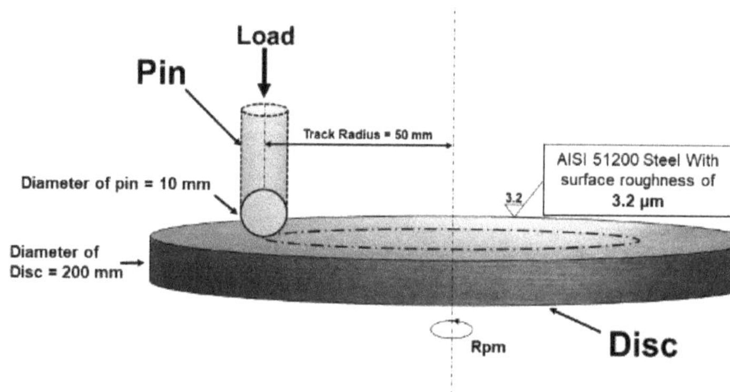

Figure 1. Pin-on-disk schematic illustrating surface parameters, geometry, and loading conditions.

The dataset array was generated for 30 experiments (number of samples) according to the values shown in Table 2. The NPs used in this study are nearly spherical in powder form, with an average size of 20 µm and 7 nm for NG and SiO_2, respectively. Dispersion of the NPs and their static stability in the oil over time is ensured based on the dispersion test. A volumetric sample is taken from each blend and examined after each hour by the naked eye for any visible sedimentation. No sedimentations were observed during the first day of sample preparation; therefore, all the tribological tests were performed on the same day as the sample preparation, which was accomplished through the signification process. Multigrade oil was used to create the NP-based blends, and the viscosity of such oils is highly dependent on the temperature. The actual temperature of the interface may vary significantly depending on the lubrication domain being experienced by the interface. This oil temperature variation is in comparison with the ambient temperature and the bulk oil temperature in the sump. Therefore, conducting a comparative analysis for variations in oil viscosities because of NP concentration requires an in-depth study considering the lubrication condition in the actual tribo-pair and is hence deliberately not presented here. This is a limitation of the present study, which will be addressed in the future.

Table 2. List of parameters involved and their values for the conducted experiments with multiple NPs.

Parameters	Minimum	Maximum	Average
SiO_2 NP concentration (wt%)	0.2	0.4	0.3
NG NP concentration (wt%)	0.2	0.4	0.3
Load (N)	30	50	40
Speed (RPM)	35	100	58
Coefficient of friction	0.02	0.3	0.16

3. Computational Models of Lubricants

The initial study to develop the computational models of NP-based lubricants was initiated by training the RFT- and SVM-based regression models. The shortcomings of these two models directed the study to create more comprehensive ANN-based regression models to cater for the non-linearity involved in the experimental data of the lubricants. Developing the ANN-based computational model for hybrid nano lubricants with optimized parameters is daunting. It is required to capture the true behavior of the lubricant's tribology, as evident from the experimental data. This study employs the Bayesian optimization (BO) method to find optimal hyperparameters for the ANN models of NP-based lubricants. Once optimized hyperparameters are known, the ANN regression models are developed accordingly to estimate the CoF for the individual- and multiple-NP-based lubricants.

3.1. Training Dataset Generation

Multiple training datasets were developed from the experimental data to train the regression models. Two datasets contained the NP concentration, load, and speed as inputs along with the response variable CoF for both the NPs, i.e., SiO_2 and NG. The third dataset contained multiple NP concentrations as input along with the other parameters. All the inputs were rescaled with min–max normalization to regularize the data for loss function and to achieve rapid convergence during training. Input normalization was applied using the *normalize* built-in function of MATLAB 9.12 (MathWorks, Natick, MA, USA) according to the following relationship:

$$X_i' = a + \frac{X_i - \min(X_i)}{\max(X_i) - \min(X_i)}(b - a) \quad (1)$$

where X_i' is the normalized value and X_i is the original value of the input i, a and b are the normalization range limits which are set as $[a\ b] = [0\ 1]$ for all the inputs.

3.2. Random Forest Trees

The initial regression model is developed using the ensemble method with bootstrap aggregating (bagging) of multiple decision trees (DT)-based regression learners. The random forest is developed at each ensemble split with a minimum leaf size of eight. A total of 30 DT learners were bagged in the ensemble with 100 learning cycles. The objective function is the MSE, which is minimized, and a threshold is set as a stopping criterion. The RFT training performance is evaluated via various regression performance assessment metrics such as root mean square error (RMSE), MSE, mean absolute error (MAE), and coefficient of determination (R-squared or R^2). All the training sessions are conducted using the 10-fold cross-validation, and the assessment metrics are calculated upon the validation results. The RFT model is implemented using the *fitrensemble* built-in function of MATLAB 9.12.

3.3. Support Vector Machines

The other initial regression model is the non-linear SVM regression learner with a radial basis function (RBF) kernel for more accurate predictions. The rapid variations in the CoF are well predicted with the fine Gaussian SVM as compared with the polynomial-based SVM models. The fine Gaussian SVM employed a Gaussian kernel RBF with the kernel scale set to $\frac{\sqrt{P}}{4}$ for P number of predictors. For three input parameters of the individual NP-based datasets, the kernel scale is set to 0.43. In model designing, the box constraint and epsilon values are calculated heuristically by gradually increasing and decreasing them. Both these parameters are fine-tuned to generate a flexible model that avoids overfitting the predictions. The 10-fold cross-validation-based model training is conducted to achieve the best RMSE, MSE, MAE, and R^2 metrics results. The SVM regression model is implemented using the *fitrsvm* function of MATLAB 9.12.

3.4. Hyperparameter Estimation with Bayesian Optimization

The well-tuned hyperparameters for all the ANN models are computed with the BO algorithm, a derivative-free optimization method for non-analytical models. The MSE is used as the objective function $f(x)$, which is minimized upon subsequent iterations of the BO with different random samples of x according to the following relationship:

$$\min_{x \in A} f(x) = \min_{x \in A}(\mathrm{MSE}) = \min_{x \in A}\left(\frac{1}{N}\sum_{i=1}^{N}(T_i - O_i)^2\right)_{x=\mathbb{R}^6} \quad (2)$$

where T_i and O_i are the actual target and predicted output values, respectively, for training sample i ranging from 1 to N number of observations, x is a random sample of six optimization variables for each iteration of the BO algorithm and always selected from the bounded domain of the structure A, containing search ranges for all the optimization variables as stated in Table 3.

Table 3. Optimized hypermeters and their search range for Bayesian optimization.

Optimization Variable (Hyperparameter)	Search Range for Optimization
Number of hidden layers	$[1 \quad 3]$
Number of neurons in 1st, 2nd, 3rd hidden layers	$[1 \quad 300]$ for each layer
L2 Regularization strength (λ)	$[1 \times 10^{-6} \quad 1 \times 10^{4}]$
Activation function	[ReLU Sigmoid Tanh None]

The selection of x from A for each iteration of BO is based upon the Gaussian distribution model, which is updated after each iteration to sample the x from the region that maximizes the acquisition function. The acquisition function (expected improvement per second plus) is used here, which is best for the global minimization of the objective function by avoiding the local minima. The local minima are avoided by the balanced exploration ratio of 0.5, which means an equal trade-off between the exploitation of already explored regions and the exploration of comparatively unexplored regions of A for sampling the new x. The maximization of the acquisition function, and hence, the convergence of the BO algorithm, is obtained by an iterative quasi-Newton numerical optimizer known as the limited-memory Broyden–Fletcher–Goldfarb–Shanno (LBFGS) algorithm. This way, the fine-tuned hyperparameters of x for global minimization of the MSE are computed using the BO method.

The additional hyperparameters are the data standardization and training iterations limit, which were not optimized and were set manually for multiple sessions of the BO algorithm for all the training datasets. As all the datasets were already normalized before the BO application, data standardization during the optimization process was set to false. All the sessions of BO were conducted with 10-fold cross-validation to find out the optimal and validated trained model. Ten sessions of the BO method were applied to each dataset to ensure optimal hyperparameter tuning. After validation, the optimization results were selected for the session with the minimum RMSE, MSE, and MAE values and the maximum R^2. The BO computational method is implemented using the *bayesopt* built-in function of MATLAB 9.12 and various optimization settings.

3.5. Design of Lubricant ANN Model

Once the optimized hyperparameters for all the lubricant models are computed, regression ANNs are developed from all three training datasets with fine-tuned hyperparameters. The general mathematical model for a single perceptron in all the ANNs is given below:

$$z_{j,i} = b_j + \sum_{i=1}^{n} w_{j,i}\, x_{j-1,i} \quad (3)$$

$$h_{j,i} = \sigma(z_{j,i}) = \frac{1}{1+e^{-z_{j,i}}} \quad (4)$$

where $w_{j,i}$ is the weight for neuron i of the layer j, b_j is the bias term for a particular layer j, x is the input value from the preceding neuron, $z_{j,i}$ is the linear output value of all the connected weighted inputs subjected to the activation function, σ is the non-linear sigmoid activation function generating the final value $h_{j,i}$ for the neuron i in layer j.

Using the tuned hyperparameters, the regression ANN models are trained with different training algorithms for neural networks. The two variants of training algorithms were tested here for multiple training sessions, the scaled conjugate gradient (SCG) and the LM backpropagation from the conjugate gradient and the quasi-Newton families. These learning algorithms were implemented in MATLAB 9.12 with the built-in functions *trainscg* for SCG and *trainlm* for LM training methods. Among the various comparative runs for both methods, the SCG showed the best validation performance compared with LM for these smaller datasets. The SCG converged to a lower MSE with fewer iterations at the expense of training time. It also performed well during the testing of the approach with a varying number of hidden neurons, as it is less sensitive to hyperparameter changes than LM. The final design models of lubricant ANNs are generated with optimized hyperparameters and an SCG backpropagation learning scheme. The convergence information for all the ANNs, along with the hyperparameters, is shown in Table 4. The optimized number of hidden layers, hidden neurons, and their activation functions are obtained from the multi-session BO application on the datasets. To further validate the BO-based hyperparameter tuning results, trial tests were conducted by changing the numbers of hidden layers and hidden neurons. It was observed that further increasing the number of hidden layers and their sizes did not significantly improve the regression performance in terms of assessment metrics. Moreover, the prediction results of such trial models showed significant deviations from the experimental data. This ensured that the best hypermeter combination was selected by the BO, which reproduced the experimental results with high accuracy and precision. The best validation results of performance assessment metrics are obtained with these hyperparameters, as shown in Table 4.

Table 4. BO-estimated hyperparameters and convergence results for all lubricant ANN models.

ANN Model	Optimized Model Hyperparameters and Convergence Results							
	Hidden Layer Size	**Activation Function**	**L2 Regularization 'λ'**	**Validation MSE at Epoch**	**Iterations**	**Training Loss**	**Gradient**	**Training Time (s)**
SiO_2 NP	10	sigmoid	0	7.81×10^{-4} at 37	43	52.31×10^{-4}	7.02×10^{-4}	213
NG NP	4	sigmoid	0	5.89×10^{-4} at 27	33	2.02×10^{-4}	10.01×10^{-4}	188
Multi-NP	2	sigmoid	0.11×10^{-4}	1.44×10^{-4} at 22	28	5.97×10^{-4}	3.96×10^{-4}	142

4. Results and Discussion

Figure 2 shows the regression plots for each NP with initial regression models, i.e., RFT and SVM. The best regression models with the most promising performance metrics are selected out of multiple training sessions with each regressor. The RFT regression plots show a significant sensitivity drift compared with the perfect predictions for both NPs, as shown in Figure 2a,b. The RFT prediction function follows the positive kurtosis with the leptokurtic distribution of predictions that can be observed from the regression plots. The RFT model failed to capture the complete variance of target data for the CoF predictions with 0.87 and 0.757 mean values for the coefficient of correlation (R) and R^2, respectively. Moreover, a slight negative skewness of prediction distribution is observed towards higher target values of the CoF. On the other hand, the SVM regression model performed better with comparatively higher R and R^2 values, as shown in Figure 2c,d for SiO_2 and NG NPs, respectively. The sensitivity drift is significantly reduced as compared with the RFT, but few predicted values of the CoF still vary significantly from the target CoF. The kurtosis of prediction distribution is significantly reduced to mesokurtic in comparison with the leptokurtic distribution of the RFT model. These attributes and performance metrics results for both models are further compared with the final design of ANN-based regression models.

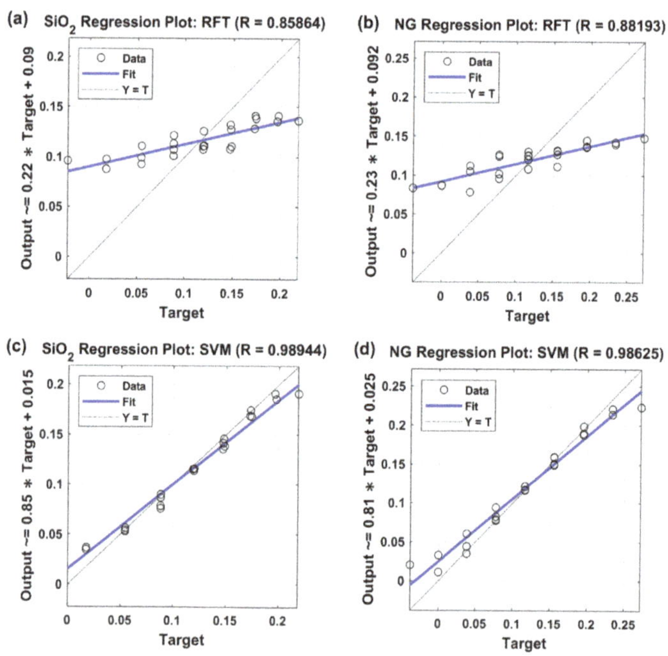

Figure 2. CoF computational efficiency for individual-NP-based lubricants with initial regression models (**a**) SiO_2 and (**b**) NG with RFT (**c**) SiO_2 and (**d**) NG with SVM.

The regression plots for the individual-NP-based and hybrid lubricant ANNs are shown in Figure 3. The regression plots for SiO_2 and NG ANN models in Figure 3a,b represent the best fit between the actual target values and ANN computed predictions of the CoF with higher R values.

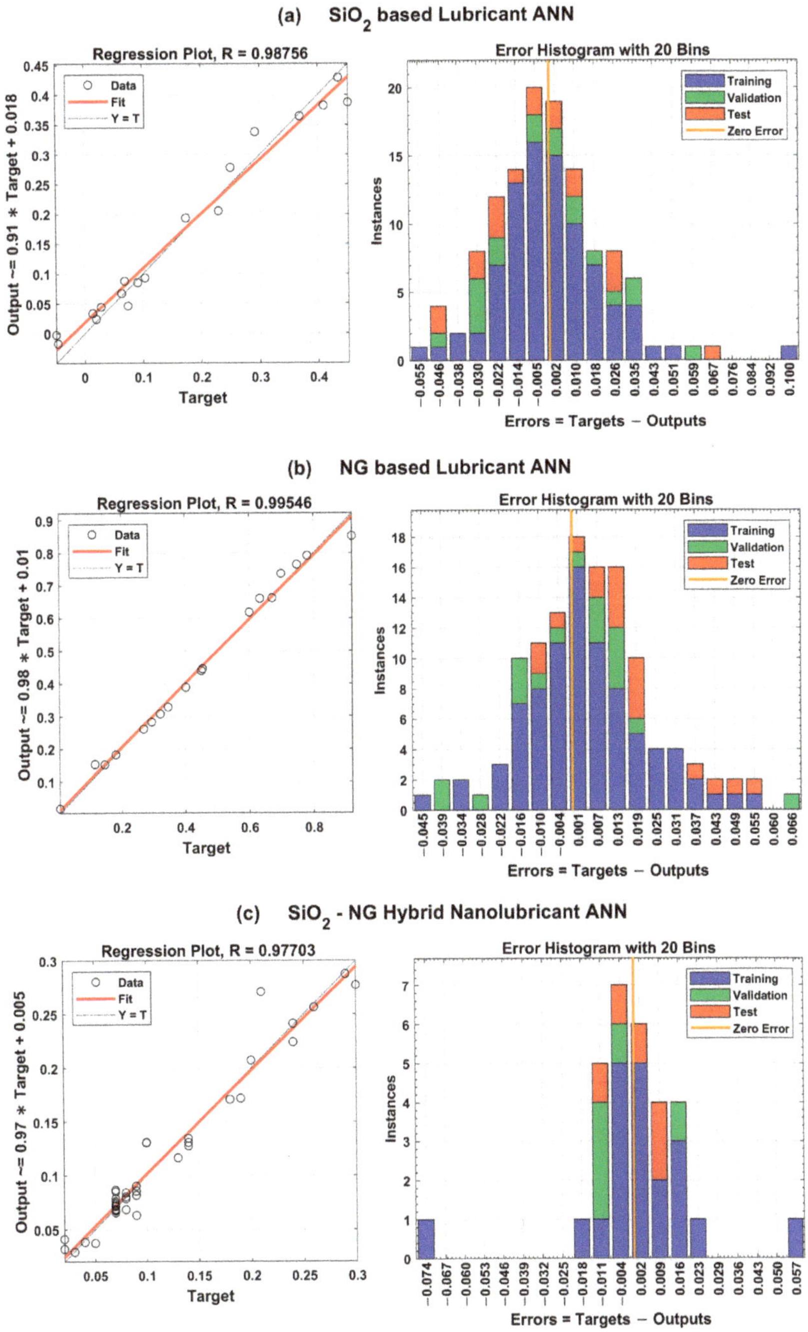

Figure 3. ANN models (regression plots, error histograms) for individual NP lubricants (**a**) SiO_2 (**b**) NG and (**c**) the multi-NP-based hybrid nano lubricant.

The R^2 is almost equal to one for all the models, representing the perfect estimation power of the designed ANNs and good confidence level in their computations. Moreover, these models are trained over a wide target range as compared with the initial models and they performed better with comparatively good sensitivity over a larger spread of the target CoF. The error histograms show the perfect Gaussian distribution of training, testing, and validation errors during ANN model convergence.

These models can be further investigated to exhibit the behavior of NP-based lubricants in terms of the CoF, with varying input parameters such as NP concentration, load, and speed. Individual-NP-based lubricants mostly exhibit limitations in achieving the required tribological characteristics. To overcome these limitations, the hybrid nano lubricant ANN model is trained to achieve the benefits of both NPs to gain the required tribological properties of the lubricants. Figure 3c represents the regression plot along with the error histogram obtained during this ANN training. The regression plot with 0.9546 R^2 shows the good computational power of this model to find out the optimum CoF against the number of observations from the multi-NP-based training dataset. Few outlier samples in the regression plot achieved the training errors, with higher magnitudes on both sides of the zero error. Despite these countering outliers, the perfect regression fit is achieved. The rest of the error histogram shows that the ANN is well trained with the SCG method and has achieved minimal errors during the training, testing, and validation phases.

Table 5 shows the validation results of four regression performance assessment metrics, i.e., RMSE, R^2, MSE, and MAE, for all the computational models (RFT, SVM, ANN) for both NPs and their hybrid (with ANN only). For SiO_2 NP, the ANN achieved the lowest RMSE, MSE, and MAE values with less difference from the SVM results and a significant difference from the RFT results. The R^2 for ANN and SVM is almost equal. For NG NP, SVM achieved slightly lower RMSE, MSE, and MAE values than the ANN, whereas the ANN achieved the best R^2 among all models.

Table 5. Performance assessment metrics results for all regression models with individual NP and hybrid nano lubricants.

Regression Model	Nanoparticle	Performance Assessment Metrics (10-Fold Cross-Validation)			
		RMSE	R^2	MSE	MAE
Random Forest Trees	SiO_2	8.8662×10^{-3}	0.7373	7.8609×10^{-5}	7.9509×10^{-3}
	NG	6.7444×10^{-2}	0.7778	4.5487×10^{-3}	6.3266×10^{-2}
Support Vector Machines	SiO_2	2.2689×10^{-3}	0.9790	5.1481×10^{-6}	2.1874×10^{-3}
	NG	3.2127×10^{-2}	0.9727	1.0321×10^{-3}	1.8183×10^{-2}
Artificial Neural Network	SiO_2	2.2181×10^{-3}	0.9753	4.9199×10^{-6}	2.1026×10^{-3}
	NG	4.2407×10^{-2}	0.9909	1.7983×10^{-3}	3.1608×10^{-2}
	Hybrid	3.6296×10^{-2}	0.9546	1.3174×10^{-3}	2.3902×10^{-2}

Hence, these ANN models can be further investigated to study the tribological behavior of computationally designed lubricants that are influenced by the individual characteristics of multiple NPs. The inherent properties of such lubricant models can be utilized to achieve better CoF values with varying NP concentrations, load, and speed trends.

During the investigation of lubricant characteristics, it was observed that the speed is a less significant input as compared with the NP concentration and load. Substantial changes in the operating speed do not considerably affect the CoF for any load and concentration, whereas changing the NP concentration significantly affects the CoF of the lubricant. Figure 4 represents the characteristics of individual-NP-based lubricants with varying speeds and concentrations at a fixed load of 50 N. In an agreement with Bhaumik et al.'s study [29], it is evident that varying the concentration (with identical speeds) varies the CoF significantly. Thus, varying loading conditions results in variation in the optimum concentration to achieve CoF minima.

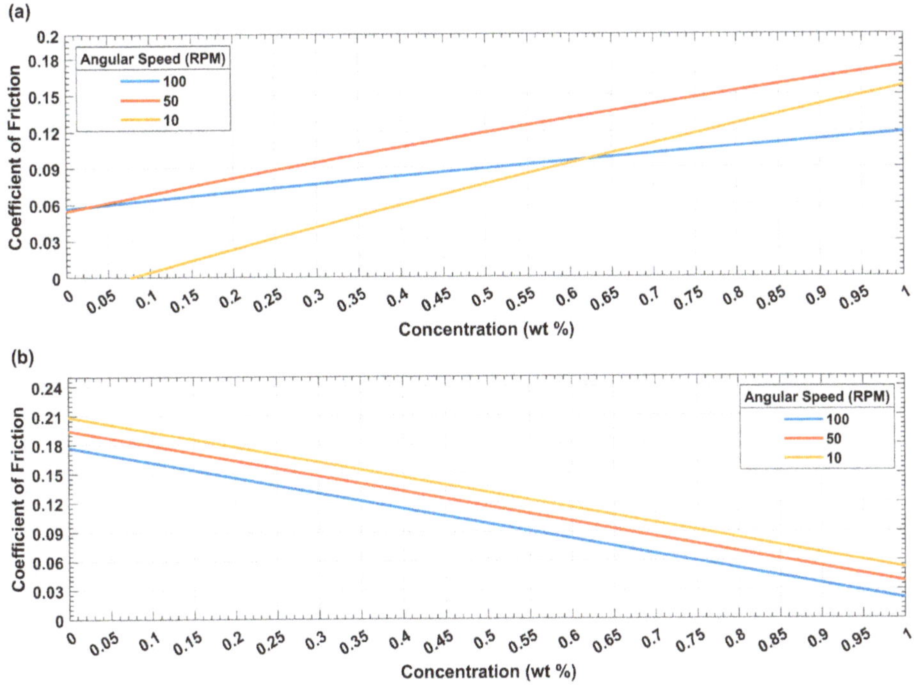

Figure 4. Significance of speed and NP concentration inputs to predict the CoF at a fixed load (50 N) for (**a**) SiO_2 and (**b**) NG nanoparticles.

Surface plots are developed to incorporate the influence of input parameters on the performance of the lubricants. Figure 5 represents the load–speed effects on the CoF with NP concentrations obtained for the individual-NP-based lubricants. These surfaces indicate that the load and NP concentrations are the influential input parameters in the NP-based lubricants and can drastically affect their tribological properties, as evident from the varying CoFs.

In Figure 5, the CoF has been plotted for varying loads and speeds for both NPs, i.e., SiO_2 and NG. Notably, regardless of the same base oil, the lubricating regimes vary for different NPs for the same loading conditions. An increasing and then decreasing CoF with increasing speeds for SiO_2 occurs, in contrast with NG, where the CoF decreases for growing speeds. This is attributed to the already fully developed lubricant oil film for the former one for identical loading conditions against the interface still experiencing mixed lubrication for the latter one. The decreasing CoF also highlights a thicker lubricant film with an increasing load because of shearing thinning for NG.

Moreover, a precise offset is evident in friction reduction with increasing concentration, regardless of the loading conditions. The percentage of each NP is different in oil to achieve an identical CoF, e.g., 1 wt% of SiO_2 and 0.1 wt% of NG results in a similar CoF at 100 RPM and counterbalance weight load conditions. This, and the above-mentioned existence of the interface in different lubricating regimes for identical loading conditions, makes it possible to develop a hybrid lubricant with more than one NP.

The magnitude of influence caused by a combination of NPs is illustrated in Figure 6 for identical speeds but at varying loads, i.e., 10–100 N. This is to observe the effect of NP combinations in different lubrication domains and how it influences the optimum concentration of the NPs to develop a composite N-enriched lubricant oil. NG facilitates the interface to reduce friction at low loads when the lubrication film thickness is well developed and the interface is experiencing pure hydrodynamic lubrication at a 10 N load, as shown in Figure 6c. In contrast, when the load is increased, e.g., 50 N, as shown in Figure 6b, the local minima for the CoF move toward a high concentration for both the NPs and keep moving until they reach an equal concentration near (1,1). Similarly, with another increase in load, e.g., at 100 N, as in Figure 6a, the SiO_2 tends to facilitate the decrease in friction more compared with the NG. This could be because of the high molecular weight of SiO_2, which increases the viscosity more than NG, or better tribofilm development caused by SiO_2. The mechanism of friction reduction, and hence the different optimum concentrations at varying lubrication domains, is a limitation of the present work and will be reported on in a future publication.

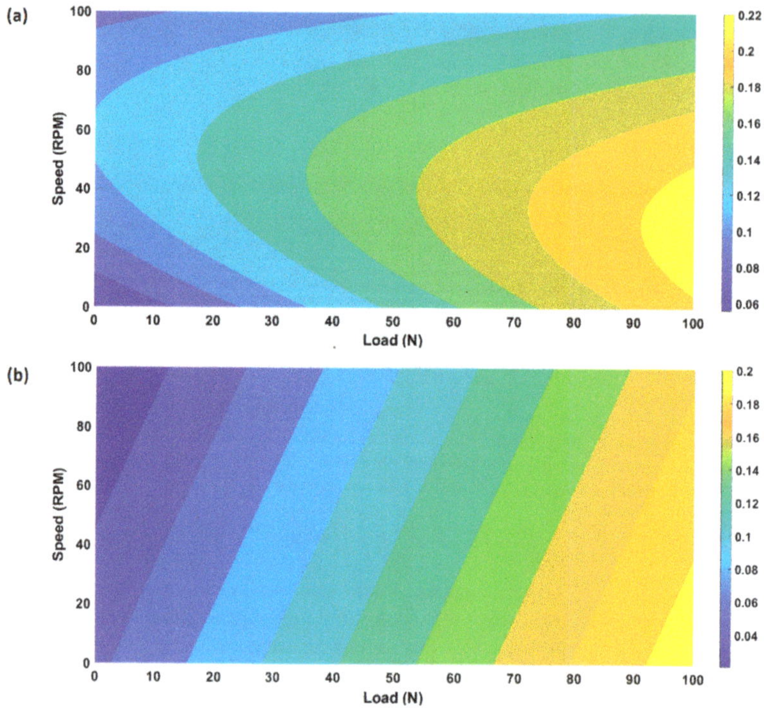

Figure 5. Speed and load effects on CoF for (**a**) SiO_2 with 1% concentration and (**b**) NG with 0.5% concentration.

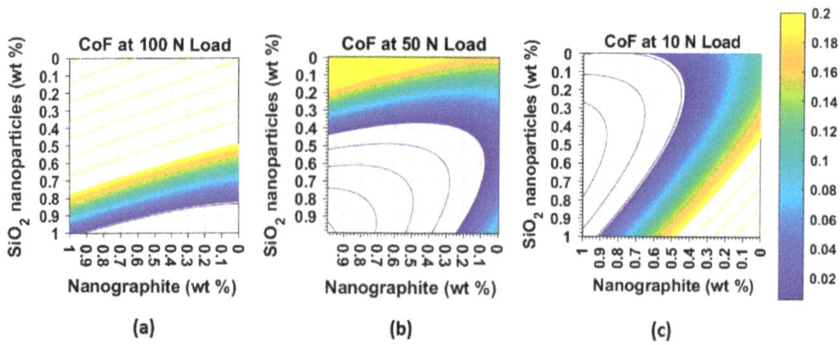

Figure 6. CoF estimation with hybrid-nano-lubricant-enriched lubricant oil containing ceramic and carbon- based NPs against varying concentrations at a constant speed (50 RPM) and three different loads; (**a**) 100 N, (**b**) 50 N, and (**c**) 10 N.

5. Conclusions

The coefficient of friction (CoF) in nano lubricants has a complex and non-linear relationship with its composition and loading conditions. Therefore, analytical models for predicting the tribological behavior of such lubricants are not available. A study is conducted utilizing machine learning (ML), and the following conclusions can be made from the observations of this study.

- The computational models given by the data-driven ML-based approaches such as random forest trees (RFT), support vector machines (SVM), and artificial neural networks (ANN) are promising solutions to predict non-linearity in such complex interactions.
- The multi-layered ANN-based regression models of lubricants having single and multiple nanoparticles (NP) are developed to examine their tribological behavior. The complex interactions of input parameters (load, speed, and NP concentration) and the output parameter (CoF) is well estimated by the ANNs when their hyperparameters are optimized.
- A better performance for ML-optimized nano lubricant models is found in decreasing the CoF between metal-to-metal interactions in sliding lubricated contact for engineering applications.
- The results have shown that the optimum concentration of NP varies with varying lubrication domains and that a composite lubricant based on multiple NPs can be beneficial to reduce frictional energy loss and improve the lubrication conditions.
- The optimum concentration of multiple NPs can be reached for interfaces that experience fluctuating loads and thus varying lubrication conditions during their service.

The future scope of this study is to examine the mechanism of friction reduction in hybrid nano lubricants with different NPs and base oil combinations. Finding out the optimum NP concentrations at varying lubrication domains is an underexplored research area requiring the further study of such ML-based applications.

Author Contributions: Conceptualization, A.U.; methodology, A.U. and S.A.; software, A.U. and S.A.; validation, A.U. and S.A.; formal analysis, A.U., S.A., M.L. and A.A.; investigation, A.U., S.A. and A.H.R.; resources, M.L. and A.A.; data curation, A.U. and A.H.R.; writing—original draft preparation, A.U. and S.A.; writing—review and editing, A.U., S.A., R.K., M.L. and A.A.; visualization, A.U., S.A. and A.H.R.; supervision, M.L. and A.A.; project administration, M.L. and A.A.; funding acquisition, M.L. and A.A. All authors have read and agreed to the published version of the manuscript.

Funding: This research was funded by the Swedish Research Council (VR) grant number 2019-04293.

Data Availability Statement: The data presented in this study are available on request from the corresponding author.

Acknowledgments: The authors acknowledge the technical and financial support provided by their organizations to complete this study.

Conflicts of Interest: The authors declare no conflict of interest.

References

1. Lim, S.C. Recent developments in wear-mechanism maps. *Tribol. Int.* **1998**, *31*, 87–97. [CrossRef]
2. Xiang, G.; Han, Y.; Wang, J.; Wang, J.; Ni, X. Coupling transient mixed lubrication and wear for journal bearing modeling. *Tribol. Int.* **2019**, *138*, 1–15. [CrossRef]
3. Xie, Z.; Jiao, J.; Yang, K. Experimental and numerical study on the mixed lubrication performances of a new bearing. *Tribol. Int.* **2023**, *182*, 108334. [CrossRef]
4. Li, W.; Wang, M.; Chen, Q.; Zhang, W.; Luo, T. A New Preparation Method of Copper oxide/Aluminium oxide Nanocomposites with Enhanced Anti-friction Properties. *ES Mater. Manuf.* **2023**, *19*, 692.
5. Jaffar, S.S.; Baqer, I.A.; Soud, W.A. Effect of Bi_2O_3 Nanoparticles Addition to Lubricating Oil on the Dynamic Behavior of Rotor-Bearing Systems. *J. Vib. Eng. Technol.* **2022**, *10*, 2005–2017. [CrossRef]

6. Zhao, Z.; Ma, Y.; Wan, H.; Ye, Y.; Chen, L.; Zhou, H.; Chen, J. Preparation and tribological behaviors of polyamide-imide/polytetrafluoroethylene lubricating coatings reinforced by in-situ synthesized CeO_2 nanoparticles. *Polym. Test.* **2021**, *96*, 107100. [CrossRef]
7. Ali, M.K.A.; Xianjun, H. Exploring the lubrication mechanism of CeO_2 nanoparticles dispersed in engine oil by bis(2-ethylhexyl) phosphate as a novel antiwear additive. *Tribol. Int.* **2022**, *165*, 107321. [CrossRef]
8. Lei, X.; Zhang, Y.; Zhang, S.; Yang, G.; Zhang, C.; Zhang, P. Study on the mechanism of rapid formation of ultra-thick tribofilm by CeO_2 nano additive and ZDDP. *Friction* **2023**, *11*, 48–63. [CrossRef]
9. Ghosh, S.K.; Miller, C.; Perez, G.; Carlton, H.; Huitink, D.; Beckford, S.; Zou, M. Effect of Cu nanoparticles on the tribological performance of polydopamine + polytetrafluoroethylene c oatings in oil-lubricated condition. *Appl. Surf. Sci.* **2021**, *565*, 150525. [CrossRef]
10. Abdel-Rehim, A.A.; Akl, S.; Elsoudy, S. Investigation of the Tribological Behavior of Mineral Lubricant Using Copper Oxide Nano Additives. *Lubricants* **2021**, *9*, 16. [CrossRef]
11. Okokpujie, I.P.; Tartibu, L.K.; Sinebe, J.E.; Adeoye, A.O.M.; Akinlabi, E.T. Comparative Study of Rheological Effects of Vegetable Oil-Lubricant, TiO_2, MWCNTs Nano-Lubricants, and Machining Parameters' Influence on Cutting Force for Sustainable Metal Cutting Process. *Lubricants* **2022**, *10*, 54. [CrossRef]
12. Taha-Tijerina, J.; Aviña, K.; Diabb, J.M. Tribological and Thermal Transport Performance of SiO_2-Based Natural Lubricants. *Lubricants* **2019**, *7*, 71. [CrossRef]
13. Cortes, V.; Ortega, J.A. Evaluating the Rheological and Tribological Behaviors of Coconut Oil Modified with Nanoparticles as Lubricant Additives. *Lubricants* **2019**, *7*, 76. [CrossRef]
14. Mirzaamiri, R.; Akbarzadeh, S.; Ziaei-Rad, S.; Shin, D.-G.; Kim, D.-E. Molecular dynamics simulation and experimental investigation of tribological behavior of nanodiamonds in aqueous suspensions. *Tribol. Int.* **2021**, *156*, 106838. [CrossRef]
15. Wu, L.; Zhong, Y.; Yuan, H.; Liang, H.; Wang, F.; Gu, L. Ultra-dispersive sulfonated graphene as water-based lubricant additives for enhancing tribological performance. *Tribol. Int.* **2022**, *174*, 107759. [CrossRef]
16. Xu, M.; Wang, X.; Wang, T.; Wang, Q.; Li, S. Ag nanoparticle decorated graphene for improving tribological properties of fabric/phenolic composites. *Tribol. Int.* **2022**, *176*, 107889. [CrossRef]
17. Wang, G.; Ruan, Y.; Wang, H.; Zhao, G.; Cao, X.; Li, X.; Ding, Q. Tribological performance study and prediction of copper coated by MoS2 based on GBRT method. *Tribol. Int.* **2023**, *179*, 108149. [CrossRef]
18. Yu, Z.; Wang, S.; Cheng, J.; Chen, J.; Chen, W.; Sun, Q.; Yang, J. Tribological behaviors of MoAlB ceramic in artificial seawater. *Tribol. Int.* **2022**, *167*, 107345. [CrossRef]
19. Pham, S.T.; Huynh, K.K.; Tieu, K.A. Tribological performances of ceramic oxide nanoparticle additives in sodium borate melt under steel/steel sliding contacts at high temperatures. *Tribol. Int.* **2022**, *165*, 107296. [CrossRef]
20. Simonovic, K.; Vitu, T.; Cammarata, A.; Cavaleiro, A.; Polcar, T. Tribological behaviour of W-S-C coated ceramics in a vacuum environment. *Tribol. Int.* **2022**, *167*, 107375. [CrossRef]
21. Xu, Z.; Zhong, M.; Xu, W.; Xie, G.; Hu, H. Effects of aluminosilicate particles on tribological performance and friction mechanism of Cu-matrix pads for high-speed trains. *Tribol. Int.* **2023**, *177*, 107983. [CrossRef]
22. Chen, F.; Yan, K.; Hong, J.; Song, J. Synergistic effect of graphene and β-Si_3N_4 whisker enables Si_3N_4 ceramic composites to obtain ultra-low friction coefficient. *Tribol. Int.* **2023**, *178*, 108045. [CrossRef]
23. Fahad, M.R.; Abdulmajeed, B.A. Experimental investigation of base oil properties containing modified TiO_2/CuO nanoparticles additives. *J. Phys. Conf. Ser.* **2021**, *1973*, 012089. [CrossRef]
24. Sharma, A.K.; Tiwari, A.K.; Dixit, A.R.; Singh, R.K.; Singh, M. Novel uses of alumina/graphene hybrid nanoparticle additives for improved tribological properties of lubricant in turning operation. *Tribol. Int.* **2018**, *119*, 99–111. [CrossRef]
25. Huang, S.; He, A.; Yun, J.-H.; Xu, X.; Jiang, Z.; Jiao, S.; Huang, H. Synergistic tribological performance of a water based lubricant using graphene oxide and alumina hybrid nanoparticles as additives. *Tribol. Int.* **2019**, *135*, 170–180. [CrossRef]
26. de Oliveira, L.R.; Rodrigues, T.A.; Costa, H.L.; da Silva, W.M., Jr. Scuffing resistance of polyalphaolefin (PAO)-based nanolubricants with oleic acid (OA) and iron oxide nanoparticles. *Mater. Today Commun.* **2022**, *31*, 103837. [CrossRef]
27. Liñeira del Río, J.M.; López, E.R.; González Gómez, M.; Yáñez Vilar, S.; Piñeiro, Y.; Rivas, J.; Gonçalves, D.E.P.; Seabra, J.H.O.; Fernández, J. Tribological Behavior of Nanolubricants Based on Coated Magnetic Nanoparticles and Trimethylolpropane Trioleate Base Oil. *Nanomaterials* **2020**, *10*, 683. [CrossRef] [PubMed]
28. Alvi, M.A.A.; Belayneh, M.; Bandyopadhyay, S.; Minde, M.W. Effect of Iron Oxide Nanoparticles on the Properties of Water-Based Drilling Fluids. *Energies* **2020**, *13*, 6718. [CrossRef]
29. Bhaumik, S.; Pathak, S.D.; Dey, S.; Datta, S. Artificial intelligence based design of multiple friction modifiers dispersed castor oil and evaluating its tribological properties. *Tribol. Int.* **2019**, *140*, 105813. [CrossRef]
30. Humelnicu, C.; Ciortan, S.; Amortila, V. Artificial Neural Network-Based Analysis of the Tribological Behavior of Vegetable Oil–Diesel Fuel Mixtures. *Lubricants* **2019**, *7*, 32. [CrossRef]
31. Haldar, A.; Chatterjee, S.; Kotia, A.; Kumar, N.; Ghosh, S.K. Analysis of rheological properties of MWCNT/SiO_2 hydraulic oil nanolubricants using regression and artificial neural network. *Int. Commun. Heat Mass Transf.* **2020**, *116*, 104723. [CrossRef]
32. Esfe, M.H.; Toghraie, D.; Amoozadkhalili, F. Optimization and design of ANN with Levenberg-Marquardt algorithm to increase the accuracy in predicting the viscosity of SAE40 oil-based hybrid nano-lubricant. *Powder Technol.* **2023**, *415*, 118097. [CrossRef]

33. Sharma, G.; Kotia, A.; Ghosh, S.K.; Rana, P.S.; Bawa, S.; Ali, M.K.A. Kinematic viscosity prediction of nanolubricants employed in heavy earth moving machinery using machine learning techniques. *Int. J. Precis. Eng. Manuf.* **2020**, *21*, 1921–1932. [CrossRef]
34. Qing, H.; Hamedi, S.; Eftekhari, S.A.; Alizadeh, S.M.; Toghraie, D.; Hekmatifar, M.; Ahmed, A.N.; Khan, A. A well-trained feed-forward perceptron Artificial Neural Network (ANN) for prediction the dynamic viscosity of Al_2O_3–MWCNT (40:60)-Oil SAE50 hybrid nano-lubricant at different volume fraction of nanoparticles, temperatures, and shear rates. *Int. Commun. Heat Mass Transf.* **2021**, *128*, 105624. [CrossRef]

lubricants

MDPI

Article

Performance Prediction Model for Hydrodynamically Lubricated Tilting Pad Thrust Bearings Operating under Incomplete Oil Film with the Combination of Numerical and Machine-Learning Techniques

Konstantinos P. Katsaros * and Pantelis G. Nikolakopoulos

Machine Design Laboratory, Department of Mechanical Engineering and Aeronautics, University of Patras, 26504 Patras, Greece
* Correspondence: k.katsaros@upnet.gr

Abstract: Pivoted pad thrust bearings are common machine elements used in rotating mechanisms in order to support axial loads. The hydrodynamic lubrication of such bearings has been a major subject of many investigations over the years. However, the majority of these investigations are based on full film lubrication models, when, in fact, incomplete oil film profiles appear during various operating conditions, such as startups and shutdowns. The lack of lubricant during operations can have severe impact on the bearing's performance, affecting its ability to carry the applied axial load. The scope of the current investigation is to combine numerical analysis and machine-learning techniques in order to create a model that predicts the thrust bearing's performance in terms of the pad's load-carrying capacity. For this purpose, the 2-D Reynolds equation is solved numerically for a variety of angular velocities and three different lubricants: SAE 20, SAE 30 and SAE 10W40. The position of the lack of lubricant within the oil film's control volume is studied and evaluated, together with the percentage of oil film coverage in the inlet of the pad. The results of the numerical analysis are used as input, in order to train and evaluate three different machine-learning models: Quadratic Polynomial Regression, Quadratic SVM Regression and Regression Trees. The results showed that the position of the film incompleteness affects the ability of the bearing to carry the axial load. At the same time as less lubricant entered the domain, the pressure drop could reach lower values, up to 93%. From the studied lubricants, SAE 10W40 was the one that showed the best performance results during incomplete oil film operation. Finally, the Quadratic Polynomial Regression model showed the best fit and 99% accuracy in predicting the pad's load-carrying capacity.

Keywords: thrust bearing; hydrodynamic lubrication; numerical analysis; machine-learning; polynomial regression; SVM; regression trees

Citation: Katsaros, K.P.; Nikolakopoulos, P.G. Performance Prediction Model for Hydrodynamically Lubricated Tilting Pad Thrust Bearings Operating under Incomplete Oil Film with the Combination of Numerical and Machine-Learning Techniques. *Lubricants* **2023**, *11*, 113. https://doi.org/10.3390/lubricants11030113

Received: 23 January 2023
Revised: 25 February 2023
Accepted: 1 March 2023
Published: 4 March 2023

1. Introduction

Over the years, hydrodynamically lubricated tilting pad thrust bearings have been widely used in many applications, such as agriculture, electrical generators, mining, naval and automotive industry. They are designed to carry axial loads of rotating machinery based on the hydrodynamic principals. A wedge created from the stationary thrust pads and the rotor, as well as the relative motion of these two friction surfaces with the lubrication film flowing in the middle, describe the fundamental principal of operation for such bearings. Many researchers have built computational algorithms in order to model the flow of the lubricant inside these mechanisms and calculate the major tribological parameters that affect the operation of the bearings [1–4]. At the same time, a wide variety of lubricants, surface profiles, texturing and coatings have been investigated in order to improve pad thrust bearings' operation targeting to maximize the load-carrying capacity with the minimum possible power losses [5–8]. The majority of these studies are based on the assumption of a full lubricant film along the pad's surface. However, in many applications, the lubricant's flow in the inlet of the pad is not sufficient enough to cover the full width, resulting in incomplete oil film operating conditions. Such operating conditions can occur in several occasions, such as in startups and shutdowns, as well as in cases of direct lubrication, regardless of the supply method. This oil film incompleteness can result in severe pressure drop inside the pad, reducing the bearing's load-carrying capacity. To begin with, Etsion et al. [9] used the finite difference technique to solve the Reynolds equation for a flat, sector-shaped pad thrust bearing with incomplete oil film. By calculating and comparing the bearing's load-carrying capacity and power loss with the results of a complete fluid film bearing, they concluded that the bearing's performance was affected by the location of the lubricant's supply. Furthermore, Heshmat et al. [10] performed a parametric study on thrust bearings with insufficient oil supply. They investigated different numbers of pads and inner and outer radii, as well as multiple degrees of starvation for tapered land bearings. The results showed that 12-pad thrust bearings with $(R_2 - R_1)/R_2 = \frac{1}{2}$ were the optimum geometry under starved conditions. Finally, Artiles and Heshmat [11] performed an analysis on starved thrust bearings that included temperature effects. They used a finite difference mesh in order to solve the 2-D temperature and pressure fields. The investigation was performed for tapered land thrust bearings for different minimum film thicknesses and levels of starvation. It was found that the effects of starvation were small when the bearing was flooded with lubricant, but accelerated rapidly below 50% of starvation level. The start of the film was mainly independent of geometric characteristics, but directly dependent on the starvation level.

Modern technological advances in the field of computer engineering and networks have already positively affected the more traditional mechanical engineering in many aspects. The so-called 4th Industrial Revolution has provided researchers with impressive computational power and digital tools, such as AI, machine-learning and IoT: enough to support more revolutionary investigations and applications. In the field of tribology, and specifically in bearings, researchers have mainly applied these tools for fault diagnosis, prognosis and residual life estimation. It was not until recently that progress was reported in applying such techniques on the design and performance prediction of bearings. First of all, A. Moosavian et al. [12] proposed a diagnostic method that can reliably separate different fault conditions for the main journal bearings of an internal combustion engine. Vibration signals of three different operating conditions were examined (normal, oil starvation and extreme wear) and then used as inputs to train two classifiers: K-nearest neighbor and artificial neural network. The artificial neural network showed better performance in journal bearing fault diagnosis compared to the K-nearest neighbor classifier. Furthermore, Alves et al. [13] presented promising results for training machine-learning algorithms with simulated data in order to perform ovalization fault diagnosis in hydrodynamic journal bearings. They built a numerical model to simulate the ovalization fault conditions; then, they used the numerical analysis results as a training data set for a deep convolutional neural network algorithm that was able to predict the fault conditions. Moreover, S. Poddar and N. Tandon [14] developed an application that takes acoustic emission data as input and diagnoses the category of faults in journal bearing operation. To do so, they used acoustic emission signals from journal bearings operating under normal conditions, cavitation, particle contamination and oil starvation. These data were then used in order to train different decision tree and K-nearest neighbor machine-learning models. The weighted k-NN classifier model showed the best prediction results and was eventually used for the application. R.L. Lorza et al. [15] proposed a combined Finite Element and Data Mining method to determine the maximum load-carrying capacity in tapered roller bearings. The FE model was run for different input loads and the corresponding contact stresses were obtained. This training data set was then used to train a regression model. Linear regression, Gaussian processes, artificial neural networks, support vector machines and regression trees were investigated in this study. The best combination of input loads was achieved by applying evolutionary optimization techniques based on genetic algorithms to the best regression models. In addition, K.P. Katsaros and P.G. Nikolakopoulos [16] proposed a combination of numerical and machine-learning techniques in order to identify optimal designs in hydrodynamically lubricated pivoted pad thrust bearings. A 2-D Reynolds-based finite difference numerical model was solved for three different lubricants and multiple operating conditions. The obtained tribological data were then used to train linear, quadratic and SVM regression models. AWS 100 was found to be the most efficient lubricant; it showed the maximum load-carrying capacity and the minimum friction force for the thrust pad. Moschopoulos et al. [17] developed a machine-learning procedure in order to predict journal bearings' performance characteristics. To this end, they recorded sound and vibration signals, applying the one-third octave filter to post process them. With this data set, they trained three ML algorithms: K-nearest neighbor, random forest classifier and gradient-boosting regressor. The investigation showed that ML algorithms that used sound signals had better prediction accuracy compared to those based on vibration signals. Finally, Zavos et al. [18] proposed a machine-learning approach, in order to design piston rings and thrust bearings with optimum coating selection. For this purpose, analytical results from the friction models of both assemblies were used as input data in order to train quadratic polynomial regression and support vector machine models. By predicting the minimum friction coefficient, the investigation showed that, in the case of piston rings, the TiN2 and TiAlN were the best design selection. On the other hand, in the case of the tiling pad thrust bearing, the DLC was the optimum coating selection.

The aim of this study is to combine numerical and machine-learning algorithms in order to create a model that predicts the performance of tilting pad thrust bearings that operate under various incomplete oil film profiles. Focusing on the load-carrying capacity of the pad as a critical performance characteristic, the pad bearing's operation is simulated for rotational velocities from 2000 up to 12,000 rpm. Three lubricants are used during the investigation: the mono-grade oils SAE 20 and SAE 30, as well as the multi-grade SAE10W40. Three different machine-learning methods (quadratic polynomial regression, support vector machine, regression trees) are applied and compared in terms of predictions accuracy. The novelty of this study lies in the fact that no similar work can be found in literature combining numerical and ML methods for incomplete oil film study and design of hydrodynamically lubricated tilting pad thrust bearings.

2. Theory

2.1. Hydrodynamic Lubrication Model

The 2-D Reynolds Equation (1) is used in the current study in order to calculate the hydrodynamic characteristics of the lubricant's flow. The pivoted pad under consideration is approximated and considered to be a center-pivoted rectangle. A schematic of the rotor–pad conjunction is presented in Figure 1. The film thickness is assumed to be small compared to the length and the width of the pad. To add to that, Newtonian, incompressible lubricants are assumed to follow a laminar and isothermal flow inside the pad- rotor conjunction. Cavitation effects, although important in specific pad geometries and high rotational velocities, are not taken into consideration for the current investigation, based on the assumption that the minimum pressure is not reaching the vapor pressure value. In the rotor-lubricant interface, the oil is assumed to gain the velocity of the wall that it comes in contact with; thus, the no-slip condition is applied [19]. Moreover, the viscosity is considered to be constant throughout the film thickness. The film thickness h is assumed to be a function of the pad's length and is calculated from equation (2), while any change in the radial direction and the corresponding misalignment issues are not taken into consideration. Normally, the inclination of the pad and the minimum film thickness are calculated at the equilibrium position, so that the pad can carry the applied load. In this study, given the specific minimum film thickness and the inclination value, the load-carrying capacity of the pad is calculated in the equilibrium position by integrating the pressure p over the bearing pad area (3). In the cases of incomplete oil film, the lubricant's width (l) is calculated based on the continuity of the flow (4).

$$\frac{\partial}{\partial x}\left(h^3\frac{\partial p}{\partial x}\right)+\frac{\partial}{\partial y}\left(h^3\frac{\partial p}{\partial y}\right)=6\mu U\frac{\partial h}{\partial x} \quad (1)$$

$$h=f(x)=h_1+\frac{x}{B}(h_1-h_0) \quad (2)$$

$$F_p=\int_A pdA=W \quad (3)$$

$$\int_0^l q_x dy=\int_0^{L_0}(q_x)_0 dy\pm\int_0^L q_y dx \quad (4)$$

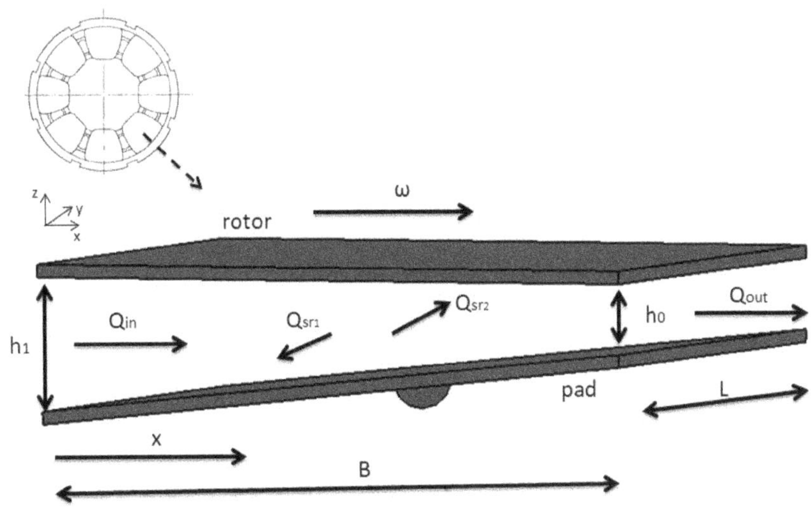

Figure 1. Pivoted pad thrust bearing schematic.

2.2. Viscosity Model

During operation, the rise in temperature leads to a decrease in the lubricant's viscosity value. As mentioned, from Nacer Tala-Ighil and Michel Fillon [20], the concept of the "effective temperature" can be considered in order to approximate the operating viscosity value without applying complex and time-consuming THD algorithms. The effective temperature value inside the lubricant's domain is calculated from Equations (5) and (6) [21]. T is the effective temperature of the lubricant, while T_0 is considered to be the inlet temperature. The constant k_e is empirical and, with a value of 0.8, gives good agreement between theory and experiment. The variation of temperature ΔT is considered to be a function of friction, rotating velocity and average axial fluid flow. The lubricant's density and specific heat capacity are also taken into consideration. To add to that, the fraction $\frac{l_{in}}{L}$ is applied, in order to define the various percentages of inlet oil coverage during the investigation. An iterative procedure is followed, in order to define the final average effective temperature for each simulation.

The Sutherland's law is used to model the viscosity variation according to temperature (7), (8). Specific coefficients are calculated as the model is adapted to fit the known dynamic viscosity values for each lubricant. A graphical representation of the dynamic viscosity variation according to temperature is shown in Figure 2.

$$T=T_0+k_e\Delta T \quad (5)$$

$$\Delta T = \frac{2FU}{\frac{l_{in}}{L} Q\rho\sigma} \tag{6}$$

$$\mu = \mu_v e^A \tag{7}$$

$$A = C_2^{\mu}\left(\frac{1}{T} + \frac{1}{C_1^{\mu}}\right) + C_3^{\mu}\left(\frac{1}{T} + \frac{1}{C_1^{\mu}}\right) \tag{8}$$

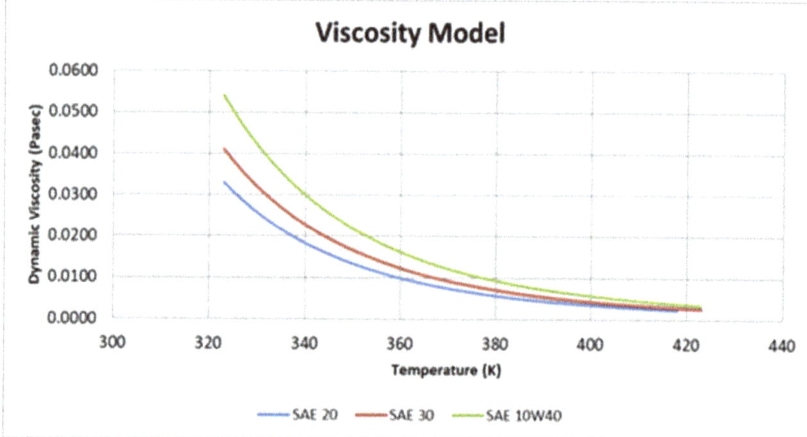

Figure 2. Dynamic Viscosity variation according to Temperature for SAE 20, SAE 30 and SAE 10W40.

2.3. Numerical Analysis

In order to numerically solve the Reynolds equation, the control domain of the lubricant inside the pad-rotor tribocouple is discretized with a typical 2-D mesh of approximately 2500 finite cells; 50 in x direction, and 50 in y direction. Spatial resolution tests showed differences in the order of 1% between typical and fine meshes. The inlet and outlet of the lubricant's control volume are assumed to be openings, and a constant pressure $P = p_{atm}$ is applied as a boundary condition. To add to that, an outflow condition is prescribed in both inner and outer pad sides: $r = R_{in}, R_{out}$. In addition, no inflow is allowed in the computational domain and the ambient pressure $P = p_{atm}$ is applied. The rotor is assumed to be moving with a constant rotational velocity ω, which corresponds to $U = \omega r_{mean}$ at the pad's mid sector. An iterative algorithm is built based on the finite differences—central differences—methodology. The Reynolds equation is adapted so that the algorithm is able to swipe over the grid and compute the corresponding pressure P_{ij} at any internal node (9). A representation of the calculation is presented in Figure 3, where *c* is the node at which the pressure is calculated and *n*, *w*, *s*, *e* are the neighboring nodes used for this calculation. Convergence to steady-state condition is verified by monitoring the computed nodal pressure based on the defined convergence criteria (10). In the cases of incomplete oil film (Figure 4), the lubricant's width limit lines *LB* (*i*), *LT* (*i*) are calculated by swiping over the nodes in the direction of the flow (11). The amount of lubricant that enters the domain l_{in} flows through the pad-rotor conjunction and adapts to the inclination of the pad. As a result, the same amount of lubricant at every step of the way through the pad (*i*) has to cover more and more of its surface until (if) it reaches the pad's sides or the end of the pad in the flow direction. Pressure $P = p_{atm}$ is then applied as a boundary condition on the area where no lubricant flows. The calculation of pressure distribution in the *y*-direction is then limited to the new boundary conditions. In addition, Case A refers to lack of lubricant on the outer part of the pad, and is modeled with *LB* (*i*) placed on the inner pad border, while *LT* (*i*) takes values within the domain. Case B refers to the lack of lubricant on the inner part of the pad. As a result, *LT* (*i*) is placed on the outer border and *LB* (*i*) runs through the fluid film domain. Finally, Case C refers to the scenario where both *LB* (*i*) and *LT* (*i*) are calculated symmetrically through the fluid film.

$$P_{i,j} = C_n P_n + C_w P_w + C_s P_s + C_e P_e + G \quad i,j = 0,\ldots,50 \tag{9}$$

$$Err_{press} = \frac{\Sigma_1^N \left| P_i^j - P_{i-1}^j \right|}{\Sigma_1^N \left| P_i^j \right|} \leq 1 \times 10^{-6} \tag{10}$$

$$l_i = l_{in} \frac{h_{in}}{h_i} \tag{11}$$

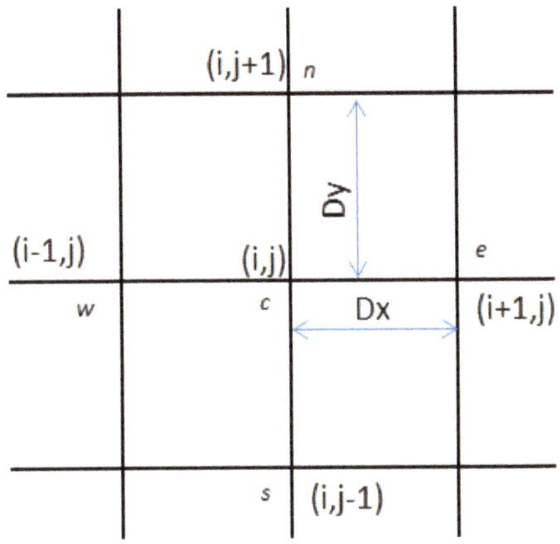

Figure 3. Finite Difference-Central Differences Computation Grid.

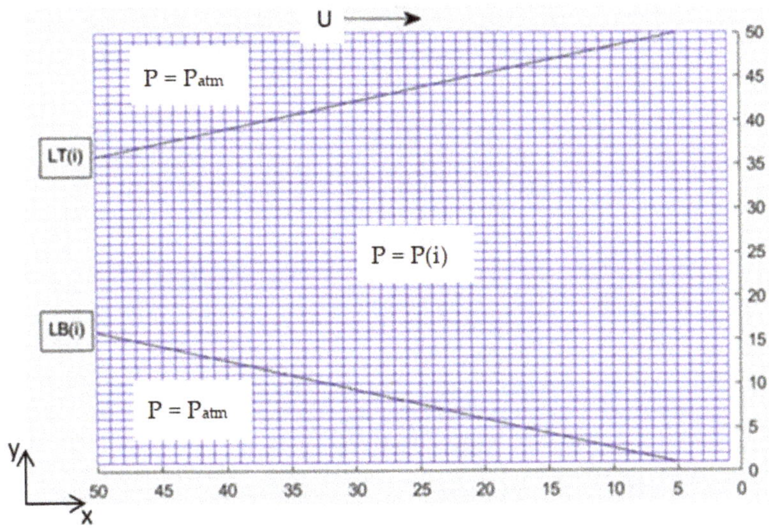

Figure 4. Computation grid, along with the incomplete oil film areas.

The hydrodynamic lubrication model is validated with experimental data obtained from the paper of Bielec and Leopard [22]. Figure 5 shows that there is a good agreement between the experimental and computed pad-specific load for different angular velocities and film thicknesses.

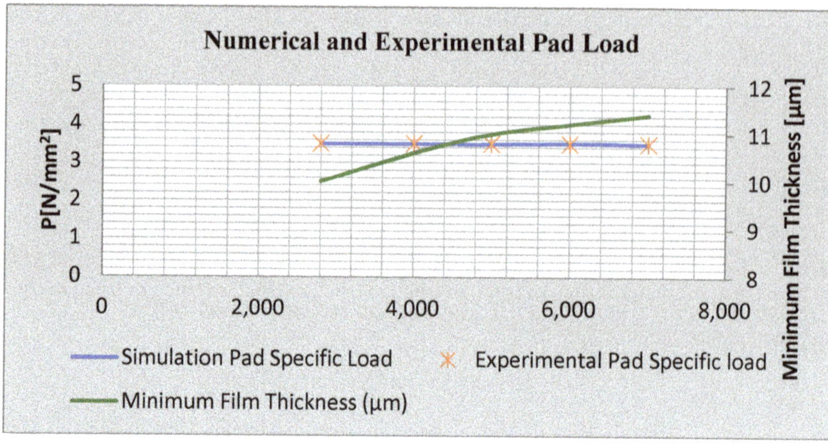

Figure 5. Numerical and Experimental Specific Pad Load- Data Validation.

2.4. Machine-Learning

For the purpose of this study, all the data obtained from the numerical simulations are used as input, in order to train and compare machine-learning models based on three different methods: the Multi-Variable Quadratic Polynomial Regression, the Quadratic Support Vector Machine and Regression Trees. These regression models are widely used in machine-learning applications, mainly due to their simplicity and accuracy to predict the corresponding response values. To begin with, the Multi-Variable Quadratic Polynomial Regression model is based on the least-squares fit methodology, in which the sum of the squares of the residuals needs to be minimized. Two independent variables, or predictors, are used x_{1i}: rotational velocity [rpm]; x_{2i}: percentage of inlet oil coverage, in order to predict the response values of one dependent variable Y: Pad's Load-carrying Capacity [N]. For a set of n-observations, Equation (12) or, in matrix form, Equation (13), is solved, in order to calculate the y-intercept: β_0 and the corresponding slopes: $\beta_1, \ldots, \beta_5$.

$$Y = XB \tag{12}$$

$$Y = \begin{bmatrix} y_1 \\ y_2 \\ \vdots \\ y_n \end{bmatrix} = \begin{bmatrix} 1 & x_{11} & x_{21} & x_{11}^2 & x_{11}x_{21} & x_{21}^2 \\ 1 & x_{12} & x_{22} & x_{12}^2 & x_{12}x_{22} & x_{22}^2 \\ \vdots & \vdots & \vdots & \vdots & \vdots & \vdots \\ 1 & x_{1n} & x_{2n} & x_{1n}^2 & x_{1n}x_{2n} & x_{2n}^2 \end{bmatrix} \begin{bmatrix} \beta_0 \\ \beta_1 \\ \beta_2 \\ \beta_3 \\ \beta_4 \\ \beta_5 \end{bmatrix} \tag{13}$$

Furthermore, the Support Vector Machine models were trained in Matlab's Regression Learner application, using the quadratic polynomial kernel function (14). In addition, with the same application, regression trees were trained and evaluated accordingly. To perform the analysis, all data were sorted in ascending order for both predictors, x_{1i} and x_{2i}. Then, all the mean squared errors were calculated separately for all the response values of both predictors (15) in each splitting candidate node t. At every iteration, the splitting node t of the regression tree was defined as the one that provided the minimum mean-squared error from all the examined data. The procedure continues repeatedly until each branch reaches the pre-defined leaf size. For the current study, a leaf size equal to 4 has been selected, as it provides the finest tree results for the Matlab's application with the optimum accuracy. In addition, the criteria chosen in the current study, in order to measure and evaluate the goodness of fit for the generated machine-learning models, is the coefficient of determination, or R^2 (16). This coefficient indicates the difference between the values of the dependent variable y_{fit} calculated from the model and the observations y_{num} obtained from the relevant numerical simulations. The higher the value of R^2, the better the model is at predicting the data. Finally, the Matlab's standard 5-fold, cross-validation procedure was applied for 5 randomly chosen partitions of the original data set. All the models where trained with 80% of the data from the data lake, while the rest 20% of the data was used for testing. Experimental data were used for the validation of the ML model as shown in [16].

$$(X, Y) = \left(c + X^T Y\right)^2 \tag{14}$$

$$MSE = \sum \frac{1}{n}(y_i - \overline{y_t})^2 \tag{15}$$

$$R^2 = 1 - \frac{\sum_1^n \left(y_{num} - y_{\hat{fit}}\right)^2}{\sum_1^n (y_{num} - \overline{y})^2} \tag{16}$$

3. Results

The simulations were performed for three different types of inlet incomplete oil profiles: Case A: where there was lack of oil on the outer radius; Case B: where there was lack of oil on the inner radius; Case C: symmetrical lack of oil from the center of the pad. Three different lubricants were examined: the mono-grade SAE 20 and SAE 30, as well as the multi-grade SAE 10W40. The simulations were run for rotational velocities, from 2000 up to 12,000 rpm, and a $k = 0.1$ inclination of the pad. The corresponding Reynolds numbers vary from $Re = 60$ up to $Re = 200$, indicating a laminar flow. The coverage of the pad's inlet with lubricant varied from 1 (full film lubrication) up to 0.4 (40% of the inlet covered with oil). The film thickness variation to rotational velocity has been considered similar to the one presented in Figure 13.3a from Bielec and Leopard [22]. All the input parameters are shown in Table 1.

Table 1. Input parameters for the simulations.

Pad's Length	32	mm
Pad's Width	28	mm
Pad's Outer Radious	62	mm
Pad's Inclination	0.1	
Pad's Pivot	center	
Rotational Velocity	2000–12,000	rpm
Percentage of Inlet Oil Coverage	0.4–1	
SAE 20 dynamic viscosity @50 °C	0.033	Pasec
SAE 30 dynamic viscosity @50 °C	0.046	Pasec
SAE 10W40 dynamic viscosity @50 °C	0.054	Pasec
SAE 20 density @40 °C	861	Kg/m^3
SAE 20 specific heat capacity	2021	J/kgK
SAE 30 density @40 °C	869	Kg/m^3
SAE 30 specific heat capacity	1950	J/kgK
SAE 10W40 density @40 °C	851	Kg/m^3
SAE 10W40 specific heat capacity	1980	J/kgK
Lubricant's Inlet Temperature	323	K

Figures 6–8 below show typical representations of the corresponding pressure profiles for the three different incomplete oil film cases studied: A, B, C, at 60% inlet coverage and 6000 rpm rotational velocity.

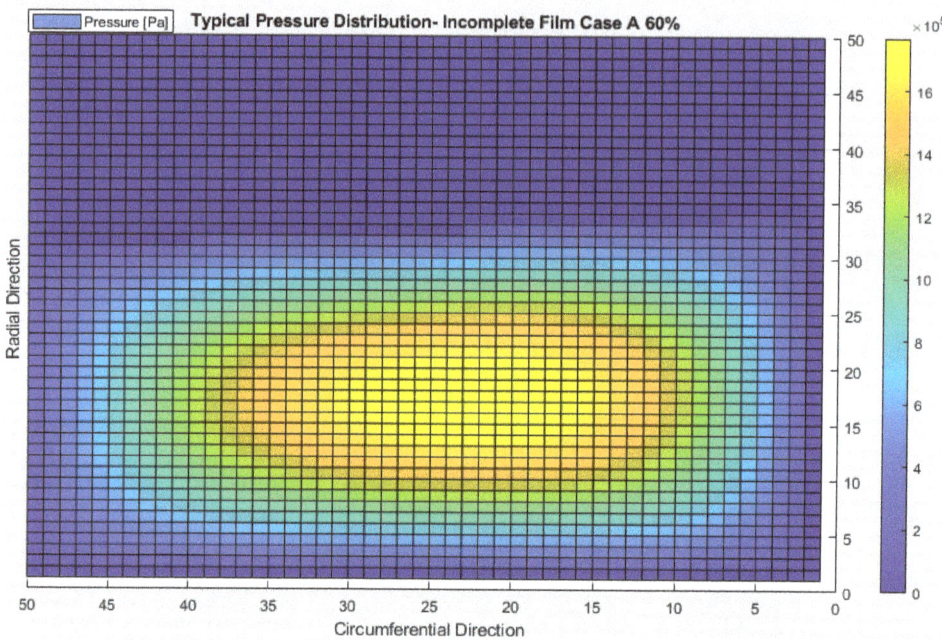

Figure 6. Typical pad's pressure distribution for the Case A incomplete oil film profile at 60% oil film coverage for the inlet of the pad.

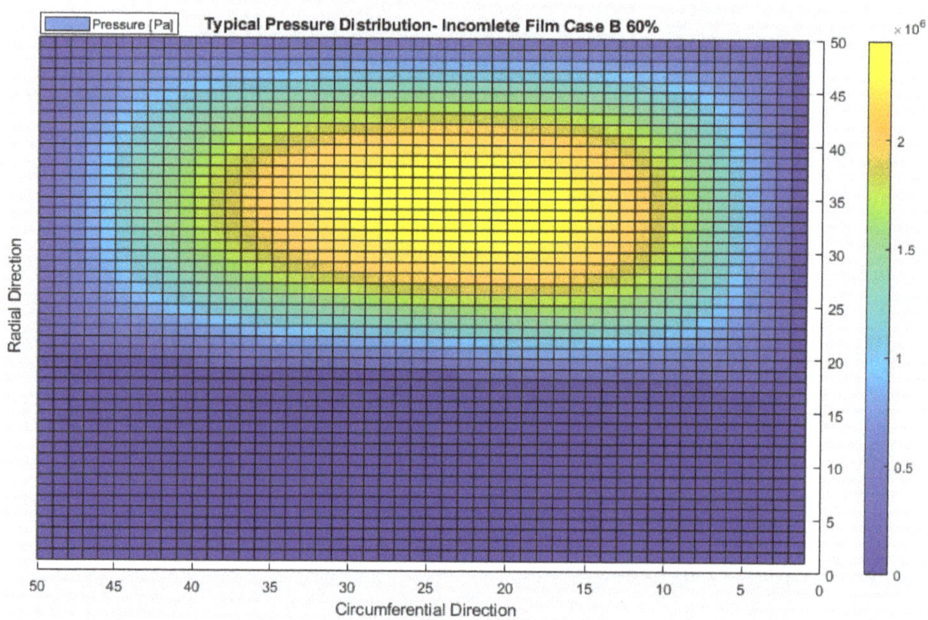

Figure 7. Typical pad's pressure distribution for the Case B incomplete oil film profile at 60% oil film coverage for the inlet of the pad.

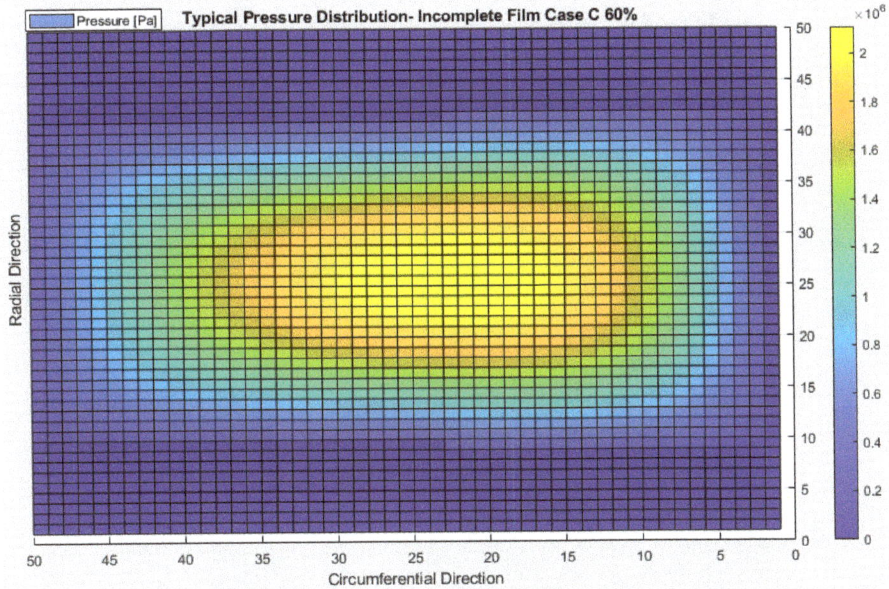

Figure 8. Typical pad's pressure distribution for the Case C incomplete oil film profile at 60% oil film coverage for the inlet of the pad.

The total amount of 2079 simulation data was used as input in order to train the machine-learning models that predict the load-carrying capacity of the pad according to rotational velocity and the percentage of oil coverage in the inlet of the pad. Table 2 shows all the Quadratic Polynomial Regression ML models, along with the corresponding R^2 values of each case:

Table 2. Quadratic Polynomial Regression models.

Case Study	ML Model	R^2
SAE 30 Case A	$y = 139.4 - 891x_1 - 0.016x_2 + 1577{x_1}^2 + 0.075x_1x_2 - 0.1 \times 10^{-5}{x_2}^2$	0.99
SAE 30 Case B	$y = 5.1 - 405.3x_1 - 0.021x_2 + 1240.7{x_1}^2 + 0.08x_1x_2 - 0.8 \times 10^{-6}{x_2}^2$	0.99
SAE 30 Case C	$y = -57.7 - 189.7x_1 - 0.02x_2 + 1087.3{x_1}^2 + 0.08x_1x_2 - 0.8 \times 10^{-6}{x_2}^2$	0.99
SAE 10W40 Case A	$y = 172.3 - 1035.4x_1 - 0.023x_2 + 1792.5{x_1}^2 + 0.09x_1x_2 - 0.1 \times 10^{-5}{x_2}^2$	0.99
SAE 10W40 Case B	$y = 101.7 - 748.3x_1 - 0.026x_2 + 1593.4{x_1}^2 + 0.09x_1x_2 - 0.8 \times 10^{-6}{x_2}^2$	0.99
SAE 10W40 Case C	$y = 23.1 - 496.5x_1 - 0.023x_2 + 1419.2{x_1}^2 + 0.09x_2 - 0.9 \times 10^{-6}{x_2}^2$	0.99
SAE 20 Case A	$y = 80.3 - 729.3x_1 - 0.01x_2 + 1409.1{x_1}^2 + 0.07x_1x_2 - 0.1 \times 10^{-5}{x_2}^2$	0.99
SAE 20 Case B	$y = -38.7 - 325.4x_1 - 0.009x_2 + 1127.1{x_1}^2 + 0.07x_1x_2 - 0.1 \times 10^{-5}{x_2}^2$	0.99
SAE 20 Case C	$y = -909 - 1418.2x_1 - 0.09x_2 + 9977.2{x_1}^2 + 0.7x_1x_2 - 0.1 \times 10^{-4}{x_2}^2$	0.99

The R^2 values in all models are close to 0.99, which means that there is a good agreement between the numerical data and the prediction models' response values. At the same time, this is also an indicator of 99% accuracy for the ML model to predict the pad's load-carrying capacity at the given predictor values.

Figures 9–11 are the graphical representations of the Quadratic Polynomial Regression ML models for all three lubricants and incomplete oil film profiles. In all cases, the load-carrying capacity of the pad decreases along with the percentage of inlet oil coverage, with the pressure drop reaching up to 93% for 40% inlet oil coverage. Furthermore, it is clearly shown that, in all cases, the lack of lubricant in the outer area of the pad—profile A—shows the minimum load-carrying capacity for the pad. On the other hand, profile C, with the symmetrical lack of lubricant, shows the maximum load-carrying capacity for the pad in all the studied cases. All three lubricants show identical response to the area of oil film incompleteness. Regardless of the angular velocity, data show a better load-carrying capacity for the profile C compared to the profile A, from 6 up to 15%, depending on the coverage of the inlet with oil. As the percentage of the lubricant's coverage decreases, the case C profile shows better and better performance for the pad of the bearing compared to the profiles A and B. For the worst studied conditions, 12,000 rpm rotational velocity and 40% of inlet oil coverage, the profile C provides up to 15% more load-carrying capacity for the pad compared to the case A profile.

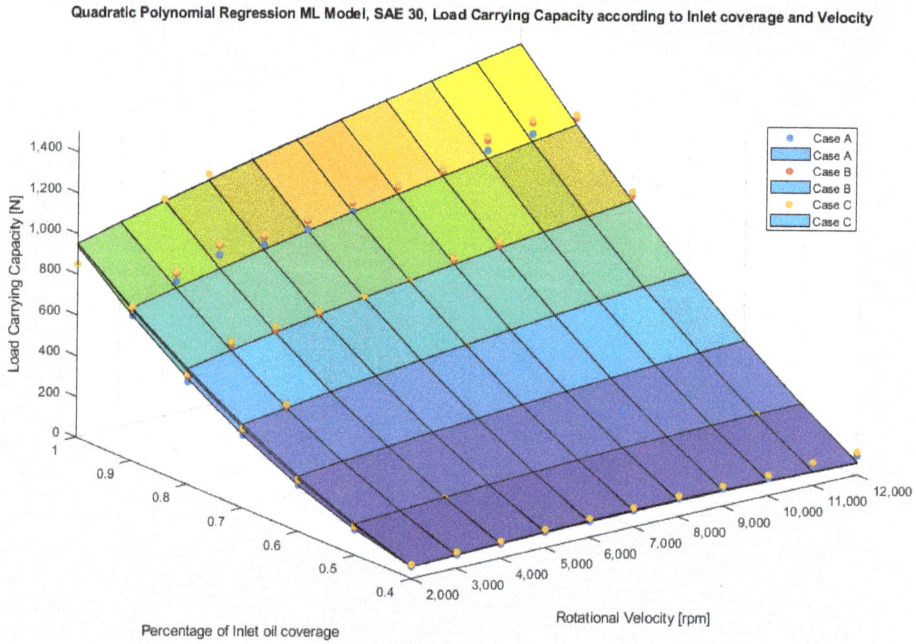

Figure 9. Quadratic Polynomial Regression model of SAE30 for all the incomplete oil film profiles. Load-carrying capacity according to percentage of inlet oil coverage and rotational velocity.

Figure 12 shows the comparison results for Case C—symmetrical oil film incompleteness—for all studied lubricants. SAE 20 shows the minimum load-carrying capacity values in comparison to SAE 10W40, which has by far the highest values in all studied conditions. This outcome is consistent with the corresponding dynamic viscosities of the lubricants. SAE 10W40 shows up to 135% better performance when studying the most extreme conditions of 12,000 rpm angular velocity and 40% coverage for the inlet of the pad.

For comparison purposes, the numerical data of the case study C (symmetrical incomplete oil film profile) were used as input, in order to train a Quadratic SVM ML model and a Binary Regression Tree model. The R^2 values, which will define the goodness of fit for all the trained models, are presented in Table 3. First of all, values of the order of 0.95 for the R^2 are, in general, accepted as very good for the fitness of the models in the data. That means that all trained models in this study have a very good response and higher than 95% accuracy to predict the load-carrying capacity of the pad. Nevertheless, in a more detailed approach, the Quadratic SVM models show better results than Regression Trees, while the Quadratic Polynomial Regression models present, in general, the best values of R^2.

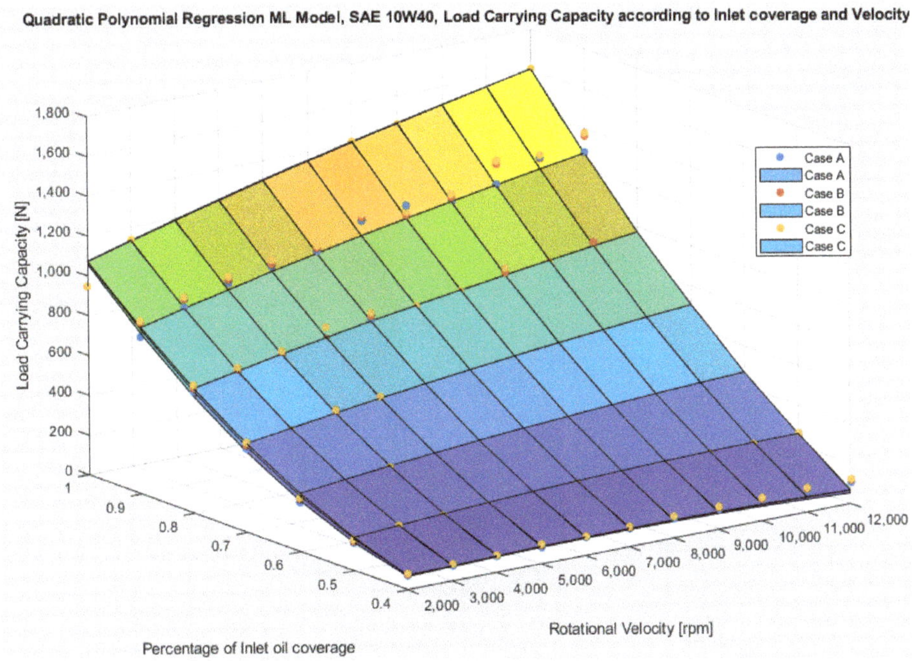

Figure 10. Quadratic Polynomial Regression model of SAE10W40 for all the incomplete oil film profiles. Load-carrying capacity according to percentage of inlet oil coverage and rotational velocity.

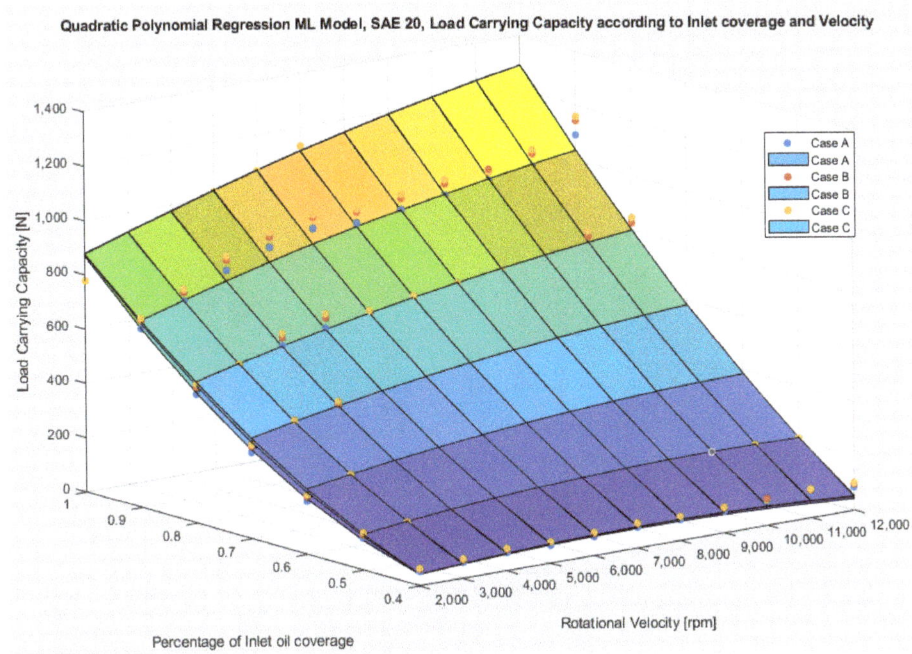

Figure 11. Quadratic Polynomial Regression model of SAE 20 for all the incomplete oil film profiles. Load-carrying capacity according to percentage of inlet oil coverage and rotational velocity.

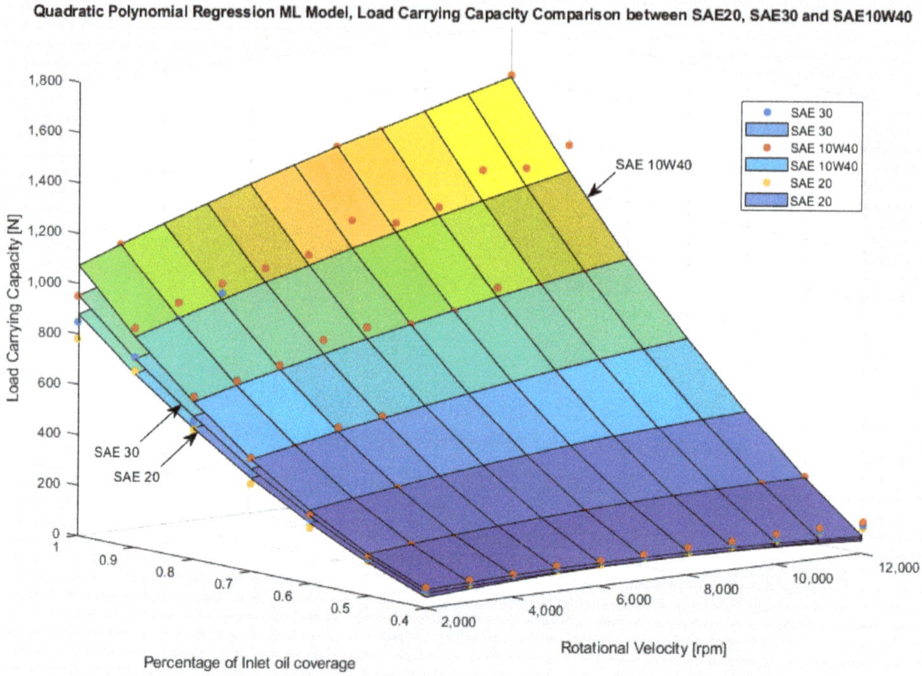

Figure 12. Quadratic Polynomial Regression model of incomplete oil film profile C for all the studied lubricants. Load-carrying capacity according to percentage of inlet oil coverage and rotational velocity.

Table 3. Quadratic SVM and Regression Tree models and their corresponding R^2.

Case Study	R^2
SAE 30 Quadratic SVM ML model	0.98
SAE 30 Regression Tree ML model	0.95
SAE 10W40 Quadratic SVM ML model	0.98
SAE 10W40 Regression Tree ML model	0.95
SAE 20 Quadratic SVM ML model	0.98
SAE 20 Regression Tree ML model	0.95

Taking a closer look at the results of case study C for the SAE 10W40, the lubricant with the optimum performance in terms of pad load-carrying capacity, one can notice that the Quadratic Polynomial Regression model has 99% accuracy in predicting the results. The Quadratic SVM model shows just 1% less accuracy with R^2 = 0.98, while the Regression Tree model has an R^2 = 0.95, which gives 4% less accuracy in load-carrying capacity prediction compared to the Quadratic Polynomial Regression model. Figure 13 is a graphical representation of the predicted versus the true response values for the Quadratic SVM and the Regression Tree models that were trained with Matlab's Regression Learner tool. It is visually verified that the Quadratic SVM model has a better fit to the results compared to the Regression Tree model, since the observations (blue markers) are gathered very close to the prediction line compared to the Regression Tree model on the right, which shows a few observations with a higher deviation from the prediction line, mainly on the upper left corner. Figures 14 and 15 (below) are the typical representation of the response plots for the SAE 10W40, case study C and Quadratic SVM model for each predictor. Similarly, Figures 16 and 17 are the typical representations of the response plots for SAE 10W40, case study C and Regression Tree model. Finally, Figure 18 is the graphical representation of the Regression Tree machine-learning model for the lubricant SAE 10W40 and case study C- symmetrical, incomplete oil film profile.

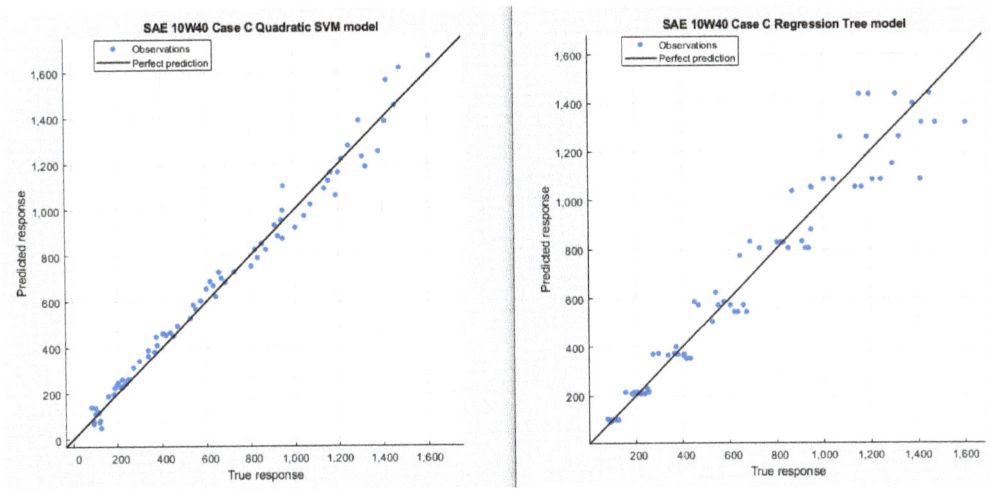

Figure 13. SVM model VS Regression Tree model- True and Prediction response plots for SAE 10W40, case C.

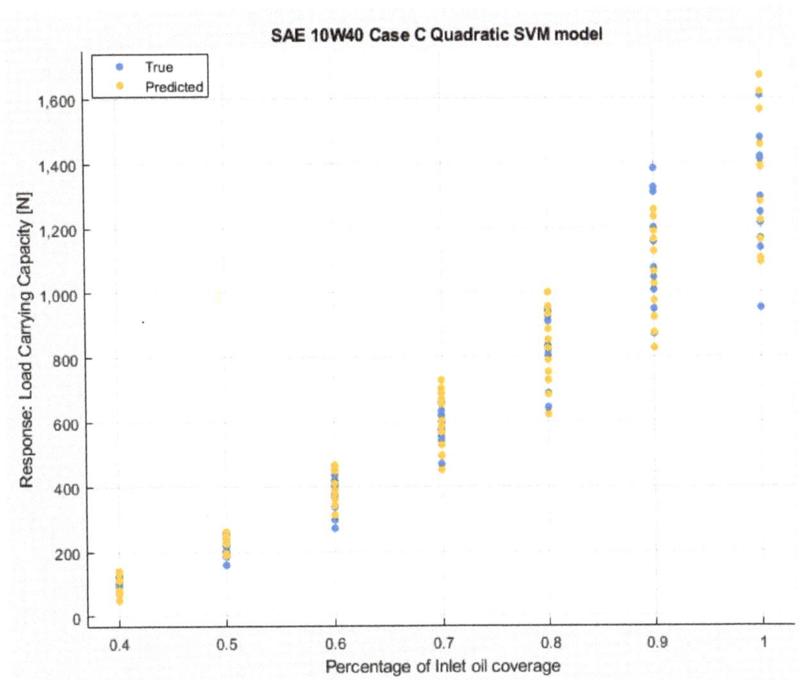

Figure 14. Typical response plot of the pad's inlet oil coverage and load-carrying capacity for the Quadratic SVM model, SAE 10W40, case C profile.

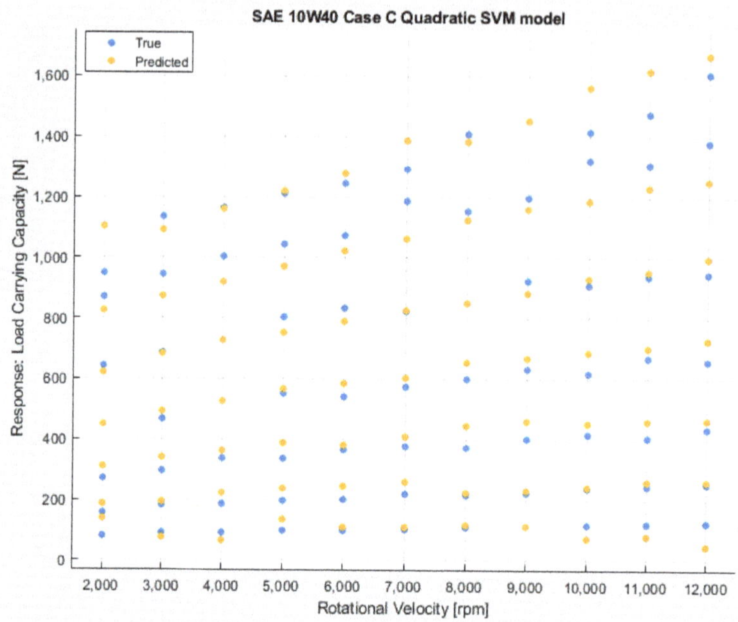

Figure 15. Typical response plot of the rotational velocity and load-carrying capacity for the Quadratic SVM model, SAE 10W40, case C profile.

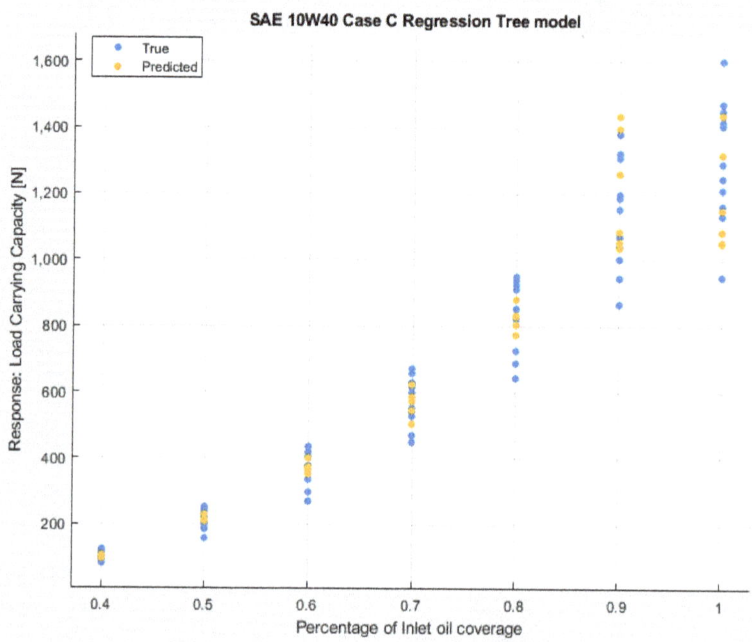

Figure 16. Typical response plot of the pad's inlet oil coverage and load-carrying capacity for the Regression Tree model, SAE 10W40, case C profile.

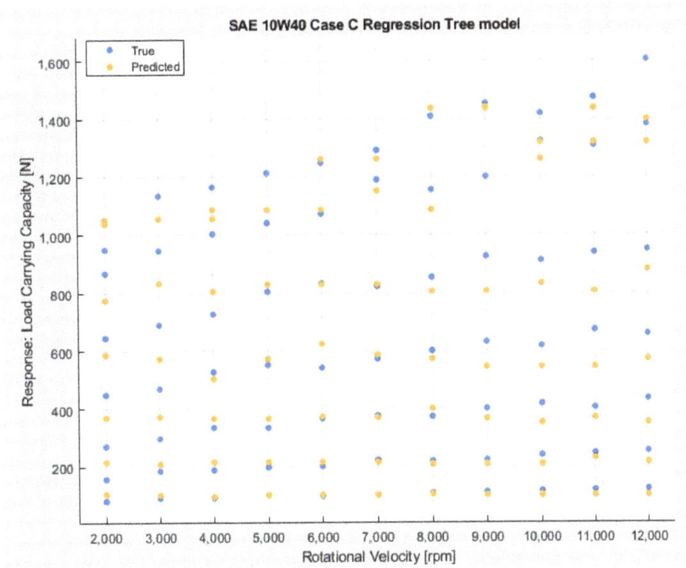

Figure 17. Typical response plot of the pad's rotational velocity and load-carrying capacity for the Regression Tree model, SAE 10W40, case C profile.

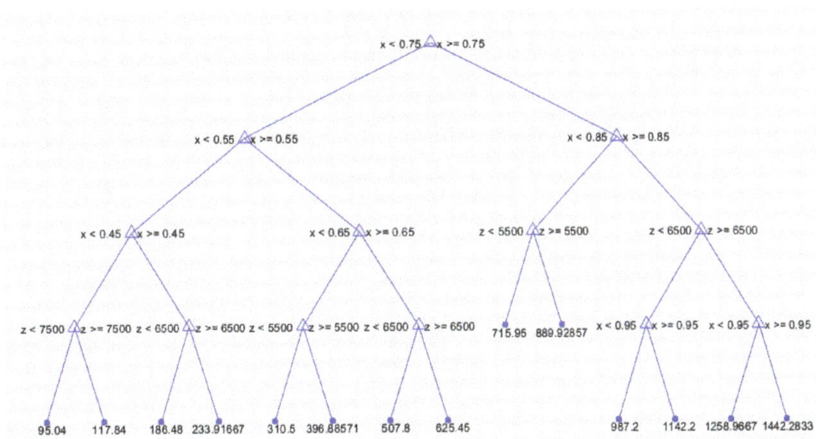

Figure 18. Graphical representation of the Regression Tree model for the SAE 10W40, case C incomplete oil film profile.

4. Conclusions

In the current paper, the performance of a tilting pad thrust bearing was investigated in terms of the pad's load-carrying capacity under various incomplete oil film profiles by combining numerical and machine-learning techniques. The 2-D Reynolds equation was solved numerically with the finite difference, central differences and method for three different lubricants: SAE 20, SAE 30 and SAE10W40. Three incomplete oil film profiles were studied, with the percentage of inlet oil coverage varying from 40% to 100%, and the rotational velocity of the rotor covering a range between 2000 and 12,000 rpm. In addition, the numerical data were used as input in order to train three machine-learning models: Quadratic Polynomial Regression, Quadratic SVM and Regression Trees. The conclusions of the investigation are summarized below:

- As less oil covers the pad's surface, the load-carrying capacity drops up to 93% for 40% of inlet oil coverage.
- The load-carrying capacity of the pad is affected by the position of the oil film incompleteness. The lack of lubricant on the outer area of the pad, profile A, shows the worst load-carrying capacity results, while the case study C profile, with symmetrical lack of lubricant, presents up to 15% better performance.
- From the studied lubricants, SAE 10W40 shows up to 135% better performance for the worst studied conditions of 12,000 rpm and 40% inlet oil coverage.
- All the machine-learning models have a good accuracy in predicting the load-carrying capacity of the pad, since all R^2 values are higher than 0.95.
- Finally, the Quadratic Polynomial Regression ML model shows 1% better accuracy compared to the Quadratic SVM model, and 4% better accuracy when compared to the Regression Tree ML model.

All in all, the chosen machine-learning model that fits the needs of the current investigation in the best possible way is the Quadratic Polynomial Regression model. The lubricant that provides the pad with the optimum load-carrying capacity when facing incomplete oil film operating conditions is the SAE 10W40, and the worst case scenario is the lack of lubricant in the outer area of the pad's surface.

Author Contributions: Conceptualization, writing—review and editing, P.G.N.; writing—original draft preparation, methodology, software and machine learning, K.P.K.; All authors carried out interpretations for the results. All authors have read and agreed to the published version of the manuscript.

Funding: This research received no external funding.

Institutional Review Board Statement: Not applicable.

Informed Consent Statement: Not applicable.

Data Availability Statement: Not applicable.

Conflicts of Interest: The authors declare no conflict of interest.

Nomenclature

A	total area of bearing pads [m^2]
B	pad length in x-direction [m]
C_1^{μ}	first viscosity coefficient—absolute temperature at which $\mu = \mu_v$ (323 K)
C_2^{μ}	second viscosity coefficient according to Sutherland's law = 3800
C_3^{μ}	third viscosity coefficient according to Sutherland's law = 30,000
$C_{n,s,w,e}$	constants for each neighbor node
h	film thickness [m]
h_0, h_1	outlet, inlet film thickness [m]
h_{min}	minimum film thickness [m]: $h_{min} = \min(h_0, h_1)$
k	convergence ratio: $k = (h_1 - h_0)/h_0$
k_e	empirical constant = 0.8 [21]
L	pad's width in y-direction [m]
p	absolute pressure [Pa]
P	absolute nodal pressure [Pa]
$q_{x,y}$	lubricant flow [m^3/h]
$Q_{in,out}$	lubricant flow in inlet and outlet area of the pad [m^3/h]
$Q_{sr1,2}$	lubricant outflow from the sides of the pad [m^3/h]
T	temperature [K]
U	linear rotor velocity [m/s]
μ	dynamic viscosity coefficient [Pas]
μ_v	nominal dynamic viscosity
x	independent variable of length along pad's width side [m]
ω	rotational velocity [rpm]

References

1. Ettles, C. The Development of a Generalized Computer Analysis for Sector Shaped Tilting Pad Thrust Bearings. *ASLE Trans.* **1976**, *19*, 153–163. [CrossRef]
2. Markin, D.; McCarthy, D.; Glavatskih, S. A FEM approach to simulation of tilting-pad thrust bearing assemblies. *Tribol. Int.* **2003**, *36*, 807–814. [CrossRef]
3. Dadouche, A.; Fillon, M.; Dmochowski, W. Performance of a Hydrodynamic Fixed Geometry Thrust Bearing: Comparison between Experimental Data and Numerical Results. *Tribol. Trans.* **2006**, *49*, 419–426. [CrossRef]
4. Papadopoulos, C.I.; Kaiktsis, L.; Fillon, M. CFD Thermohydrodynamic Analysis of 3-D Sector-Pad Thrust Bearings with rectangular dimples. In Proceedings of the ASME Turbo Expo 2013: Turbine Technical Conference and Exposition, GT2013, San Antonio, TX, USA, 3–7 June 2013.
5. Wasilczuk, M. Friction and Lubrication of Large Tilting-Pad Thrust Bearings. *Lubricants* **2015**, *3*, 164–180. [CrossRef]
6. Fouflias, D.G.; Charitopoulos, A.G.; Papadopoulos, C.I.; Kaiktsis, L.; Fillon, M. Performance comparison between textured, pocket, and tapered-land sector-pad thrust bearings using computational fluid dynamics thermohydrodynamic analysis. *Proc. Inst. Mech. Eng. Part J J. Eng. Tribol.* **2014**, *229*, 376–397. [CrossRef]
7. Gropper, D.; Harvey, T.J.; Wang, L. A numerical model for design and optimization of surface textures for tilting pad thrust bearings. *Tribol. Int.* **2018**, *119*, 190–207. [CrossRef]
8. Katsaros, K.; Bompos, D.A.; Nikolakopoulos, P.G.; Theodossiades, S. Thermal-Hydrodynamic Behavior of Coated Pivoted Pad Thrust Bearings: Comparison between Babbitt, PTFE and DLC. *Lubricants* **2018**, *6*, 50. [CrossRef]
9. Etsion, I.; Barkon, I. Analysis of a Hydrodynamic Thrust Bearing with Incomplete Film. *J. Lubr. Technol.* **1981**, *103*, 355–359. [CrossRef]
10. Heshmat, H.; Artiles, A.; Pinkus, O. Paper IV(ii) Parametric study and optimization of starved thrust bearings. *Tribol. Ser.* **1987**, *11*, 105–112. [CrossRef]
11. Artiles, A.; Heshmat, H. Analyis of Starved Thrust Bearings Including Temperature Effects. *J. Tribol.* **1987**, *109*, 395–401. [CrossRef]
12. Moosavian, A.; Ahmadi, H.; Tabatabaeefar, A.; Khazaee, M. Comparison of Two Classifiers; K-Nearest Neighbor and Artificial Neural Network, for Fault Diagnosis on a Main Engine Journal-Bearing. *Shock. Vib.* **2013**, *20*, 263–272. [CrossRef]
13. Alves, D.S.; Daniel, G.B.; de Castro, H.F.; Machado, T.H.; Cavalca, K.L.; Gecgel, O.; Dias, J.P.; Ekwaro-Osire, S. Uncertainty quantification in deep convolutional neural network diagnostics of journal bearings with ovalization fault. *Mech. Mach. Theory* **2020**, *149*, 103835. [CrossRef]
14. Poddar, S.; Tandon, N. Classification and detection of cavitation, particle contamination and oil starvation in journal bearing through machine learning approach using acoustic emission signals. *Proc. Inst. Mech. Eng. Part J J. Eng. Tribol.* **2021**, *235*, 2137–2143. [CrossRef]
15. Lorza, R.L.; Garcia, R.E.; Martinez, R.F.; Cueva, M.I.; MacDonald, B.J. Using the finite element method and data mining techniques as an alternative method to determine the maximum load capacity in tapered roller bearings. *J. Appl. Log.* **2016**, *24*, 4–14. [CrossRef]
16. Katsaros, K.P.; Nikolakopoulos, P.G. On the tilting-pad thrust bearings hydrodynamic lubrication under combined numerical and machine learning techniques. *Lubr. Sci.* **2021**, *33*, 153–170. [CrossRef]
17. Moschopoulos, M.; Rossopoulos, G.N.; Papadopoulos, C.I. Journal Bearing Performance Prediction Using Machine Learning and Octave-Band Signal Analysis of Sound and Vibration Measurements. *Pol. Marit. Res.* **2021**, *28*, 137–149. [CrossRef]
18. Zavos, A.; Katsaros, K.P.; Nikolakopoulos, P.G. Optimum Selection of Coated Piston Rings and Thrust Bearings in Mixed Lubrication for Different Lubricants Using Machine Learning. *Coatings* **2022**, *12*, 704. [CrossRef]
19. Aurelian, F.; Patrick, M.; Mohamed, H. Wall slip effects in (elasto) hydrodynamic journal bearings. *Tribol. Int.* **2011**, *44*, 868–877. [CrossRef]
20. Tala-Ighil, N.; Fillon, M. A numerical investigation of both thermal and texturing surface effects on the journal bearings static characteristics. *Tribol. Int.* **2015**, *90*, 228–239. [CrossRef]
21. Stachowiak, G.W.; Batchelor, A.W. Thermal Effects in Bearings. In *Engineering Tribology*; Paragraph 4.6; Butterworth-Heinemann Elsevier Ltd.: Oxford, UK, 2014; Chapter 4.
22. Bielec, M.K.; Leopard, A.J. Paper 13: Tilting Pad Thrust Bearings: Factors Affecting Performance and Improvements with Directed Lubrication. *Proc. Inst. Mech. Eng. Conf. Proc.* **1969**, *184*, 93–102. [CrossRef]

MDPI AG
Grosspeteranlage 5
4052 Basel
Switzerland
Tel.: +41 61 683 77 34

Lubricants Editorial Office
E-mail: lubricants@mdpi.com
www.mdpi.com/journal/lubricants

www.ingramcontent.com/pod-product-compliance
Lightning Source LLC
Chambersburg PA
CBHW041017270926
51850CB00024B/1344